Catalytic Hydrogenation for Biomass Valorization

RSC Energy and Environment Series

Series Editors:
Laurence Peter, *University of Bath, UK*
Heinz Frei, *Lawrence Berkeley National Laboratory, USA*
Roberto Rinaldi, *Max Planck Institute for Coal Research, Germany*
Tim S. Zhao, *The Hong Kong University of Science and Technology, Hong Kong*

Titles in the Series:

How to obtain future titles on publication:
A standing order plan is available for this series. A standing order will bring delivery of each new volume immediately on publication.

For further information please contact:
Book Sales Department, Royal Society of Chemistry, Thomas Graham House, Science Park, Milton Road, Cambridge, CB4 0WF, UK
Telephone: +44 (0)1223 420066, Fax: +44 (0)1223 420247
Email: booksales@rsc.org
Visit our website at www.rsc.org/books

Catalytic Hydrogenation for Biomass Valorization

Edited by

Roberto Rinaldi
Max Planck Institute for Coal Research, Mülheim an der Ruhr, Germany
Email: rinaldi@kofo.mpg.de

THE QUEEN'S AWARDS
FOR ENTERPRISE:
INTERNATIONAL TRADE
2013

RSC Energy and Environment Series No. 13

Print ISBN: 978-1-84973-801-9
PDF eISBN: 978-1-78262-009-9
ISSN: 2044-0774

A catalogue record for this book is available from the British Library

Published by The Royal Society of Chemistry,
Thomas Graham House, Science Park, Milton Road,
Cambridge CB4 0WF, UK

Registered Charity Number 207890

For further information see our web site at www.rsc.org

Printed and bound by CPI Group (UK) Ltd, Croydon, CR0 4YY

Preface

Catalytic hydrogenation and hydrogenolysis are key reaction classes in the upgrading of biomass molecules into platform chemicals and synthetic fuels. Over the last two decades, the catalytic community has devoted a considerable amount of experimental and computational work to develop new methodologies and processes, and to understand the intricacies of hydrogenation and hydrogenolysis performed on bioderived molecules. The main objective of this book is to bring together and synthesize the scattered information from around the world through the perspective of leading researchers in the field. This book is intended to serve as an excellent source of information in addition to providing a platform for critical discussion connecting the multitude of fields related to catalytic hydrogenation. To facilitate a well-balanced discussion pertaining to the several subfields, the overlap among the chapters was kept to a minimum. To guide the reader through the array of opinions commonly found in this field, related topics are clearly linked throughout this book. Finally, I present this book with my sincere gratitude to the authors for their contributions. Without their valuable insight, the creation of this book would not have been possible. Furthermore, I am indebted to Katrina Hong and Marco Kennema for discussions and a careful review of this work. These students provided me with valuable comments and constructive criticism, helping me to edit a book worthy of being accessible key literature to even young undergraduate students in the field.

Roberto Rinaldi
Mülheim an der Ruhr

RSC Energy and Environment Series No. 13
Catalytic Hydrogenation for Biomass Valorization
Edited by Roberto Rinaldi
© The Royal Society of Chemistry 2015
Published by the Royal Society of Chemistry, www.rsc.org

Contents

RSC Energy and Environment Series No. 13
Catalytic Hydrogenation for Biomass Valorization
Edited by Roberto Rinaldi
© The Royal Society of Chemistry 2015
Published by the Royal Society of Chemistry, www.rsc.org

Chapter 7   Catalytic Hydrotreatment of Fast Pyrolysis Oils Using
Supported Metal Catalysts   151
Agnes Retno Ardiyanti, Robertus Hendrikus Venderbosch,
Wang Yin and Hero Jan Heeres

Chapter 8   Hydrodeoxygenation of Biomass-Derived Liquids over
Transition-Metal-Sulfide Catalysts   174
Barbara Pawelec and Jose Luis Garcia Fierro

CHAPTER 1

Hydrogen: Economics and its Role in Biorefining

FERDI SCHÜTH

MPI für Kohlenforschung, Kaiser-Wilhelm-Platz 1, 45470 Mülheim, Germany
Email: schueth@kofo.mpg.de

1.1 Introduction

Hydrogen is perhaps one of the most promising energy carriers of the future. In renewable energy systems with high fractions of intermittent supply (*e.g.* wind power and solar thermal energy), potential surplus electricity could be converted into hydrogen through water electrolysis. This hydrogen can be used in a wide variety of applications. The most often discussed option, the reconversion of hydrogen into electricity, be it by gas turbines or by fuel cells, appears to be rather unattractive, due to the low round-trip efficiencies. Electrolysis – based on the process scale – can be estimated to have an efficiency of about 60% (if higher efficiencies are given, they are typically relative to the cell level). A recent NREL analysis, based on questionnaires given to manufacturers, indicate a mean efficiency value of 53% for the system.[1] Considering that the fuel-cell efficiency on the systems' level and gas turbines (not available for hydrogen yet) is estimated at about 50–60%, the overall round-trip efficiency is thus reduced to slightly above 30%. It will certainly be possible to improve this figure to some extent, but substantial losses in the round trip from electricity to electricity will invariably always be present. Therefore, the use of "renewable" hydrogen – not for the reconversion into electricity, but rather as a feedstock for the chemical industry, in oil refineries, or in biorefineries – appears to be promising. For

RSC Energy and Environment Series No. 13
Catalytic Hydrogenation for Biomass Valorization
Edited by Roberto Rinaldi
© The Royal Society of Chemistry 2015
Published by the Royal Society of Chemistry, www.rsc.org

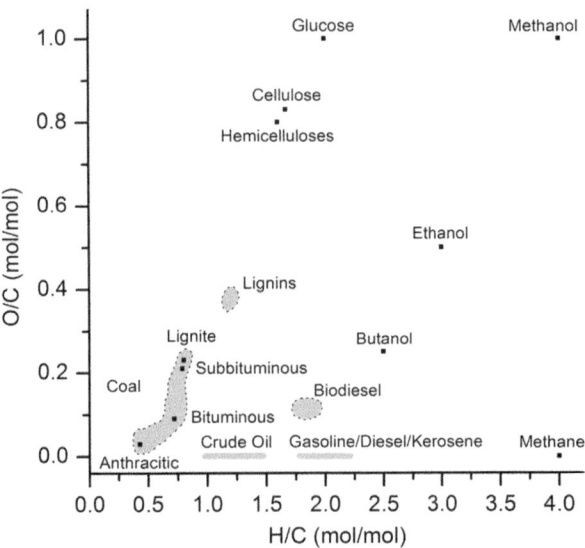

Figure 1.1 O/C and H/C molar ratios of biomass components, fossil energy resources and fuels derived from them. Oil biomass has approximately the
same composition as biodiesel (fatty acid methyl esters).
Reproduced with permission from Ref. 2.

biomass upgrading, a substantial need for hydrogen undoubtedly exists due
to the high oxygen content present in biogenic molecules.

Figure 1.1 plots the chemical composition of different energy carriers in
an O/C *vs.* H/C diagram.[2] The typical biomass constituents contain much
more oxygen than potential target molecules. In addition, the hydrogen
content often needs to be increased. Decarboxylation and decarbonylation
pathways are one of the possibilities for the reduction of the O/C ratio, but
they alone are insufficient for this purpose. Accordingly, for further oxygen
removal, hydrogen is often required as a reducing agent in order to convert
biogenic molecules into less-oxygenated target compounds. In order to increase the H/C ratio, hydrogen is needed directly either in the hydrogenation
and hydrogenolysis pathways, or indirectly after dehydration, since the dehydration leads to unsaturated compounds that are often undesired intermediates, as they are very reactive, and thus may undergo side reactions,
decreasing overall product yields. The various process options that hydrogen
is used for in order to convert biomass into chemical intermediates and end
products will be briefly discussed at the end of this chapter.

1.2   Conventional Routes for Hydrogen Production and Corresponding Costs

The vast majority of hydrogen is currently produced from fossil fuels (estimated at 49% from natural gas, 29% from liquid hydrocarbons, either

directly from naphtha or related feedstocks, or indirectly by converting residues in refineries or as off-gases from chemical or refinery processes, 18% from coal, and 4% from electrolysis).[3] Most of today's production is intended for further processing in the chemical and refinery industries, and is thus not traded on the market. It is estimated that only *ca.* 10% of the produced hydrogen is traded (*i.e.* merchant hydrogen); the rest is produced and directly used onsite (*i.e.* captive hydrogen).[4] Due to this fact, there are various figures available, as the production levels are difficult to assess. From the estimates published for different years and projected growth, current global production is about 60 million metric tons per year, with wide margins of error.

For the purposes of this chapter, we will consider processes rendering hydrogen as the main product (*i.e.* hydrogen made on purpose). Thus, typical refinery processes (*e.g.* coking and visbreaking) are not further discussed. Also disregarded are petrochemical processes, such as steam cracking for lower olefins production, since here the olefin is the main product, although the hydrogen produced contributes to the overall profitability of the process. Some processes can be considered as borderline cases, such as cracking, in which the amount of produced hydrogen can be adjusted by the processes conditions, and can thus be tuned to the hydrogen requirements of the refineries.

1.2.1 Steam Reforming/Autothermal Reforming/Partial Oxidation of Fossil Feedstocks

There are three main processes for the production of hydrogen from carbon-containing feedstocks: catalytic steam reforming (SR), autothermal reforming (AR) and partial oxidation (PO), as well as other configurations, which contain various aspects of any of the aforementioned processes. The selection of the reforming technology depends on many factors, such as the intended use of the hydrogen, acceptable impurity level, pressure level of downstream processes, price and availability of hydrocarbon and fuel, investment and operational costs, catalyst price, and several others secondary factors.[5] Overall, the reforming technologies are intimately connected to the chemical reaction networks that govern hydrogen formation, making them perhaps best to be discussed altogether.

The basic reaction of steam reforming is given as eqn (1.1) for the example of methane,

$$CH_4 + H_2O \leftrightarrows 3\,H_2 + CO \qquad \Delta H_R^0 = 206\ \text{kJ mol}^{-1} \qquad (1.1)$$

however, any other carbon-containing feedstock can be converted by a similar process, although this may require changes in process technology. Moreover, the syngas composition is dependent on the elemental composition (*i.e.* H/C/O ratios) of the feedstock.

Partial oxidation can be described by eqn (1.2), again formulated for methane as feed:

$$2\,CH_4 + O_2 \leftrightarrows 2\,CO + 4\,H_2 \qquad \Delta H_R^0 = -36\ kJ\ mol^{-1} \qquad (1.2)$$

However, it is difficult to achieve full selectivity to syngas as expressed in eqn (1.2), since partial oxidation is always competing against total oxidation:

$$CH_4 + 2\,O_2 \leftrightarrows CO_2 + 2\,H_2O \qquad \Delta H_R^0 = -802\ kJ\ mol^{-1}\ (\text{water in gas phase})$$
$$(1.3)$$

Moreover, several other equilibria are relevant in all of these systems, such as:

$$CO + H_2O \leftrightarrows CO_2 + H_2 \qquad \Delta H_R^0 = -90\ kJ\ mol^{-1} \qquad (1.4)$$

$$C + O_2 \leftrightarrows 2\,CO \qquad \Delta H_R^0 = -222\ kJ\ mol^{-1} \qquad (1.5)$$

$$C + H_2O \leftrightarrows CO + H_2 \qquad \Delta H_R^0 = 131\ kJ\ mol^{-1} \qquad (1.6)$$

In addition, methane can be cracked into hydrogen and carbon; for higher hydrocarbons, cracking reactions also come into play, and heteroatoms, which are almost invariably present in the feedstocks, react as well under the conditions of the hydrogen-generating reactions. While all of these reactions can occur in all of the systems, it is helpful to conceptually separate them, and discuss the three processes as prototypes. SR is described by eqn (1.1) and PO by eqn (1.2). As the overall reaction network contains exothermic and endothermic reactions, the feed can be composed in such a manner that the resulting enthalpy becomes nearly zero. This option is called autothermal reforming (AR).

In SR, the hydrocarbon, methane for the majority of the produced hydrogen, and steam react over a nickel/γ-Al_2O_3 catalyst at 1073–1173 K under a pressure of 1.5 to 3 MPa in a tubular reactor, producing syngas. One of the key problems is the supply of the heat to the reactor, since the reaction is highly endothermic (206 kJ mol^{-1}). Typically, the heat is integrated into the system *via* a firebox, in which part of the feed gas is combusted. Methane and other fuel gases can be used in the firebox. The product gas after the steam-reforming reactor has the equilibrium composition at the exit temperature of the reformer, containing H_2, CO, CO_2, and CH_4 in the case of natural gas and naphtha as feedstock;[6] in turn, with oil or coal, nitrogen and sulfur compounds can be present, as well. In order to maximize hydrogen yields, the product gas is first exposed to a high temperature shift reaction (1.4) over iron-based catalysts at 623–723 K. Depending on the plant configuration, the final hydrogen product is obtained after purification *via* pressure swing adsorption (standard in modern plants)[7] or by a low-temperature shift stage at 493 K over Cu/ZnO/Al_2O_3 catalysts, followed by CO_2 scrubbing, with a final removal of CO_x by catalytic methanation. Instead of providing heat for the reforming reaction by an external heating of the reformer tubes, the required energy can be supplied to the process by partial combustion of the feed gas (autothermal reforming). In this case, the feed is mixed with oxygen before it enters the catalyst bed. In sequence, the gas is combusted in the entrance section of the bed, thus providing the heat for the reforming reaction in the later section of the unit. Such designs are advantageous if high temperatures are desired in order to minimize methane concentration, to allow high pressures, or to directly provide nitrogen for ammonia synthesis (in which air is thus used instead of oxygen).[8]

Conventional partial oxidation processes for hydrogen production are used for heavy feeds (*e.g.* heavy residues or oil fractions with high sulfur or metal contents). The PO reaction takes place in a flame in an empty furnace with substoichiometric amounts of oxygen. For the production of hydrogen, CO needs to be shifted to hydrogen, which is advantageously carried out in the presence of a catalyst at temperatures below 773 K after the steam addition to the feed. The purification of hydrogen proceeds by one of the many available scrubbing processes. Partial oxidation has also been reported to be suitable for syngas production from methane with very high selectivity and yields.[9] At residence times of milliseconds, methane can be partially oxidized with oxygen rendering *ca.* 90% yields of CO and H_2. Key for this development is the very short contact time with noble-metal catalysts supported on ceramic foam monoliths. Despite the promising results, this technology does not seem to have been commercially implemented, probably due to the potential hazards of the gas mixture that is in the middle of the explosion regime.

1.2.2 Cost Analysis for Hydrogen Production

Cost analysis for hydrogen production is not a simple task, and data provided in the literature have to be interpreted with caution. Should just production costs be analyzed, centralized facilities have clear advantages over distributed production, as the former can provide significant economies of scale. However, this advantage can be easily lost upon including the cost for delivery (*i.e.* liquefaction/pressurization and transport).[10] An analysis for specific boundary conditions gave a cost for the dispensed hydrogen of 3.3 \$/kg H_2 for a distributed production. Costs were only half of that in a centralized plant, but liquefaction and delivery cost of 3.5 \$/kg H_2 have to be added to the pure production costs. However, in other studies, the additional cost for transportation has been estimated at only 1.02 \$/kg H_2.[11] Transportation costs are obviously dependent on the mode of hydrogen transportation (pressurized, liquefied, truck, pipeline, *etc.*). In the following, if no other information is given, the hydrogen cost is referred to as the production costs alone. The studies quoted are also from different years, and thus for a direct comparison, inflation would need to be taken into account. However, most of the studies quoted are from the last five to ten years, and thus the error in neglecting inflation is minor when compared to strongly fluctuating prices of raw materials. Finally, the differences in exchange rates

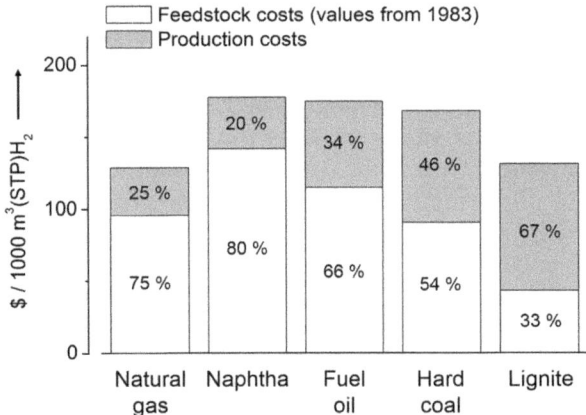

Figure 1.2   Cost for different hydrogen production routes (values from 1983), with a separation in feedstock costs and production costs. Adapted from Ref. 6.

will also affect the figures. For the purpose of this contribution, the values are given in the currency that had been used in the original publication; when currencies are converted, an exchange rate of 1.38 $/€ is utilized.

Due to the broad range of feedstocks that can be used for production of hydrogen and the different process options available, the production costs greatly depend on local conditions. A detailed analysis from 1983 is available for hydrogen production in Germany.[12] The situation, especially with respect to feedstock prices, has changed significantly since then, but some of the features, especially the split between feedstock and investment costs, are probably still valid. On the one hand, in the case of methane as a feedstock, investment costs for the reformer are the lowest, while feed/fuel costs are intermediate. On the other hand, a reformer operating on lignite shows the highest investment costs; however, due to the low price of lignite, the lowest feed cost. Most unfavorable are naphtha- and fuel-oil-based reformers. In the case of fuel oil, the feed cost is much higher than the investment costs so that the overall cost is one of the highest (Figure 1.2). Altogether, the pathway starting with natural gas is currently the most favored, as it is associated with the lowest overall costs.[10]

In a centralized methane reforming, the natural gas price dominates the cost for hydrogen production. Natural gas prices are influenced by regional economic conjunctures, since it is traded globally only to a limited extent, despite the availability of liquefied natural gas terminals in addition to the steady expansion of their capacity over the last decades. Considering the natural gas price at 5 $/GJ, the cost for hydrogen has been estimated at approximately 1 $/kg H_2 gate price for a large-scale plant with a production capacity of 427 tons per day.[13] In the United States, due to the shale gas boom, natural gas prices are currently below 5 $/GJ (4.3 $/MMBTU,[14] corresponding to about 4.1 $/GJ). Accordingly, the cost for hydrogen should be *ca.* 0.9 \$/kg H_2 or 0.65 €/kg. In Europe, natural gas prices are about 22.5 €/MWh (*i.e.* corresponding to approximately 8.5 \$/GJ) at the European Energy Exchange (March 24th 2014). The EU feed cost leads to a hydrogen production cost of about 1.2 €/kg H_2 (calculation based on the figures given in Ref. 13). Nonetheless, with the increasing supply of natural gas traded globally, it can be expected that also the European price of natural gas will eventually fall within the next few years, and thus hydrogen production costs at around 1 €/kg H_2 may also be estimated as a base-case scenario, against which other hydrogen production technologies will have to compete. In addition to the raw materials costs, it is worth mentioning that such estimates are dependent on other boundary conditions, *e.g.* the process scale (hydrogen production is at substantially higher cost on small-scale plants), price of possible coproducts (*e.g.* oxygen in electrolyzers), to mention just a few.

Importantly, hydrogen production from natural gas is very CO_2 intensive. This problem is even more serious for hydrogen production from feedstocks rich in carbon (*e.g.* lignite). Therefore, the price of CO_2 emission certificates – by either traded emission certificates or taxes on emissions – comes at the costs to other parts of the world. Depending on plant size and specific conditions, CO_2 equivalent emissions for hydrogen produced from natural gas are given as 11.88 kg CO_2/kg H_2.[15] Considering the current European market prices for the CO_2 emission certificates (*ca.* 5 €/t CO_2), there is no considerable change in the economics of the process yet. However, when the European Emission Trading System was implemented, target prices were at 40 €/t (this amount had to be paid during the first trading period between 2005 and 2007, if an emitter did not have the corresponding certificate), this price corresponds to additional costs of *ca.* 0.5 €/kg H_2, if the full certificate price would have to be paid.

The cost situation is not much different for hydrogen production by coal gasification in centralized plants.[10,16] In 2008, an analysis of different studies revealed that the cost of hydrogen from methane reforming and coal gasification differed by less than 10% – again, this is strongly dependent on the regional context, since prices for gas and coal differ worldwide. However, coal would have a strong disadvantage, if CO_2 emission certificate prices would rise substantially in the future. Coal gasification has about twice the CO_2 footprint (*i.e. ca.* 25 kg CO_2/kg H_2) compared to centralized methane reforming. Accordingly, for the coal-based hydrogen, a CO_2 emission certificate at a cost of 40 €/t would lead to a cost increase of more than 1 €/kg H_2.

1.3 Future Routes for Hydrogen Production

1.3.1 Reforming of Biomass

Biomass can be used for the production of hydrogen *via* reforming reactions. As for the conversion of fossil carbon-containing materials, there are also different general methods for the production of hydrogen from biomass.[17] For instance, biomass can directly be gasified by partial oxidation at high temperatures of 1073 to 1173 K. This process resembles the gasification of solid fossil feedstocks. Gasification is possible with oxygen, air, or steam, with differences in the process layout and the heating value of the produced gas. In addition to hydrogen, the product gas contains CO, CO_2, water, and hydrocarbons. Water and carbon monoxide can be converted into hydrogen and CO_2 by the water-gas-shift (WGS) reaction (1.4).

Alternatively, plant biomass can be pyrolyzed in the absence of air. The standard process operates at around 773 K, and results in the formation of gases, pyrolysis oil, and solid char residue (as presented in Chapter 7). In turn, the pyrolysis oil can be converted into hydrogen by steam reforming, in a similar manner to that for the reforming of liquid fossil feedstocks. The major advantage of first producing a pyrolysis oil and then reforming it, as compared to direct gasification, is the marked reduction in volume provided by decentralized, small pyrolysis plants.[18] Compared to plant biomass, pyrolysis oil is more efficiently transported to a centralized, large facility for the gasification. In this case, such plants are characterized by substantial economies of scale.

The pyrolysis can be carried out at higher temperatures of 973–1073 K in the presence of a catalyst, directly rendering a hydrogen-rich gas. Such biomass utilization schemes typically target liquid transportation fuels, which can be produced from syngas. When hydrogen production is the target, the advantage of the relatively easier transportation of pyrolysis oil might be lost, since the transportation of hydrogen is costly, and thus hydrogen production facilities are advantageous when strategically located directly at the point of use.

The conversion of biomass into hydrogen, be it by direct gasification or *via* pyrolysis schemes, seems to be the most cost-competitive pathway of all hydrogen-production routes from renewable energy. As in hydrogen production of natural gas and other fossil resources, the cost depends on many factors, such as local availability of biomass, transportation cost, exact plant layout, nature of the biomass, to mention just a few. Thus, there is a substantial spread of cost estimates for the production of hydrogen from biomass.

In a comparative study of different hydrogen-production technologies, biomass-based routes were identified as the only ones that are able to economically compete against natural gas reforming and coal gasification under the current market conditions.[10] The same conclusion was reached in a recent publication, where again natural gas reforming, coal gasification and biomass gasification were identified as the most cost competitive methods at a level of approximately 1.50 \$/kg H_2, based on an average of a number of reported figures.[16]

For two selected sources of biomass, waste forest residues and straw, the costs were analyzed for a 2000 $t_{dry\ biomass}$/d biomass plant (resulting in a hydrogen production capacity of approximately 170 t/d). Depending on the exact feedstock and the design of the plant (atmospheric or pressurized operation), the costs were estimated at 1.17 $/kg H_2 and 1.33 $/kg H_2.[19] Again, these figures are in the same range as those found for the afore-mentioned reports. This publication also provides a sensitivity analysis, *i.e.* the costs are broken down into different components in a transparent manner. Interestingly, also the transportation costs, which are substantial for biomass, are carefully discussed. This information is useful in order to assess the optimum plant size, where the cost benefit due to the economy of scale is offset by the additional transportation distance – and thus cost – for a larger plant. Although the optimum plant size depends again on feedstock and gasifier technology, the optimum volume production is estimated at about 5000 $t_{dry\ biomass}$/d. Nonetheless, it is important to point out that the curve becomes rather flat beyond 3000 $t_{dry\ biomass}$/d.[24]

As will be discussed later in this chapter, many of the routes for trans-forming biomass into fuels and chemicals require hydrogen as a reducing agent. Since the infrastructure for transportation and handling of biomass is already well established in a biorefinery, synergies can be expected between the biomass exploitation to give the starting materials for fuel and chemical production, and biomass as a feedstock for hydrogen production. On a biorefinery site, gasification of biomass could thus become economically the most attractive method for hydrogen production.

1.3.2 Water Electrolysis

Currently, a small fraction (around 4%)[17] of the world demand of hydrogen is supplied by water electrolysis. However, under most circumstances, electrolysis for the production of hydrogen is economically not optimal. It is often argued that this situation will change in systems with a high fraction of intermittent renewable electricity, such as from wind power or solar energy, since the price for electricity goes down substantially at times of high availability. However, one has to keep in mind that the capital expenditure of electrolyzers is sub-stantial. Consequently, electrolysis plants running only during the (currently) relatively short periods of low electricity prices are not economical yet. Our analyses will be discussed further below, after the technologies available for hydrogen production by electrolysis have been introduced.

There are two technologies, alkaline electrolyzers and PEM (polymer electrolyte membrane or proton exchange membrane) electrolyzers, which are commercially available, albeit on different scales. High-temperature electrolysis is a third option. Such systems, however, are still being de-veloped and heavily studied, and thus, no commercial units are yet available. A very useful survey on the state of the art of such technologies is given by Smolinka *et al.*[20] or in a book edited by Stolten.[21]

When considering efficiency of electrolyzers, which is one important cri-terion to judge their performance, one has to be careful in checking how the efficiency is defined. Generally, efficiency is defined as energy output (as given by the energy content of the hydrogen produced) divided by energy input. Typically, electrolyzers are just driven by the electricity supplied, without additional heat supply. Thus, the energy input is directly provided by the electrical power consumed by the electrolyzer.

As for an energy output in the form of hydrogen, it is important to consider whether the higher heating value (HHV; condensation enthalpy of water included, 39.4 kWh/kg H_2 or 3.52 kWh/m^3 H_2) or the lower heating value (LHV; combustion product gaseous water 33.3 kWh/kg H_2 or 2.96 kWh/m^3 H_2) should be used. On the one hand, should hydrogen be converted into another form of energy, it is appropriate to use the lower heating value, but on the other hand, should hydrogen be utilized as a chemical raw material, the higher heating value is more suitable for the analysis.

One can also assess the energy efficiency by considering the thermodynamic minimum voltage required for water splitting divided by the voltage applied under operation conditions. The equilibrium (reversible) potential for the water splitting reaction is 1.23 V, based on ΔG_R^0. However, under adiabatic conditions, in order to provide the required heat for the reaction, the potential has to be higher at 1.48 V (thermoneutral voltage at 298 K). Due to the overpotentials and the Ohmic losses in electrolyzers, this heat is typically available in any case, and heat supply is not the reason to use higher potential than 1.23 V. Thus, 1.23 V is probably the most useful value for analyzing efficiency based on voltage. Finally, one has to check whether efficiency is based on the cell itself or on the electrolyzer plant, *i.e.* including balance-of-plant energy needs.

In order to compare technical systems, efficiencies are often not used, but absolute values, *i.e.* total energy requirement for the hydrogen production in kWh/kg or m^3 of the H_2 produced. In 2008, based on company questionnaires and a literature search, a study considered an energy requirement value of 62.8 kWh/kg H_2 as the base case. This estimate corresponds to an efficiency of 53% (LHV) or 63% (HHV).[1] In another survey, energy requirements between 4 and 5 kWh/m^3 H_2 are quoted.[4] In turn, in Ref. 13, a value of 4.3 kWh/m^3 was proposed (based on information from one manufacturer). In Ref. 20, values between 4.1 and 6.3 kWh/m^3 are given for alkaline electrolyzers, where the lower values correspond to large atmospheric units and the higher values to very small pressurized units. However, in most of these publications, it is not fully clear which balance-of-plant components are included in the analyses. Such a differentiation is available for PEM electrolyzers.[20] Stack productivities of 4–5 kWh/m^3 H_2 are quoted, rather independent of size (it has to be noted that several of the systems considered were precommercial). On the systems level, the smallest units (less than 1 m^3/h) had an energy consumption of up to 8 kWh/m^3, whereas the large systems (of several 10 m^3/h) showed values ranging from 4.5 to 6 kWh/m^3. One should note, however, that the PEM technology is not as mature as alkaline electrolyzer technology. Alkaline electrolyzers, which are about one order of magnitude larger than PEM systems, are commercially available.

While alkaline electrolysis is already well advanced, there is still a high level of development work going on (for a recent survey, see Ref. 22). The electrolyte is a concentrated (20–40%) KOH solution, cathodes (where hydrogen is evolved) are typically made of corrugated steel sheets on which catalysts are typically deposited in order to improve efficiency. Unlike PEM electrolyzers, the alkaline electrolysis does not utilize noble metals as catalysts, but instead, nickel or nickel compounds are typically employed. Nickel, an active catalyst for electrolysis, or nickel alloys are used as electrodes on the anode side in the form of sheets, which are structured in a way to maximize surface area. There are also electrolyzers where both electrodes are made of nickel or nickel alloys. Most commercial electrolyzers are operated with electrodes mounted in series (bipolar plates) so that overall high voltages can be employed.

For most applications, be it for direct use or for intermediate storage, hydrogen would be required under high pressure. Compression is energy intensive, and it would thus be advantageous to carry out electrolysis at higher pressure. Thermodynamically, this would require higher reversible and thermoneutral potentials, but this is often offset by reduction of the overpotential caused by small gas bubbles formed under high pressure.[23] As a result, almost identical efficiencies are found, on the cell level, for both atmospheric and pressure electrolyzers, especially for large-scale units.[20,24]

On the one hand, high-pressure electrolyzers save the investment and energy required for postelectrolysis compression of the hydrogen produced, but on the other hand, this advantage is partly offset by the more expensive and complex layout. Most alkaline pressure electrolyzers operate under moderate pressures between 0.2 and 1.0 MPa, but systems are available also in the range of 1 MPa up to several MPa.[24] Alkaline electrolyzers have one disadvantage when combined with intermittent renewable energy. They typically have a lower operation threshold of approximately 20% of the maximum capacity rating and a slower dynamic behavior.[13] They thus cannot be shut down completely without significant repercussions for the overall performance.

In contrast, PEM electrolyzers (the state-of-the-art has recently been surveyed in Ref. 25) have a low operation threshold of 0% of the power rating. Practically, however, one would require approximately 5% of the rating to supply the electricity to the auxiliary components.[20] On the high-power side, such systems can also be operated appreciably above the nominal rating. For a Siemens prototype, short-term operation at three-fold nominal rating is reported.[26] In addition, power can be ramped up and down with much higher transients than for alkaline electrolyzers.

PEM electrolyzers became possible when proton conducting polymer membranes were available as solid electrolytes. Compared to alkaline electrolyzers, they have the main advantages of higher possible current densities, lower hydrogen crossover, and better dynamic behavior. Moreover, they are apt to operate under high pressure, since hydrogen crossover is less problematic. PEM electrolyzers operate under acidic conditions (around pH 2), which poses substantial corrosion problems for the system components operating at high potentials. This holds true especially for the catalysts, which are, in most cases, iridium on the anode side (for oxygen evolution) and platinum at the cathode. Since these elements are scarce and expensive, research efforts are underway to replace them by cheaper and more abundant alternatives. However, currently the noble metals still have to be considered the state-of-the-art, contributing appreciably to the investment costs for the stack.

Hydrogen production *via* electrolysis, be it by using alkaline or PEM electrolyzers, is currently not cost competitive. Two extreme cases can be discerned with respect to the mode of operation: (i) hydrogen could only be produced by electrolysis during times of very low electricity prices, which would lead to a high effect of the capital expenditure (capex) on hydrogen price, or (ii) the electrolyzers could be operated basically all the time in order to compensate for the high capex. The latter process mode would translate into high operation costs, since electricity would also be bought at peak costs. Altogether, both strategies are certainly inappropriate choices. Accordingly, an optimized operation scenario has to be found that minimizes total production costs by operating the electrolyzer as much as possible at moderate overall electricity costs.

Mansilla *et al.*[13] have analyzed the cost of hydrogen from electricity for France, Germany and Spain, taking the real spot price electricity data into account for the years 2010, 2011 and 2012. The authors assumed continuous electrolyzer operation, and compared this with an operation mode in which the electrolyzer was only operated when the electricity price was below a certain threshold. Calculated prices for all three countries and years were – with one exception – between 3 and 3.5 €/kg H_2, with only very little difference between the two cases discussed. Should a reduction of electrolyzer investment of 50% be assumed, prices drop by approximately 0.5 €/kg H_2, and operation at low electricity cost only become somewhat more favorable, although the effect upon it was still below 5%.

Similar costs for hydrogen from off-peak power by electrolysis (for a capacity of 10 tons per day) were estimated for the United States.[27] The hydrogen costs calculated by Saur and Ainscough[28] for hydrogen from wind electricity ranged between 3.74 and 5.86 $/kg H_2 (dollars at purchasing power for the year 2007), which is in a similar range as the values calculated for the European countries quoted above. An interactive online tool is available to perform cost analyses for different sites in the USA using different assumptions.[29]

Smolinka *et al.*[20] estimate hydrogen-production costs for several cases, concerning electrolyzer technology, size of the plant, and yearly operation hours. In most cases for hydrogen production in the renewable energy systems, costs were estimated between approximately 3 and 4 €/kg H_2. High costs of about 9 €/kg H_2 were calculated for the case of a small PEM electrolyzer with a capacity of 30 m^3/h, operation under 2.5 MPa, plant utilization of 75% at electricity costs of 0.09 €/kWh. At high utilization, fixed costs do not impair the overall hydrogen cost. However, at utilization below 50%, fixed costs dominate the cost of the hydrogen produced. In this study, it was also estimated that costs of about 2 €/kg H_2 should be possible at high plant utilization, assuming additional development of technology. Investment cost reduction probably needs to concentrate on the stack cost, since the stack contributes about 46% to the total cost for PEM electrolyzers and 57%, for alkaline electrolyzers.[1]

In summary, the available data suggest that costs of hydrogen produced by electrolysis are not competitive with other methods for hydrogen production. However, electrolyzers can relatively easily be decentralized and deliver hydrogen at the point of use. These factors speak in favor of hydrogen production by electrolysis. Nevertheless, it is important to consider the high investment costs for the construction of small-sized plants, which accounts for the high price of hydrogen from a small PEM electrolyzer, as reported in Ref. 20.

1.3.3 High-Temperature Processes

In addition to the more conventional routes for hydrogen production, the gas can also be synthesized *via* thermochemical cycles. These cycles were originally developed for use in connection with high-temperature nuclear power plants, but several of these processes are also suitable for use in connection with concentrating solar power plants.[16] In total, more than 200 such cycles are claimed.[27] However, commercial processes have not yet been implemented. For nuclear heat, the sulfur–iodine (S–I) and the copper–chlorine (Cu–Cl) cycles appear to be among the systems most intensively explored. The S–I cycle relies on the following three prototypical reactions, the temperature levels and specifics may vary to some extent:[27]

$$I_2 + SO_2 + 2\,H_2O \leftrightarrows 2\,HI + H_2SO_4 \qquad \text{at 393 K} \qquad (1.7)$$

$$H_2SO_4 \leftrightarrows H_2O + SO_2 + 1/2\,O_2 \qquad \text{at 1123 K} \qquad (1.8)$$

$$2\,HI \leftrightarrows I_2 + H_2 \qquad \text{at 723 K} \qquad (1.9)$$

The Cu–Cl cycle is again available in several embodiments, but also here prototypical reactions describe the system:[27]

$$2\,Cu_{(s)} + 2\,HCl_{(g)} \leftrightarrows 2\,CuCl_{(melt)} + H_{2(g)} \qquad \text{at 723 K} \qquad (1.10)$$

$$4\,CuCl_{(aq.)} \leftrightarrows 2\,Cu_{(s)} + 2\,CuCl_{2(aq.)} \qquad \text{electrolysis at 303–353 K} \qquad (1.11)$$

$$CuCl_{2(aq)} + n_f H_2O_{(l)} \leftrightarrows CuCl_2 \cdot n_h H_2O_{(s)} + (n_f - n_h)H_2O \qquad \text{drying} \qquad (1.12)$$

$$2\,CuCl_2 \cdot n_h H_2O_{(s)} + H_2O_{(g)} \leftrightarrows CuOCuCl_{2(s)} + 2\,HCl_{(g)} + n_h H_2O_{(g)}$$
$$\text{hydrolysis at 648 K} \qquad (1.13)$$

$$CuOCuCl_{2(s)} \leftrightarrows CuCl_{(melt)} + 1/2\,O_{2(g)} \qquad \text{decomposition (800 K)} \qquad (1.14)$$

In both cases, the net reaction is the splitting of water into hydrogen and oxygen.

Thermal efficiencies for these thermochemical cycles are reported to be as high as 43% for the Cu–Cl cycle and 52% for the S–I cycle.[27] In the various cost estimates, the Cu-Cl and the S–I cycle are rather similar, with the S–I cycle estimated at about 1.5 \$/kg H_2 (2003 basis) and about 2 \$/kg H_2 for the Cu–Cl cycle.[27] In another study, the cost was estimated depending on plant size, where the hydrogen price from a 200 t/d plant is given as 2 \$/kg H_2, while the price increases to 3.49 \$/kg H_2 for a 2 t/d plant.[30] The cost is also compared to that of steam gasification of methane, which is given as 2.67 \$/kg H_2 for a 10 t/d plant. This value, however, takes CO_2 costs into account, and relatively high costs for distribution; in turn, for the thermochemical cycle, only about half of the costs for distribution are assumed. Therefore, these results should be considered with caution.

The published data suggest that hydrogen can be produced by thermochemical cycles at comparable costs to those published for the steam-reforming processes. Solar heat from concentrating solar power plants could be used to drive such cycles. For solarthermal plants, the Cu–Cl or the S–I systems, however, are not the center of the attention. Research focus is on the Zn/ZnO system,[31] for which costs of 0.14–0.15 \$/kWh is estimated (4.66 to 5 \$/kg H_2), or based on solarthermal reduction of CeO_2 and reoxidation with water vapor, by which hydrogen is released.[32]

1.4 Hydrogen Infrastructure and Storage

Regarding the utilization of hydrogen in upgrading processes performed on biomass, the hydrogen should be supplied at the biorefinery site, or at least at a hydrogenation plant, where *e.g.* pyrolysis oil would be hydrogenated to stabilize it prior to its transportation and storage. Ideally, hydrogen is provided onsite, reducing the considerable costs associated with its transportation, and thus, being conducive to the economics of the entire chain of process for biomass conversion.

In general, hydrogen can be transported and stored either under high pressure or in liquefied form.[33] In addition, hydrogen can also be transported *via* pipelines. Other means are possible, but have not reached the same level of maturity and shall not be discussed here; a survey is given in Ref. 34. For large amounts of hydrogen, the transport and storage of liquefied hydrogen is the best approach. Usually, the higher the amount transported or stored, the lower the problems with reducing boil-off losses are, since with higher amounts the surface-to-volume ratio of the cryogenic vessels becomes more favorable. However, liquefaction typically entails a high loss of energy. It is estimated that at up to 30% of the energy content of the hydrogen is lost upon H_2 liquefaction.[35] Compression is less energy intensive, although also here appreciable losses are encountered. Typical pressures for transport are 20 MPa; for onboard storage for fuel cell powered vehicles, 70 MPa are the state-of-the-art. For pressures of 70 MPa, the energy requirement for compression is about 15% of the energy content of the hydrogen.[34] Compressed hydrogen is transported in steel cylinders or bundles of cylinders. In fuel-cell-powered cars, the pressure vessel is a fiber-reinforced polymer composite. Since the force scales with the area of the container, the scaling behavior is the opposite of that for the liquefied hydrogen: the larger the container, the more serious the problems. Thus, the strength of the container material has to be markedly increased.

Geological formations, often salt domes, are the best storage option for large amounts of hydrogen. Such domes are in operation at several sites with a high concentration of refineries and chemical production plants. In Texas, there are two large salt domes (the Clemens dome and the Moss Bluff dome) connected by a pipeline system in a refinery region close to Houston. The Clemens dome has a volume of 580 000 m^3, and a storage capacity of 5400 t. Leak rates as low as 0.01% per year are reported.[36] Three caverns of 70 000 m^3 each are operated in Teesside, UK. Above ground, pressurized hydrogen is the best option for short times and small amounts; for larger volumes, liquefied hydrogen is the preferred alternative.[37]

Considering the demand of hydrogen for the conversion of biomass, several conclusions with respect to the production and infrastructure technology can be drawn. Nonetheless, a detailed analysis is certainly required for each potential site. In order to avoid the transportation cost, onsite production is favored. In effect, the cheapest alternative seems to be biomass reforming, since the infrastructure for transporting and handling of biomass is available already for the biomass itself, since the cost of hydrogen from biomass reforming does not appear to be substantially higher than natural-gas reforming. Moreover, carbon-containing waste streams from biomass processing may be used for gasification, which could potentially further decrease the costs. Should fossil carbon resources be available at the site where the hydrogen is needed, reforming or co-reforming of fossil feedstocks could also be an attractive option. If this is not the case, electrolysis might be suitable, even if the production cost is still relatively high, because electrolytic processes can respond to the demand for hydrogen highly dynamically. For electrolysis, the production cost is at least partially offset by the fact that the expensive transportation of hydrogen is not required. In addition, hydrogen could be supplied directly at the pressure level needed. Thermochemical cycles using renewable energy do not seem to be suitable to provide hydrogen for biomass conversion, since solarthermal plants are typically located in arid regions with a high fraction of direct irradiation and little rainfall. Logically, these regions are not conducive to energy crops, and so high costs for the hydrogen transportation would result.

1.5 Hydrogen Use in Different Routes of Biomass Transformation

Depending on the kinds of feedstocks and the initial conversion processes, the details of the processes used for upgrading of the initial products change. Gasification of biomass typically results in a syngas with a H_2/CO

ratio of approximately one. Such a syngas is highly suitable for the direct one-step production of dimethylether (DME),[38] an interesting fuel molecule for compression ignition engines.[39] However, any other syngas conversion process requires the addition of hydrogen to the biomass-derived syngas. Methanol production needs a H_2/CO ratio of two, the Fischer–Tropsch process requires essentially the same ratio, and methanation even requires a ratio of three. Thus, either gasification needs to be coupled with a water-gas-shift stage in order to increase the hydrogen content of the syngas, or hydrogen has to be supplied externally. Since hydrogen production from biomass seems to be almost competitive with hydrogen production from fossil sources, changing the CO/H_2 ratio by the shift reaction is probably the economic option, since the gasification unit is available in any case.

The situation could be somewhat different in two-step thermochemical processes, where first a pyrolysis oil is generated in a pyrolysis step, which is later converted into syngas in a centralized plant. The pyrolysis oil has a number of limitations, including poor chemical stability, and thus it typically needs to be subjected to upgrading treatments.[40] The upgrading serves the purpose to reduce the oxygen content and increase the H/C ratio. Hydrodeoxygenation is an approach to accomplish this. However, the removal of the high oxygen content of crude pyrolysis oil should consume very high amounts of hydrogen. Combination with other processes, which stabilize the pyrolysis oil, is thus advisable. Nevertheless, facilities for production of pyrolysis oil will probably always require some hydrogen, and the question of how this is best produced must be answered.

The production of hydrogen from biomass appears to be less favorable than in the case of direct biomass gasification, since the gasifier would need to be built up separately. Separation of decentralized generation of pyrolysis oil and centralized gasification and downstream processes is a key advantage of pyrolysis/processing routes starting with biomass. Integration of other complex equipment, such as a gasifier for hydrogen production, at the pyrolysis site may jeopardize the attractiveness of the overall scheme. Of some advantage is the fact that the full infrastructure for biomass handling (transportation, storage, preprocessing) is available at the pyrolysis plant. Whether this makes the gasification onsite more attractive is questionable, though. Overall, the viability of hydrogen generation from biomass at the pyrolysis site is most probably a question of scale on which the hydrogen would be needed. For low demand, one would purchase hydrogen at market price; if large amounts of hydrogen are required, a separate gasifier unit might become attractive.

Yet another situation is encountered, if biomass enters into the bio-refinery stream *via* the hydrolytic pathway.[41] In addition, the oxygen content of the primary products (sugars from cellulose and hemicellulose, aromatic alcohols from the lignin fraction) has to be reduced and the C/H ratio needs to be increased in order to obtain valuable chemical products or fuels. Most pathways described in the literature starting from sugars use externally supplied hydrogen to optimize C/H and C/O ratios. This holds, for instance, for the pathways leading to furan derivatives (Figure 1.3).[42]

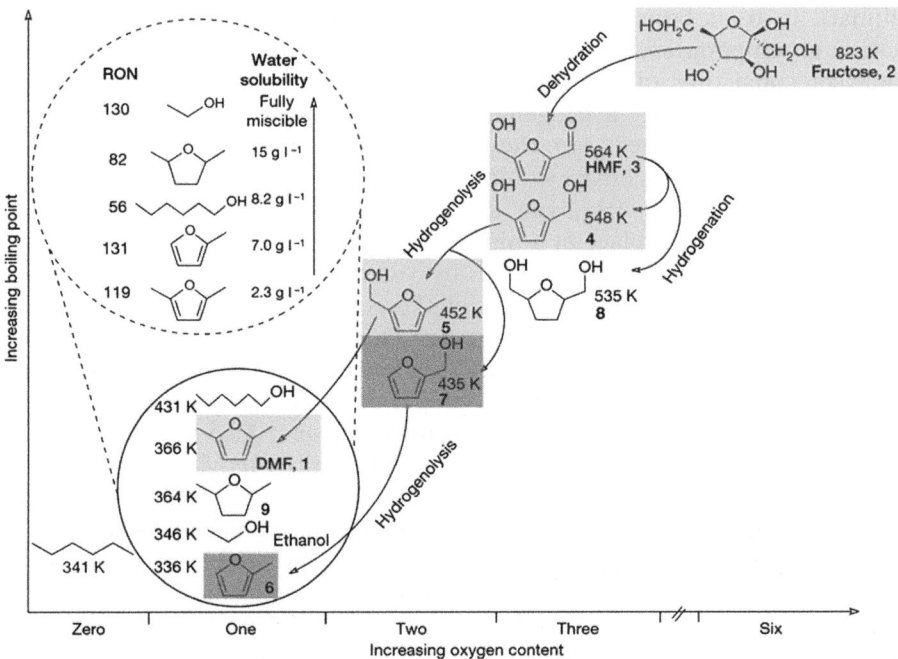

Figure 1.3 Boiling points of products derived from hexoses by successive removal of oxygen atoms and transformation pathways from hexose to corresponding compound.
Reproduced with permission from Ref. 42.

Glucose is first isomerized to fructose and then by three dehydration reactions to 5-hydroxymethylfurfural, which is in turn hydrogenated to 2,5-dimethylfuran (DMF), a compound with excellent fuel properties.[43] This reaction is ideal, since it requires the minimum amount of hydrogen – however, this is still three moles of hydrogen per mole of DMF; further hydrogenation of the ring and/or hydrogenolysis of the methyl groups should be avoided. Thus, there are highly attractive pathways in which the hydrogen is generated *in situ* by partial catalytic aqueous phase reforming of the sugars. The formed hydrogen is used in the conversion of the sugars into monofunctional alcohols, ketones, or carboxylic acids, which undergo further conversion rendering alkanes.[44] If such processes could be sufficiently well controlled on a commercial scale, they would most probably be advantageous over the hydrogen production in a dedicated separate plant or externally sourced. However, the partial catalytic aqueous phase reforming of the sugars needs to be assessed in detail for specific process layouts.

Finally, hydrogen is also required, if the lignin fraction is to be upgraded to liquid compounds – nowadays the lignin is mostly burnt in order to generate process heat or electrical energy in biomass processing units. Due to the complexity of lignin and the relative inertness of phenols against dehydroxylation, conversion of the aromatic alcohols into aromatic or aliphatic hydrocarbons is rather difficult. This reaction is possible, for instance, at temperatures of 473 K at hydrogen pressure of 5 MPa over mixed catalysts consisting of H-ZSM-5 and Pd/C.[45] This method leads to deoxygenation and ring hydrogenation, so that substantial amounts of hydrogen are consumed. For the generation of this hydrogen, probably gasification of biomass onsite would be the economic option, since it is close to competitive in any case, and the facilities for biomass handling and storage are already in place at a biorefinery site.

An alternative to hydrogenation by externally supplied hydrogen is transfer hydrogenation from an alcohol or olefin to yield deoxygenated products. This approach has been proven successfully in a recent study by Wang and Rinaldi.[46] In this study, a mixture of Raney nickel and H-beta-zeolite was used as catalysts, and 2-propanol as an H-donor, for the low-severity hydrodeoxygenation of phenol into benzene. Also with real feeds, such as lignin and pyrolysis oil, high hydrogenation activity and good selectivity to arenes were observed in this study, which reduces hydrogen consumption, as the ring is not hydrogenated. This transfer hydrogenation would be especially favorable, if the aliphatic alcohols – such as cyclohexanols obtained from full saturation of lignin – are used as the H-donors. This process option does not only overcome the need for hydrogenation of acetone and its recycle in the process, but also works as an effective H-recovery process. Accordingly, the molecular hydrogen stored in the cyclohexanols could be used for the conversion of the phenolics into arenes through catalytic transfer hydrogenation performed at a second-stage process.

1.6 Conclusion

Due to the high O/C and low H/C ratio of biomass, as compared to traditional fuels and chemicals, large amounts of hydrogen are required in biorefinery schemes, regardless of whether the thermochemical or hydrolytic pathway is chosen for the conversion of lignocellulose biomass. There are several routes available for hydrogen production. Currently, the available data from technoeconomic analyses suggest that onsite hydrogen production is more advantageous over externally sourced hydrogen due to savings in H_2 transportation, but this economy will certainly depend on the amounts used at a site. Of the different hydrogen production routes, gasification methods seem to be economically advantageous, regardless of the feedstock used (*i.e.* gas, coal and biomass). Since a site for biomass conversion, irrespective of the detailed processes implemented, will need all the installations for transportation and storage of biomass in any case, there are most probably synergies between biomass gasification for hydrogen production and the other processes operated at a specific site. Nevertheless, in order to find the most economical solution, a detailed analysis will be necessary for each site in question. This analysis is certainly complicated by the fact that the boundary conditions in this field (*e.g.* CO_2 emissions cost, feedstock prices, availability of nonexpensive electricity, electrolyzer costs, and legal boundary conditions, to name only a few) are constantly changing.

Acknowledgements

Work in this field is funded by the Cluster of Excellence "Tailor-made fuels from biomass", in addition to the basic funding provided by the MPG.

References

1. G. Saur, *Wind-to-Hydrogen Project: Electrolyzer Capital Cost Study*, Technical Report, National Renewable Energy Laboratory, NREL/TP-550-44103, Golden, CO 2008.
2. R. Rinaldi and F. Schüth, *Energy Environ. Sci.*, 2009, **2**, 610.
3. B. Suresh, R. Gubler, Y. Yamaguchi and X. He, Hydrogen, IHS Chemical Economics Handbook, 2013, http://chemical.ihs.com/CEH/Public/Reports/743.5000, accessed on September 9, 2014.
4. P. Häussinger, R. Lohmüller and A. M. Watson, *Hydrogen, in: Ullmann's Encyclopedia of Industrial Chemistry*, Wiley-VCH, Weinheim, Online Edition, 2012, Vol. 18, 354.
5. M. A. Pena, J. P. Gómez and J. L. G. Fierro, *Appl. Catal. A – Gen.*, 1996, **144**, 7.
6. H. Hiller, R. Reimert and H.-M. Stönner, *Gas Production, 1. Introduction, in: Ullmann's Encyclopedia of Industrial Chemistry*, Wiley-VCH, Weinheim, Online Edition, 2012, Vol. 16, 407.
7. P. Häussinger, R. Lohmüller and A. M. Watson, *Hydrogen, in: Ullmann's Encyclopedia of Industrial Chemistry*, Wiley-VCH, Weinheim, Online Edition, 2012, Vol. 18, 267.
8. P. Häussinger, R. Lohmüller and A. M. Watson, *Hydrogen, in: Ullmann's Encyclopedia of Industrial Chemistry*, Wiley-VCH, Weinheim, Online Edition, 2012, Vol. 18, 269.
9. D. A. Hickman and L. D. Schmidt, *Science*, 1993, **259**, 343.
10. R. Guerrero Lemos and J. M. Martinez Duarte, *Int. J. Hydrogen Energy*, 2010, **35**, 3929.
11. S. Sarkar and A. Kumar, *Trans. ASABE*, 2009, **52**, 519.
12. J. Schulze and H. Gaensslen, *Chem. Ind.*, 1984, **36**, 135 and *ibid.*, 202.
13. C. Mansilla, J. Louyrette, S. Albou, C. Bourasseau and S. Dautremont, *Energy*, 2013, **55**, 996.
14. NYMEX Herny Hub future for April 2014, as of March 24th 2014.
15. P. L. Spath and M. K. Mann, *Life Cycle Assessment of Hydrogen Production via Natural Gas Steam Reforming*, NREL/TP-570-27637 (2001).
16. C. Acar and I. Dincer, *Int. J. Hydrogen Energy*, 2014, **39**, 1.
17. B. L. Salvi, K. A. Subramanian and N. L. Panwar, *Renew. Sustain. Energy Rev.*, 2013, **25**, 404.
18. N. Dahmen and E. Dinjus, *Chem. Ing. Tech.*, 2010, **82**, 1147.
19. S. Sarkar and A. Kumar, *Energy*, 2010, **35**, 582.

20. T. Smolinka, M. Günther and J. Garche, NOW-Studie (Kurzfassung) "Stand und Entwicklungspotenzial der Wasserelektrolyse zur Herstellung von Wasserstoff aus regenerativen Energien", Freiburg 2011.
21. D. Stolten (ed.) *Hydrogen and Fuel Cells*, Wiley-VCH, Weinheim 2010.
22. K. Zeng and D. Zhang, *Prog. Energy Combust.*, 2010, **36**, 307.
23. P. Häussinger, R. Lohmüller and A. M. Watson, *Hydrogen, in: Ullmann's Encyclopedia of Industrial Chemistry*, Wiley-VCH, Weinheim, Online Edition, 2012, Vol. 18, 274.
24. D. Stolten and D. Krieg, *Alkaline Electrolysis – Introduction and Overview in: Hydrogen and Fuel Cells*, D. Stolten (ed.), Wiley-VCH, Weinheim 2010, pp. 243–268.
25. M. Carmo, D. L. Fritz, J. Mergel and D. Stolten, *Int. J. Hydrogen Energy*, 2013, **38**, 4901.
26. http://www.siemens.com/innovation/apps/pof_microsite/_pof-spring-2011/_html_en/electrolysis.html, accessed on April 22nd, 2014.
27. Z. L. Wang, G. F. Naterer, K. F. Gabriel, R. Gravelsins and V. N. Daggupati, *Int. J. Hydrogen Energy*, 2010, **35**, 4820.
28. G. Saur and C. Ainscough, *U.S. Geographic Analysis of the Cost of Hydrogen from Electrolysis*, Technical Report, National Renewable Energy Laboratory, NREL/TP-5600-5264, Golden, CO 2011.
29. http://www.nrel.gov/hydrogen/production_cost_analysis.html, accessed April 27th 2014.
30. G. F. Naterer, M. Fowler, J. Cotton and K. Gabriel, *Int. J. Hydrogen Energy*, 2008, **33**, 6849.
31. A. Steinfeld, *Int. J. Hydrogen Energy*, 2002, **27**, 611.
32. W. C. Chue, C. Falter, M. Abbott, D. Scipio, P. Furler, S. M. Haile and A. Steinfeld, *Science*, 2010, **330**, 1797.
33. P. Häussinger, R. Lohmüller and A. M. Watson, *Hydrogen, in: Ullmann's Encyclopedia of Industrial Chemistry*, Wiley-VCH, Weinheim, Online Edition, 2012, Vol. 18, 341–352.
34. U. Eberle, M. Felderhoff and F. Schüth, *Angew.Chem. Int. Ed.*, 2009, **48**, 6608.
35. J. Wolf, *MRS Bull.*, 2002, **27**, 684.
36. F. Crotogino, S. Donadei, U. Bünger and H. Landinger, *Presentation at World Hydrogen Energy Conference*, Essen, May 20th 2010.
37. S. Dunn, *J. Int. Hydrogen Energy*, 2002, **27**, 235.
38. H. Jiang, H. Bongard, W. Schmidt and F. Schüth, *Micropor. Mesopor. Mater.*, 2012, **164**, 3.
39. B. L. Salvi, K. A. Subramanian and N. L. Panwar, *Renew. Sustain. Energy Rev.*, 2013, **25**, 404.
40. T. P. Vispute, H. Zhang, A. Sanna, R. Xiao and G. W. Huber, *Science*, 2010, **330**, 1222.
41. R. Rinaldi and F. Schüth, *ChemSusChem*, 2009, **2**, 1096.
42. Y. Roman-Leshkov, C. J. Barrett, Z. Y. Liu and J. A. Dumesic, *Nature*, 2007, **447**, 982.

43. S. H. Zhong, R. Daniel, H. M. Xu, H. Zhang, D. Turner, M. L. Wyszynksi and P. Richards, *Energy Fuel*, 2010, **24**, 2891.
44. E. L. Kunkes, D. A. Simonetti, R. M. West, J. C. Serrano-Ruiz, C. A. Gärtner and J. A. Dumesic, *Science*, 2008, **322**, 417.
45. C. Zhao and J. A. Lercher, *ChemCatChem*, 2012, **4**, 64.
46. X. Wang and R. Rinaldi, *Angew. Chem. Int. Ed.*, 2013, **52**, 11499.

CHAPTER 2

General Reaction Mechanisms in Hydrogenation and Hydrogenolysis for Biorefining

NING LI, WENTAO WANG, MINGYUAN ZHENG AND
TAO ZHANG*

State Key Laboratory of Catalysis, Dalian Institute of Chemical Physics,
Chinese Academy of Sciences, Dalian 116023, China
*Email: taozhang@dicp.ac.cn

2.1 Introduction

Among various processes for biorefining, hydrogenation and hydrogenolysis are of particular significance for the biomass upgrading, and constitute a great challenge. Generally, hydrogenation is performed on biomass platform molecules in order to saturate C=C and C=O bonds, whereas hydrogenolysis describes an overall chemical transformation in which a carbon–carbon or carbon–heteroatom bond is broken by insertion of hydrogen atoms. In effect, the biomass upgrading requires a substantial amount of hydrogen. As discussed in Chapter 1, the utilization of hydrogen as an energy carrier of the future is associated with substantial losses in the round trip from electricity–hydrogen–electricity. Therefore, the use of "renewable" hydrogen – not for the reconversion to electricity, but rather as a feedstock in biorefineries – appears to be a very promising alternative. Over the last decade, significant progress in understanding the mechanistic aspects of hydrogenation and hydrogenolysis of biomass-derived molecules has been achieved. The aim of

RSC Energy and Environment Series No. 13
Catalytic Hydrogenation for Biomass Valorization
Edited by Roberto Rinaldi
© The Royal Society of Chemistry 2015
Published by the Royal Society of Chemistry, www.rsc.org

this chapter is to cover the various mechanistic aspects of hydrogenation and hydrogenolysis present in the current research. Better understanding of the mechanistic aspects involved in the upgrading of biogenic molecules is key for improving efficiency (*i.e.* conversion and selectivity) in addition to the hydrogen-atom economy of the processes.

2.2 Hydrogenation of α,β-Unsaturated Carbonyl Compounds

Biomass-derived α,β-unsaturated aldehydes and ketones (such as furfural, cinnamaldehyde, citral, crotonaldehyde, mesityl oxide and isophorone) are important chemical intermediates for the synthesis of fine chemicals. The unsaturated alcohols obtained from the selective hydrogenation of α,β-unsaturated carbonyl compounds can be used in the production of pharmaceuticals, cosmetics, and food, while the saturated aldehydes or ketones, from the selective hydrogenation of the C=C bond occurring in α,β-unsaturated carbonyl compounds, are also very important in the manufacture of solvents, perfumes and medicines. Thermodynamically, C=C bonds should be easier to be hydrogenated than C=O bonds. However, due to the conjugation in α,β-unsaturated carbonyl compounds, the C=O bonds are sometimes easier to undergo hydrogenation compared to the C=C bonds. In recent years, noble metals (*e.g.* Pt, Pd, Ir, Rh, Ag, Au), base metals (*e.g.* Cu, Ni, Co, Sn) and alloy catalysts have been extensively applied to the hydrogenation of α,β-unsaturated carbonyl compounds.[1–5]

Among noble-metal catalysts, Pd is the most selective for the C=C bond hydrogenation.[1,6] In turn, Os and Ir are considered the most selective catalysts for the hydrogenation of the carbonyl group.[1,2] To investigate the intrinsic reasons for the different performances, Delbecq and Sautet[7] studied the adsorption geometries of various biomass-derived α,β-unsaturated aldehydes on the surfaces of Pt and Pd by semiempirical extended Hückel calculations.

Six adsorption modes were proposed (Scheme 2.1). It was found that a di-σ mode was preferred on Pt(111); a $\pi_{C=C}$ mode was preferred on Pt(110) and on the steps; and a planar η^4 mode was preferred on Pt(100) and Pd(111). These preferred adsorption modes are proposed to account for the behaviors of Pt and Pd catalysts in the hydrogenation of the α,β-unsaturated carbonyl compounds. According to the calculations by Delbecq and Sautet,[7] the C=C bond is predicted as more difficult to be adsorbed onto metals with larger radial expansion of the *d* orbitals or the broader *d*-band width. Considering that Os and Ir have larger *d*-band widths compared to Pt,[1,8] the adsorption *via* C=O should then predominate on Os or Ir catalyst, explaining the high chemoselectivity of the reduction towards unsaturated alcohols.

Zhang and coworkers found that nickel phosphide is also chemoselective for the C=C bond hydrogenation,[3] while Ag is more selective than Au and Pt for the C=O bond hydrogenation.[4] According to density functional theory

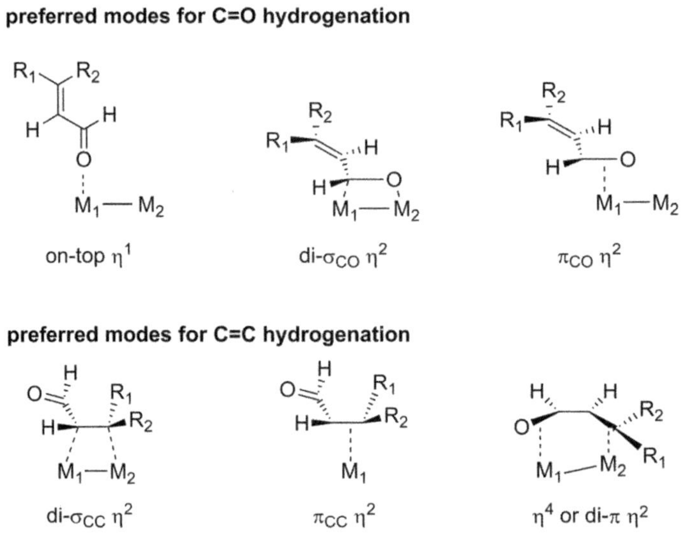

preferred modes for C=O hydrogenation

on-top η^1 di-σ_{CO} η^2 π_{CO} η^2

preferred modes for C=C hydrogenation

di-σ_{CC} η^2 π_{CC} η^2 η^4 or di-π η^2

Scheme 2.1   Adsorption modes of α,β-unsaturated aldehydes on metal catalysts. Adapted from ref. 7.

(DFT) calculation on selected M_{19} clusters and the M(111) surfaces (M: Ag, Au and Pt), the di-σ_{CO} η_2 adsorption is predicted to be the most favorable for Ag, thus contributing to the C=O bond activation.

The molecular structure of α,β-unsaturated carbonyl compounds also has a considerable influence on their adsorption on metal catalysts. Due to steric hindrance, the adsorption of the carbonyl group of an α,β-unsaturated ketone on the catalyst surface is more difficult than that of an α,β-unsaturated aldehyde. Therefore, the selective hydrogenation of a carbonyl group in α,β-unsaturated ketone is challenging.[9,10] Analogously, better selectivity to C=O hydrogenation is observed for substrates having high substitution on the C=C bond. The acrolein adsorption on Pt(111) surface is preferred by the C=C bond, but this mode becomes unstable when a methyl group is present at the terminal position of the C=C bond.[11,12] Therefore, the C=O mode adsorption becomes the most favorable one. This phenomena can be explained by the weakened C=C bond adsorption by "steric-like" Pauli repulsion between the methyl groups on terminal carbon and the *d* electrons of the metal atoms on the catalyst surface.

The metal particle size also exerts a very important effect on the hydrogenation of α,β-unsaturated carbonyl compounds. Larger particles can pose problems for the C=C bond adsorption. Therefore, larger particles can lead to higher selectivity to C=O hydrogenation. Gallezot and coworkers[13] studied the hydrogenation of cinnamaldehyde in the presence of Rh and Pt catalysts. At 50% cinnamaldehyde conversion, they found that the selectivity to cinnamyl alcohol improved (from 18 to 42%) with the increase in average size of Rh particles (from 2.5 to 7 nm) in a Rh/graphite catalyst. Analogously, the selectivity to cinnamyl alcohol improved (from 83 to 98%) with the increase in the average size of Pt particles (from 1.3 to 5 nm) in a Pt/graphite catalyst.

Alternatively, the unsaturated alcohol selectivity can be improved by increasing the interaction between carbonyl group and catalyst surface. For example, acid sites can activate the carbonyl group, thus decreasing the energy level of the acceptor molecular orbital π^*_{CO}. In a recent work by Luo and coworkers,[14] crotonaldehyde was hydrogenated to crotyl alcohol in the presence of an Ir/ZrO$_2$ catalyst. The formation of crotyl alcohol is claimed to take place on the interfacial region and to involve both Ir and Lewis acid sites.

Jacobs and coworkers[15] studied the hydrogenation of α-ionone and β-ionone over Ir catalysts loaded on different supports, such as β-zeolite (H-β), sodium-exchanged β-zeolite (Na-β), active carbon, CaCO$_3$, Al$_2$O$_3$ and TiO$_2$. Among these catalysts, Ir/H-β exhibited the highest selectivity to unsaturated alcohols. The selectivity to unsaturated alcohols with Ir/Na-β was lower than that observed with Ir/H-β. This result indicated that Brønsted acid is also important for the selective hydrogenation of the carbonyl group.

The chemoselective hydrogenation of the carbonyl group in an α,β-unsaturated carbonyl compound can be improved by the presence of a second, more electropositive metal, *e.g.* iron[16] or tin.[17–20] The utilization of oxide supports (*e.g.* CeO$_2$,[21–23] TiO$_2$,[24] ZnO,[21,25] Fe$_3$O$_4$,[10] Ga$_2$O$_3$,[26] or SnO$_2$[27]) which can interact with the metal catalysts, contributes to the hydrogenation of carbonyl groups. The improved selectivity can be explained by electronic effects, the formation of an alloy, or the strong metal–support interactions (SMSI).[28]

In a recent work by the Tomishige group,[29] it was found that some biomass-derived unsaturated aldehydes can be selectively hydrogenated with an Ir–ReO$_x$/SiO$_2$ catalyst. An evident synergistic effect was observed between the closely contacted Ir and ReO$_x$ species. According to the experimental results, they suggested that the partially reduced ReO$_x$ oxide clusters promote the adsorption of unsaturated aldehyde and the heterolytic dissociation of hydrogen (into H$^+$ and H$^-$) on the Ir metal surface. Both effects led to the high activity and selectivity of the studied Ir–ReO$_x$/SiO$_2$ catalyst.

2.3 Conversion of Organic Acids and Esters

Conversion of biomass-derived organic acids and esters into alcohols, alkanes or (partially) saturated triglycerides is one of the most important reactions in biomass conversion for several reasons:

1. Compared to the direct production of ethanol from fermentation, hydrogenation of organic acids produced in fermentation could result in a 50% increase in theoretical yield of ethanol (glucose fermentation to ethanol: 2 mol of ethanol/mol of six-carbon sugar; glucose conversion by hydrogenolysis: 3 mol of ethanol/mol of six-carbon sugar).[30]

2. Many carboxylic acids obtained from biological or chemical treatment of biomass can be converted into value-added chemicals (*e.g.* polyols and fatty alcohols).[31–36]

3. Pyrolysis oils show high contents of organic acids, which make them corrosive. Hence, pyrolysis oils cannot be directly used as fuels without hydrotreating (as discussed in Chapter 7). Moreover, the hydrogenation of organic acids is regarded as one of the slowest reactions in the hydrotreatment of pyrolysis oils.[37]

4. The direct conversion of vegetable oils or animal fats into linear or branched alkanes is the core technology for the production of the second generation of biodiesel,[38–40] as presented in more detail in Chapters 8 and 9.

5. Finally, hydrogenated vegetable oils and fats are extremely important for the food and chemical industries. The selectivity of the catalytic process determines the chemical and physical properties of the hydrogenated products, defining the application potential and therefore the product value, as discussed indepth in Chapter 10.

Huber and coworkers[41] investigated the aqueous phase hydrogenation of acetic acid in the presence of a series of transition-metal catalysts. Among them, Ru/C exhibited the highest activity and selectivity to ethanol. The turnover frequency of acetic acid over various catalysts increased as follows: $Cu < Ni < Ir \approx Pd < Pt \approx Rh < Ru$. Ethyl acetate, acetaldehyde, alkanes, and CO_2 were also identified as the byproducts. DFT calculations were performed to shed light on the factors accounting for the activity and selectivity difference of various catalysts. The DFT predictions suggest that the conversion should initiate with the C–O bond cleavage in acetic acid or acetate. In sequence, the acetyl intermediate is hydrogenated to ethanol. The cleavage of C–O bond was found to be the rate-determining step on the studied catalysts.[41] Exceptionally, in the presence of a Cu catalyst, the formation of a hemiacetal intermediate is, however, the rate-determining step in the hydrogenation of acetic acid.

Zhu and coworkers[42] studied the hydrogenation of carboxylic acids over Ru/ZrO_2, Ru/C and Ru/Al_2O_3 catalysts. Based on the kinetic data and diffuse reflection infrared Fourier transform (DRIFT) spectra, it was found that acidic support in combination with Ru metal could favor the C=O hydrogenation of carboxylic group. In turn, decarbonylation is favored at high temperatures (463–483 K) and high metal loading (4–6 wt%). Furthermore, alkanes could also be produced by the deep hydrogenolysis of alcohols. According to these observations, the reaction mechanism was proposed (Scheme 2.2).

Hardacre and coworkers[33] explored the hydrogenation of stearic acid with Pt and Pt–Re catalysts supported on SiO_2, Al_2O_3, CeO_2, ZrO_2, $CeZrO_4$ or TiO_2. Relatively higher stearic acid conversion was observed with the Pt/TiO_2 catalyst. The modification of Pt/TiO_2 with Re increased its catalytic activity, but decreased the selectivity to stearic alcohol. In contrast, the TiO_2

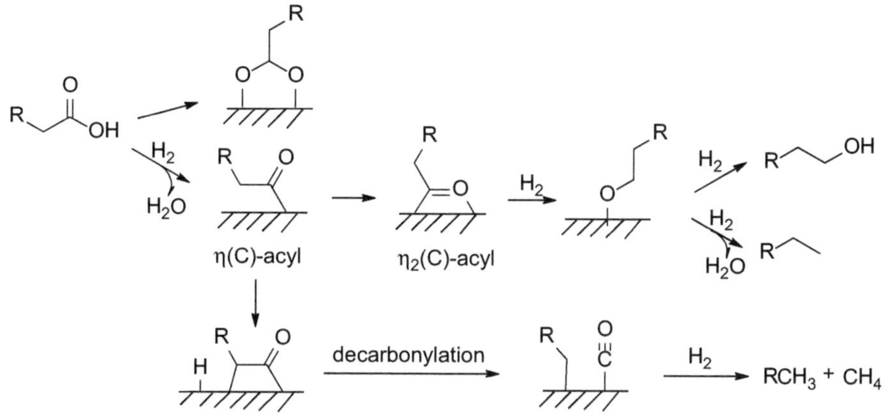

Scheme 2.2 Reaction mechanism for the hydrogenation of carboxylic acid over Ru catalyst.
Adapted from ref. 42.

support and Re/TiO$_2$ catalyst were inactive for the reaction. To elucidate the reaction pathway, they compared the catalytic performance of Pt–Re/TiO$_2$ under a hydrogen or nitrogen atmosphere. When stearic acid was used as a substrate, the formation of *n*-heptadecane was slower under a nitrogen atmosphere than that under a hydrogen atmosphere. It was also noticed that the Pt–Re/TiO$_2$ catalyst was very active for the conversion of stearic alcohol under a hydrogen atmosphere, with *n*-heptadecane as the only product. In contrast, no conversion of stearic alcohol was detected over Pt–Re/TiO$_2$ catalyst under a nitrogen atmosphere. The higher activity of TiO$_2$-supported Pt catalyst was explained by the interaction between the carbonyl oxygen in the stearic acid and the support metal ion/oxygen vacancy created by low-temperature reduction on the reducible oxide (Route 1 in Scheme 2.3).

Such an interaction of the carbonyl group with the catalyst surface weakens the C=O and promotes its hydrogenation, and the following C–O bond cleavage. The promotion by Re was rationalized by the enhanced oxyphilicity of the surface associated by Re cations, which interact with the carbonyl group (Route 2 in Scheme 2.3). The interaction of the adsorbed carboxylic acid molecule with Re on the surface of catalyst promotes the formation of alkanes and CO$_2$ (Route 3 in Scheme 2.3).

Tomishige and coworkers[34] reported a rather different trend on the selective hydrogenation of fatty acids to fatty alcohols. They found that ReO$_x$/SiO$_2$ catalyzes the reduction of stearic acid to stearic alcohol. The modification of ReO$_x$/SiO$_2$ with noble metals showed that the best promotion effect occurring in the presence of Pd. The products of the consecutive hydrogenolysis of stearic alcohol and the decarboxylation of stearic acid were *n*-octadecane and *n*-heptadecane. From the noticeably high selectivity to *n*-octadecane, the main side reaction is inferred to be the consecutive

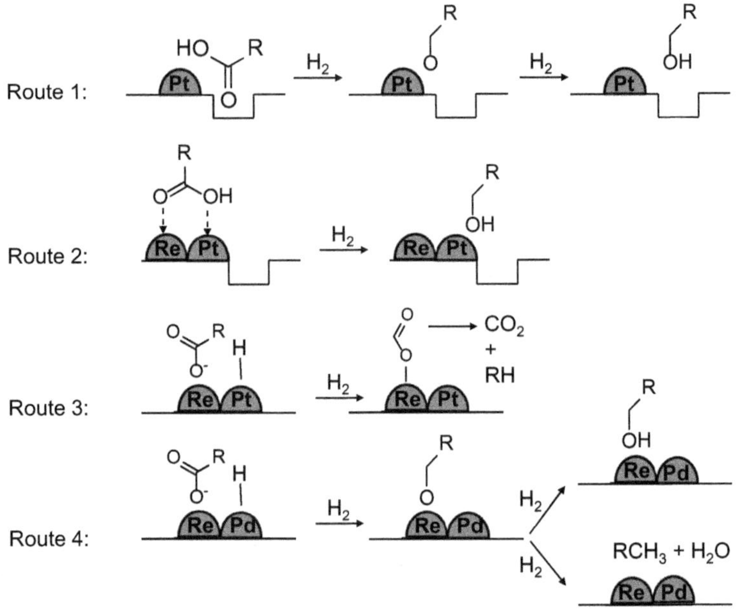

Scheme 2.3 Reaction pathways for the hydrogenation of carboxylic acid over Re-promoted Pt and Pd catalysts. Adapted from refs. 33 and 34.

hydrogenolysis of stearic alcohol instead of the decarboxylation of stearic acid. In contrast, as reported by Rooney and coworkers,[33] the experiments performed in the presence of Pt–Re/TiO$_2$ catalyst showed a high selectivity to *n*-heptadecane, indicating that the consecutive decarboxylation of stearic acid should be the main side reaction.

Tomishige and coworkers[34] compared the HDO of stearic acid, 1-hexadecanol and their mixture. It was found that the adsorption of the carboxylic acid is more favorable than that of the alcohol over ReO$_x$–Pd/SiO$_2$ catalyst.[34] Therefore, *n*-octadecane should be formed by the hydrogenolysis of stearic alcohol, which occurs before it is desorbed from the catalyst (Route 4 in Scheme 2.3). Tomishige and coworkers[34] also investigated the role of Pd in the reaction. Under the same conditions, it was found that Pd/SiO$_2$ is inactive for the hydrogenation of stearic acid. This result indicates that ReO$_x$ species is the active site in ReO$_x$-Pd/SiO$_2$ catalyst.[34] From the results obtained from temperature-programmed reduction (TPR), the addition of Pd enhances the reduction of ReO$_x$ species supported on SiO$_2$. Therefore, the promotion effect of Pd on ReO$_x$/SiO$_2$ catalyst was attributed to the activation of hydrogen species, while the higher selectivity of ReO$_x$–Pd/SiO$_2$ catalyst was rationalized by the lower hydrogenolysis activity of Pd.[34]

Pinel, Besson, and coworkers studied the hydrogenation of succinic acid and levulinic acid with noble-metal catalysts.[32,43,44] Over Pt, Pd or Ru catalysts, γ-butyrolactone or γ-valerolactone was identified as the major product, respectively. After modification of these catalysts with Re, the main product shifted to 1,4-butanediol or 1,4-pentanediol. This may be explained by the higher activities of bimetallic catalysts for the hydrogenation of the carboxylic group.

In the recent works by Li and coworkers,[35,36] the modification of Ir/SiO$_2$ and Rh/SiO$_2$ catalysts with Mo significantly improved their activities in the hydrogenation of levulinic acid, which was explained by the synergism effects of Ir or Rh and the partial reduced MoO$_x$ species. In the hydrogenation of levulinic acid with a Rh-MoO$_x$/SiO$_2$ catalyst, γ-valerolactone was detected as the intermediate between levulinic acid and 1,4-pentanediol.

In the previous work by Corma and coworkers[40] and Lercher and coworkers,[38–40] the reaction mechanism for the hydrogenolysis of vegetable and microalgae oils to C$_{15}$–C$_{18}$ alkanes was explored. The transformation of vegetable and microalgae oils to C$_{15}$–C$_{18}$ alkanes is a four-step reaction. In the first step, the C=C bond in the alkyl chain is hydrogenated on the metal sites, generating a saturated triglyceride. In the second step, the C–O bond in the saturated triglyceride is selectively hydrogenolyzed to fatty acids and propane. In the third step (the rate-determining one), the carboxylic group of the fatty acid is hydrogenated to the corresponding aldehyde. This step can be catalyzed either by metal itself or by the synergism of a metal with reducible oxide supports (*via* a ketene intermediate). As the final step, the aldehyde is hydrogenated to alcohol, which can be further dehydrated and hydrogenated, rendering alkane products.

Bifunctional catalysts, that is, metal supported on acidic solids (*e.g.* Ni/H-β) are the preferred catalyst type for the hydrogenolysis of vegetable and microalgae oils to linear alkanes.[38–40] Notably, the alkanes can also be further hydroisomerized or hydrocracked on the acidic sites. As another option, the aldehyde generated in the third step can also undergo decarbonylation on the metal sites and produce alkanes and carbon monoxide. Under the process conditions, CO and H$_2$ may react, which lead to the generation of methane and water. This overall pathway is favored on metal catalysts supported on oxides (*e.g.* Ni/ZrO$_2$).

2.4   Hydrogenolysis of Glycerol

Glycerol is the byproduct of biodiesel produced by the transesterification of animal fat or vegetable and microalgae oils. With the rapidly annual growth of biodiesel production, the market price of purified glycerol dropped significantly.[45] The catalytic conversion of glycerol into some value-added products has thus drawn increasing attention.[46–50] 1,2-Propanediol is a chemical widely used as an antifreeze agent, as a monomer for polyester resins, as well as in cosmetics, liquid detergents, and food additives. 1,3-Propanediol is an important monomer in the synthesis of biodegradable polypropylene terephthalate (PPT), which has also great potential applications in the production of textiles. In the past years, both noble-metal catalysts (Pt, Pd, Rh and Ru) and transition-metal catalysts (Ni and Cu) have been successfully used in the hydrogenolysis of glycerol to propanediols.[46,51] Based on the experimental observations, three kinds of reaction mechanisms were proposed in the literature.

2.4.1 Dehydration–Hydrogenation Mechanism

Recently, Tomishige and coworkers found that the presence of acid (*e.g.* Amberlyst 15, sulfonated zirconia, zeolites or H_2SO_4) led to an increase in glycerol conversion and 1,2-propanediol selectivity over Ru/C catalyst.[52,53] The formation of 1,2-propanediol is promoted by the metal-acid bifunctional catalyst system, as explained by the dehydration–hydrogenation mechanism (Route 1 in Scheme 2.4). The acid catalyzes the dehydration of glycerol to hydroxyacetone, which is then hydrogenated to 1,2-propanediol by the metal catalyst. The preferable formation of hydroxyacetone instead of 3-hydroxypropanal in a dehydration step can be explained by the higher thermodynamic stability of hydroxyacetone.

Under the same reaction conditions, the 1,3-propanediol conversion is comparable to that of glycerol, while the 1,2-propanediol conversion is much lower than that of glycerol. Moreover, the substrate has significant influence on the selectivities of 1-propanol and 2-propanol.[54] When the experiment was performed on glycerol or 1,3-propanediol, the selectivity ratios of 1-propanol to 2-propanol were evidently higher (almost 1 : 1), compared to that obtained from the experiment with 1,2-propanediol. According to this result, both 1-propanol and 2-propanol are mainly generated from 1,3-propanediol. The sum of 1-propanol, 2-propanol and 1,3-propanediol yield from the hydrogenolysis of glycerol over Ru/C was not affected by the addition of Amberlyst 15. In contrast, the TOF obtained from the conversion of glycerol into 1,2-propanediol with Ru/C evidently increased by the addition of Amberlyst 15, indicating that there is a synergism between Ru/C and Amberlyst 15 in the catalytic conversion of glycerol into 1,2-propanediol.

2.4.2 Glyceraldehyde-Based Mechanism

Bases can also act as cocatalysts for glycerol hydrogenolysis. In a work by Chen *et al.*,[55] an evident promotion effect of sodium or lithium base was reported for the glycerol hydrogenolysis over Ru/TiO$_2$. The addition of a lithium or sodium base (*i.e.* LiOH, NaOH, Li$_2$CO$_3$ or Na$_2$CO$_3$) promoted the conversion of glycerol into 1,2-propanediol, and suppressed the formation of ethylene glycol. These findings were explained by the glyceraldehyde-based mechanism (see Route 2 in Scheme 2.4). In the first step, glycerol undergoes dehydrogenation to glyceraldehyde. A base can accelerate this step. Moreover, a lithium or sodium base can also catalyze the dehydration of glyceraldehyde to 2-hydroxyacrolein. Finally, the 2-hydroxyacrolein is hydrogenated to 1,2-propanediol on Ru particles. Under basic conditions, the decrease in ethylene glycol selectivity was rationalized by the suppression of retro-aldol cleavage of glyceraldehyde.

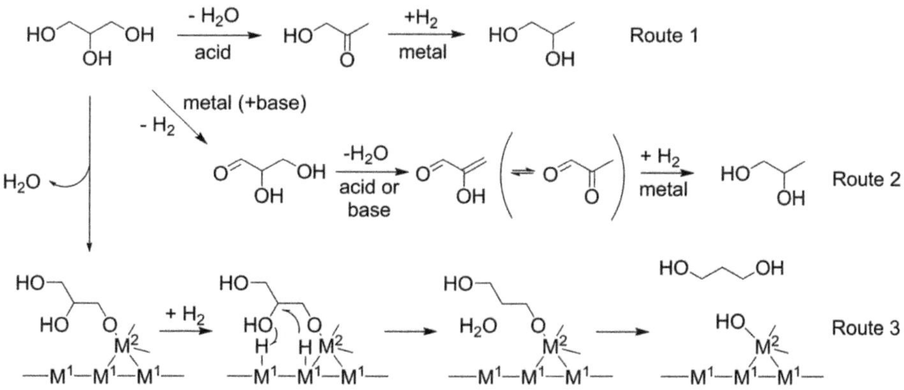

Scheme 2.4 Reaction pathways for the hydrogenolysis of glycerol to propanediols. Adapted from ref. 46.

2.4.3 Hydride-Attack Mechanism

The hydrogenolysis of glycerol to 1,2-propanediol[56,57] and 1,3-propanediol[57–61] in the presence of a Rh (or Ir) catalyst is improved by the modification of Re, Mo and W. Among these promoters, Re was demonstrated to exert the best promotion effect on the selective conversion of glycerol into 1,3-propanediol. Low conversion of glycerol was achieved when methanol, ethanol or butanol was used as a solvent. This result was explained by the competitive adsorption of alcohol and glycerol to the catalyst surface. Furthermore, it was also found that a Rh–ReO$_x$/SiO$_2$ catalyst was also more active than Rh/SiO$_2$ for the hydrogenation of hydroxyacetone to 1,2-propanediol.[62] The characterization of catalysts indicated the formation of partially reduced ReO$_x$ clusters on the surface of Rh or Ir particles.[62,63] Apparently, ReO$_x$ clusters facilitated the adsorption of glycerol by specific interactions with the substrate hydroxyl groups, while the Rh or Ir promoted the formation of hydride that is able to attack the carbocation generated by the protonation of second alcohol.

Based on these results, a synergism between ReO$_x$ species and Rh or Ir particles was proposed (Route 3 in Scheme 2.4). The addition of mineral acid (*e.g.* H$_2$SO$_4$) or solid acid (such as H-ZSM-5) was found to be favorable for the activity, stability and 1,3-propanediol selectivity over an Ir–ReO$_x$/SiO$_2$ catalyst. The acid may protonate the surface of ReO$_x$ clusters and produce more hydroxorhenium sites, which are helpful for glycerol activation through the formation of glyceride surface species.[58]

In recent works by Zhang and coworkers, mesoporous Ti–W oxides[64] and mesoporous WO$_3$[65]-loaded Pt catalysts were used for the hydrogenolysis of glycerol and exhibited high selectivity to 1,3-propanediol. It was suggested that a strong interaction between the Pt particles and the partially reduced

WO$_x$ species promotes the heterolytic dissociation of hydrogen into H$^+$ and H$^-$, which are both very important in the selective hydrogenolysis of glycerol to 1,3-propanediol.

2.5 Hydrogenolysis of Furfural and 5-Hydroxymethylfurfural

1,5-Pentanediol and 1,6-hexanediol are two important monomers in polyesters and polyurethanes production. With the increase of oil price, the preparation of 1,5-pentanediol and 1,6-hexanediol through a nonpetroleum-based route is attracting increasing attention.[66] Furfural is produced by the hydrolysis–dehydration of hemicellulose on an industrial scale.[62] 5-Hydroxymethylfurfural (HMF) can also be produced by the hydrolysis of inulin, Jerusalem artichoke tuber (JAT), and cellulose, followed by dehydration.[67,68] In recent years, several routes have been proposed for the production of 1,5-pentanediol or 1,6-hexanediol with furfural or HMF as a feedstock. Accordingly, some research work was performed on the mechanistic aspects of these reactions.[62,69]

2.5.1 Hydrogenolysis of Furfural to 1,5-Pentanediol

Tetrahydrofurfuryl alcohol (THFA) is the full saturation product of furfural. The initial technology for the synthesis of 1,5-pentanediol starting with THFA is composed of three steps (Scheme 2.5): 1) Dehydration of THFA to 3,4-dihydro-2*H*-pyran; 2) hydration of 3,4-dihydro-2*H*-pyran to 5-hydroxypentanal; 3) hydrogenation of 5-hydroxypentanal to 1,5-pentanediol. The disadvantages of such a reaction system are the many separation and purification steps and the low overall yield of 1,5-pentanediol (< 70%).[70]

Tomishige and coworkers[71,72] found that a Re-modified Rh/SiO$_2$ (Rh-ReO$_x$/SiO$_2$) was active for the conversion of THFA into 1,5-pentanediol by selective hydrogenolysis. In comparison, the unmodified Rh/SiO$_2$ catalyst showed low activity for the same reaction, yielding 1,2-pentanediol as the major final product instead. The modification of the Rh/SiO$_2$-based catalyst with Re, W, Mo increased the THFA conversion. Most importantly, the major product was also shifted from 1,2-pentanediol to 1,5-pentanediol.

To lend insights into the factors accounting for the promotion effect of Re species, they also investigated the catalytic performance of ReO$_x$/SiO$_2$

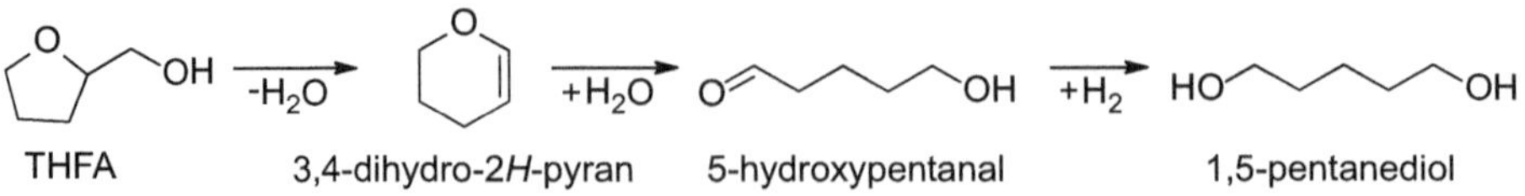

| THFA | 3,4-dihydro-2*H*-pyran | 5-hydroxypentanal | 1,5-pentanediol |

Scheme 2.5   Initial pathway for the synthesis of 1,5-pentanediol from THFA. Adapted from ref. 66.

and the catalyst mixture of Rh/SiO$_2$ and ReO$_x$/SiO$_2$.[71] Low THFA conversions were achieved in these experiments. This result indicates that the synergism between Rh particles and closely contacting ReO$_x$ species are very important for the hydrogenolysis of THFA. They also explored the effect of the reactant on the catalytic performance. The Rh-ReO$_x$/SiO$_2$ catalyst exhibited excellent performance for the hydrogenolysis of THFA and tetrahydropyran-2-methanol to 1,5-pentanediol and 1,6-hexanediol. However, when they used tetrahydrofuran as a substrate, the Rh-ReO$_x$/SiO$_2$ catalyst exhibited very poor activity. These results suggest that the introduction of Re to Rh/SiO$_2$ increased the adsorption of THFA through interactions involving the alcohol group. These interactions may be responsible for the promotion effect of Re species.

Dumesic and coworkers[69] explored the reaction mechanism for the hydrogenolysis of THFA in the presence of a Rh-ReO$_x$/C catalyst. Based on the results obtained from NH$_3$-TPD and DFT predictions, the hydroxyl group on ReO$_x$ contacting with Rh particles was proposed to be acidic. The selective hydrogenolysis of THFA to 1,5-pentanediol catalyzed by Rh–ReO$_x$/C was thus proposed to be facilitated by the acid-catalyzed hydrolysis (or ring opening) and dehydration reactions followed by hydrogenation occurring on the metal sites (Route 1 in Scheme 2.6).

According to the distribution of products from the hydrogenolysis of THFA and 2-isopropoxyethanol, different reaction pathways (hydride–proton mechanism) were proposed by Tomishige and coworkers.[62] It was suggested that the hydrogenolysis of THFA catalyzed by Rh–ReO$_x$ included the generation of anion (by hydride transfer) and proton species by the heterolytic activation of H$_2$ (Route 2 in Scheme 2.6).

Although high yield and selectivity to 1,5-pentanediol have been achieved by the hydrogenolysis of THFA, the direct hydrogenolysis of furfural to 1,5-pentanediol is still preferred in practical applications. Recently, Wang and coworkers[73] found a Pt/Co$_2$AlO$_4$ catalyst for the conversion of furfural into 1,5-pentanediol under mild conditions with ethanol as a solvent. From the analysis of liquid-phase products, it was found that the hydrogenation of

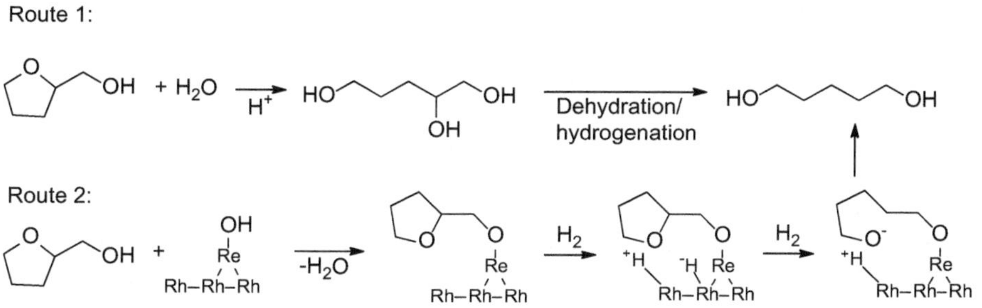

Scheme 2.6   Reaction pathways for the chemoselective hydrogenolysis of THFA to 1,5-pentanediol in the presence of a Rh-ReO$_x$ catalyst. Adapted from refs. 69 and 71.

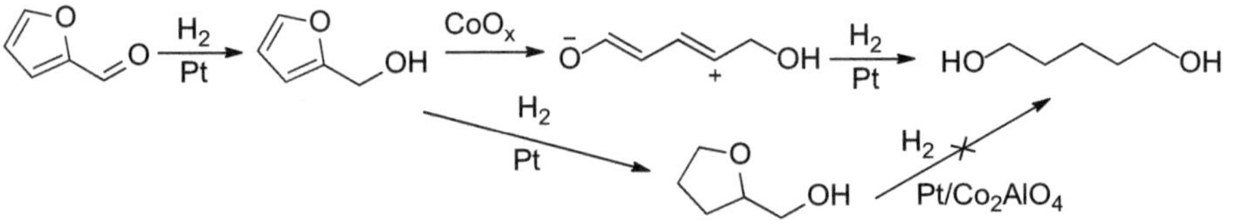

Scheme 2.7   Reaction mechanism for the direct hydrogenolysis of furfural to 1,5-pentanediol over a Pt/Co$_2$AlO$_4$ catalyst. Adapted from refs. 66 and 73.

C=O group in furfural takes place first and rapidly. Then, the intermediate furfuryl alcohol undergoes hydrogenolysis and produces 1,5-pentanediol and THFA, as the two major final products.

In order to clarify the pathway for the generation of 1,5-pentanediol, Wang and coworkers[73] also investigated the hydrogenolysis of THFA on Pt/Co$_2$AlO$_4$ under the same reaction conditions. No 1,2-pentanediol, 1,5-pentanediol or 2-methyltetrahydrofuran was detected. This result indicated that the 1,5-pentanediol was generated from the hydrogenolysis of the furfuryl alcohol other than THFA. A high yield of 1,5-pentanediol up to 30.8% was also obtained in the presence of a Pt/Co$_3$O$_4$ catalyst prepared by coprecipitation. Therefore, they suggested that CoO$_x$ (Co^{3+} ions, especially) is responsible mainly for the absorption of C=C and furan ring-opening reaction, while Pt catalyzes hydrogenation of the intermediate. Scheme 2.7 shows the reaction pathway proposed for the transformation of furfural into 1,5-pentanediol.

2.5.2   Hydrogenolysis of 5-Hydroxymethylfurfural to 1,6-Hexanediol

Vries and coworkers[74] reported the selective hydrogenolysis of HMF to 1,6-hexanediol, which can be further converted into ε-caprolactam, the monomer for Nylon-6. 2,5-Tetrahydrofuran-dimethanol (THFDM) obtained by the full saturation of HMF was used as a substrate. The hydrogenolysis of THFDM was performed with a Rh-Re/SiO$_2$ catalyst. At low temperatures (*e.g.* 353 K), the main product was 1,2,6-hexanetriol. At higher temperatures and longer reaction duration, the cogeneration of 1,6-hexanediol and 1,5-hexanediol was observed. On the Brønsted acid sites, 1,2,6-hexanetriol can be converted into tetrahydro-2*H*-pyran-2-ylmethanol (2-THPM) at very high yield (>99%). As demonstrated by Tomishige and coworkers,[75] 2-THPM is a suitable starting molecule for the production of 1,6-hexanediol by selective hydrogenolysis. Vries and coworkers[74] showed that the hydrogenolysis of THFDM, in the presence of a Rh-Re/SiO$_2$ catalyst and Nafion SAC-13 as a solid acid catalyst, produces 1,6-hexanediol at yields up to 86% at 393 K. Hence, the reaction pathway for the one-pot hydrogenolysis of THFDM to 1,6-hexanediol was proposed as shown in Scheme 2.8.

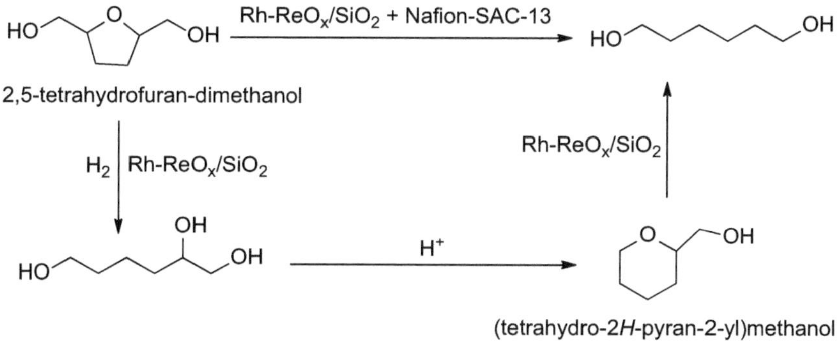

Scheme 2.8   Reaction pathways for the selective hydrogenolysis of THFDM to 1,6-hexanediol.
Adapted from ref. 74.

2.6   Hydrogenolysis of Cellulose, Hemicellulose and Inulin to Sugar Alcohols

Cellulose and hemicellulose as two major components of agriculture waste and forest residue are nonedible, nonexpensive and abundant raw materials. Inulin is the main component of Jerusalem artichoke tuber (JAT). This perennial plant grows rapidly, needs less pesticide, fertilizer, water, compared with other crops. In addition, JAT can be cultivated on barren land (*e.g.* desert and saline-alkali soil). In recent years, the direct conversion of cellulose, hemicellose and JAT into sugar alcohols has attracted considerable attention.

Sorbitol, mannitol and xylitol are important chemicals widely used as sweetener in the food and pharmaceutical industries.[76,77] These sugar alcohols can also be used as intermediates in the synthesis of useful chemicals (*e.g.* 1,4-sorbitan, isosorbide, glycols, glycerol, lactic acid and vitamin C). Sugar alcohols are also important intermediates in the conversion of carbohydrates into alkanes,[78–80] as will be discussed in Section 2.9. Currently, sugar alcohols are mainly produced by the hydrogenation of sugar or starch. However, to fulfil the need of sustainable development, the production of sugar alcohols with no edible part of biomass has greater significance. Key aspects concerning the conversion of cellulose into sugar alcohols in addition to the process modeling are also addressed in Chapter 5 and Chapter 12, respectively.

In an early work by Peters and coworkers,[81] a one-pot process was developed for the hydrogenolysis of inulin to D-mannitol and D-sorbitol with a bifunctional Ru/C catalyst. In the first step, the inulin was hydrolyzed to fructose and glucose, which was catalyzed by an acidic carbon support prepared by preoxidation of active carbon. Fructose and glucose were then hydrogenated to D-mannitol and D-sorbitol over Ru particles.

Unlike inulin, cellulose is protected by its crystalline structure formed by extensive networks of intra- and intermolecular hydrogen bonding.[82–84] In a recent work by Fukuoka and Dhepe,[85] it was found that cellulose can be selectively hydrogenolyzed to sorbitol and mannitol under the catalysis of Pt/Al_2O_3. As they suggested, the acidity of the support is very important for the conversion of cellulose.

Palkovits *et al.*[86] investigated the hydrogenolysis of cellulose to hexitols through the cocatalysis of H_3PO_4 or H_2SO_4 with carbon-loaded Pt, Pd and Ru at 433 K. Besides mineral acids, a heteropolyacid, $H_4SiW_{12}O_{40}$, was also found to be effective for the hydrogenolysis of cellulose to sorbitol when it was used in combination with a noble-metal catalyst.[87,88]

Wang and coworkers studied the hydrogenolysis of cellobiose in the presence of ruthenium supported on carbon nanotubes (CNT).[89] They found that both the acidity and the Ru particle sizes were crucial for the transformation of cellobiose into sorbitol. Acidic sites and large Ru particles (8.7–12 nm) favored the production of sorbitol, while catalysts comprising less-acidic sites and small Ru particles (2.4–6.8 nm) produced 3-β-D-glucopyranosyl-D-glucitol. Wang and coworkers[89] suggested that cellobiose was hydrogenated to 3-β-D-glucopyranosyl-D-glucitol, which was then hydrogenolyzed to sorbitol by acid-catalyzed hydrolysis followed by hydrogenation over the metal catalyst.

Wang and coworkers[90] also showed that Ru-loaded Keggin-type polyoxometalate ($Cs_3PW_{12}O_{40}$) is a highly efficient catalyst for the hydrogenolysis of cellobiose and cellulose to sorbitol at low temperatures ($\leq$ 433 K). The Brønsted acid sites generated from H_2 spillover are responsible for the excellent performance of $Ru/Cs_3PW_{12}O_{40}$ catalyst. The hydrogenolysis of cellulose to hexitols was realized with a Ni_2P/C catalyst.[91,92] The acidity formed by the excess phosphorus on the surface of Ni_2P catalyst was proposed to account for the hydrolysis of the 1,4-β-glucans, leading to glucose that was then hydrogenated to sorbitol.

Fukuoka and coworkers[93] found that Ru/C catalysts were active for the hydrogenolysis of cellulose to sugar alcohols using 2-propanol as a hydrogen donor. The active species for the transfer hydrogenation is cationic Ru species. In turn, Zhang *et al.*[94] reported the one-pot catalytic conversion of hemicellulose (xylan) into xylitol through catalytic transfer hydrogenation. Under mild conditions (413 K, 3 h), xylitol yields up to 80% were achieved in the presence of Ru/C and an acid catalyst. The reaction pathways for the production of sugar alcohols from inulin, cellulose and hemicellulose are shown in Scheme 2.9.

In recent work by Rinaldi and coworkers,[95,96] a high yield of hexitol (94%) was obtained, in the presence of a commercial Ru/C catalyst, from water-soluble oligosaccharides previously produced by mechanocatalytic depolymerization of H_2SO_4-impregnated cellulose.

Altogether, based on these works, the metal-acid bifunctional mechanism was proposed for the hydrogenolysis of cellulose to sorbitol (Scheme 2.9). Generally, the   acid   site   was   considered   the   active   site   for   the

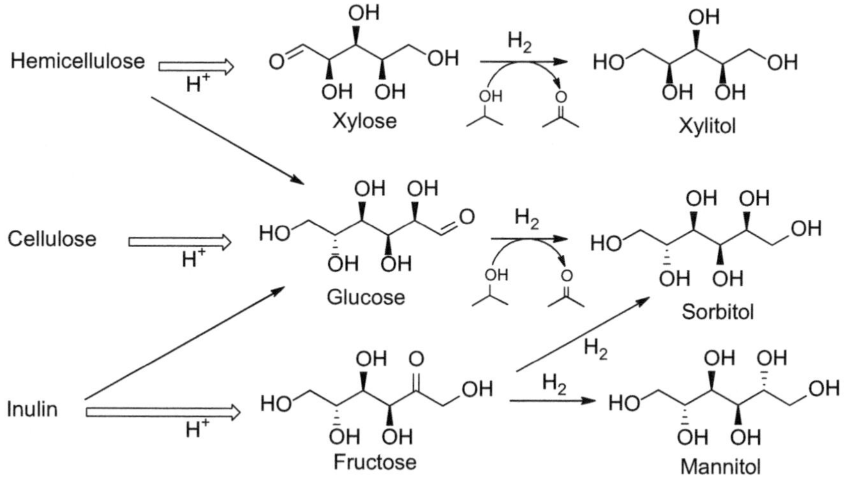

Scheme 2.9   Reaction pathways for the hydrogenolysis of cellulose, hemicellulose and inulin to sugar alcohols.
Adapted from refs. 82, 93 and 94.

depolymerization of cellulose to glucose, which is, in turn, hydrogenated to sorbitol on the metal sites. However, it was found that metals could also promote cellulose hydrolysis.[97,98] The acid catalyst for the hydrolysis of cellulose can be a mineral acid, a solid acid, or even generated from water or heterolytic cleavage of H_2.

2.7   Hydrogenolysis of Cellulose, Hemicellulose and Inulin to C_2 and C_3 diols

Ethylene glycol (EG), 1,2-propylene glycol (1,2-PG) and other diols are very important bulk chemicals currently produced predominantly from petroleum. Compared with sugar alcohols, EG and 1,2-PG have higher demand and prices.[99–101]

Zhang and coworkers reported the first example on the direct hydrogenolysis of cellulose to EG and 1,2-PG under hydrothermal condition.[101,102] They demonstrated that cellulose is fully converted in the presence of a tungsten carbide catalyst supported on active carbon (W_2C/AC), rendering an EG yield up to 27% and a 1,2-PG yield up to 6%. With the addition of a small amount of Ni, or by using mesoporous carbon as the support, the EG yield can be further improved to 61% or 73%, respectively.[101,103] However, the 1,2-PG yield was kept almost unchanged under these conditions. Similar catalytic performances were obtained by experiments performed in the presence of tungsten phosphide or Ni-promoted tungsten phosphide catalysts.[104]

The Ni-W_2C/AC catalyst also showed promising results in the hydrogenolysis of whole lignocellulosic biomass (*e.g.* corn stalk, birch wood, jerusalem artichoke stalk and miscanthus),[105–108] resulting in high yields of EG, 1,2-PG in addition to phenols. EG is produced from the hydrogenolysis of cellulose and hemicellulose. 1,2-PG is obtained mainly from the hydrogenolysis of hemicellulose. In turn, phenols are generated from the hydrogenolysis of lignin. Notably, a higher 1,2-PG yield than that of EG was obtained in the presence of the Ni-W_2C/AC catalyst when JAT was used as the feedstock.[109] From the hydrogenolysis of xylitol by the concurrent use of Ca(OH)$_2$ and Ru/C catalysts, Sun and Liu [100] obtained high EG and 1,2-PG yields. Mu and coworkers showed that EG and 1,2-PG can be produced simultaneously by the cellulose hydrogenolysis in the presence of Ni/ZnO[110] or Ni-Cu/ZnO[111] catalysts.

The reaction mechanism for the production of EG, 1,2-PG and other diols from the hydrogenolysis of cellulose, hemicellulose and JAT was proposed by the Zhang group.[102,107,109,112] First, it was found that a high cellulose conversion and high yield of EG can also be achieved by W-containing bimetallic catalysts.[113] Compared with monometallic catalysts, a significantly higher EG yield was reached in the presence of the bimetallic catalysts. This observation indicates that there is an apparent synergism between W catalyst and Ni, Pd, Pt, Ru, Ir catalysts. Tungsten plays an important role in the production of glycolaldehyde by the retro-aldol cleavage of glucose. This intermediate is then hydrogenated by Ni, Pd, Pt, Ru, Ir catalysts, leading to EG.

Liu *et al.*[114] investigated the structural changes of W, W_2C, WO_3, and WO_2 during the hydrogenolysis of cellulose to EG by XPS and Raman spectroscopy. Irrespective to the initial tungsten species, WO_3 species was always the dominant one on the catalyst surfaces. According this result, WO_3 was considered as the real active sites for the C–C cleavage occurring on glucose. However, these characterizations were not carried out under *in situ* conditions. Therefore, they may not fully account for the actual oxidation state of the tungsten species during the reaction.

In recent work by the Zhang group,[102,112,115] tungstic acid (H_2WO_4) was utilized as a temperature-controlled phase-transfer catalyst in combination with a Ru/AC (or Raney Ni) catalyst. These catalytic systems exhibited good catalytic performances for the hydrogenolysis of cellulose to EG. The combination of Ru/AC with other Brønsted acids (*e.g.* HCl or H_3PO_4) and Lewis acids (*e.g.* ZnCl$_2$ and H_3BO_3) was found to be ineffective for the hydrogenolysis of cellulose to EG. In contrast, high yields of EG were obtained when metallic tungsten, tungsten compounds, *e.g.* tungsten carbide (W_2C), tungsten oxide (WO_3) and heteropolyacid ($H_3PW_{12}O_{40}$), were used in combination with Ru/AC.

It is clear that tungsten and its compounds are unique for the catalytic hydrogenolysis of cellulose to EG. For the aforementioned tungsten-containing compounds, dissolved H_xWO_3 species was always formed under hydrothermal conditions. These dissolved H_xWO_3 species should be the genuinely active component for the conversion of cellulose into glycolaldehyde that is then hydrogenated to EG.[116]

It is noteworthy that the main product from the hydrogenolysis of sugar is also determined by its structure.[102] Subjecting glucose to the W-based catalytic system results in EG as the major product, while the conversion of fructose leads to 1,2-PG under the same reaction conditions. These results can be explained by the different positions of the carbonyl group. In the hydrogenolysis of cellulose, glucose is the major hydrolysis product.[117–119] Therefore, the main product from the hydrogenolysis of cellulose is EG. When JAT is used, the major hydrolysis product is fructose. As the result, 1,2-PG is obtained as the major product. Glucose may also undergo isomerization to fructose under the process conditions applied to hydrolysis of cellulose. As a result, the cogeneration of 1,2-PG is also observed in the hydrogenolysis of cellulose. The detailed pathways for the hydrogenolysis of cellulose, hemicellulose and JAT to EG or 1,2-PG are proposed in Scheme 2.10.

2.8 Hydrogenolysis of Lignin to Phenols and Fuels

Lignin is a highly branched, three-dimensional, phenolic polymer composed of various methoxy- and hydroxy-substituted phenylpropane units (as shown in Figure 3.1). Lignin accounts for 20–30% weight fraction and about 40% energy content of lignocellulose.[120] Therefore, its hydrogenolysis to fuel and chemicals has received increasing attention.[121,122] However, due to its complex polymeric structure, the depolymerization of lignin into small molecules is very difficult.[123–125]

In a recent work by Kou and coworkers,[126] it was found that lignin could be hydrogenolyzed to cyclic alkanes by a two-step process under mild conditions. In the first step, a series of active-carbon-supported noble-metal catalysts were used for the cleavage of ether linkages occurring in lignin, in the presence of hydrogen. Four main monomers (namely, guaiacylpropane, syringylpropane, guaiacylpropanol, syringylpropanol) and two dimers were identified by GC-MS in the product mixture. Among the studied catalysts, Pt/C exhibited the best catalytic performance. The catalytic performance of Pt/C could be further improved by the addition of H_3PO_4 or organic solvents (*e.g.* ethyleneglycol monoethyl ether and dioxane). In the second step, the monomers and the dimers generated in the first step were hydrogenolyzed into alkanes and methanol in the presence of Pd/C and H_3PO_4.

Zhao and Lercher examined a series of catalysts (*e.g.* mixtures of Pd/C and H-ZSM-5,[127] the combination of Raney Ni and Nafion/SiO$_2$ catalysts, and Ni/H-ZSM-5).[128] These materials exhibited excellent catalytic performance for the hydrogenolysis of lignin-derived phenols into cycloalkanes. Moreover, using guaiacol as a model compound, Lercher and coworkers investigated the reaction mechanism for the low-temperature (423 K) hydrogenolysis of phenol derivatives through the cocatalysis of Pd/C and H_3PO_4.[129,130] At the beginning of the reaction, 2-methoxycyclohexanone was identified as the major product. This compound was then hydrogenated to 2-methoxy-cyclohexanol. According to this result, the hydrogenation of aromatic ring

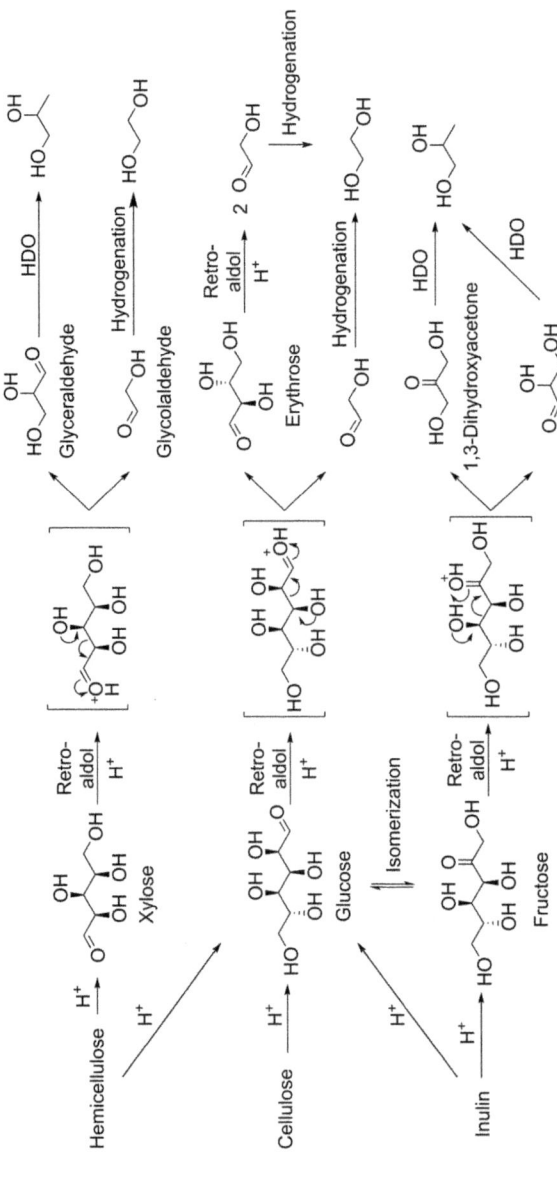

Scheme 2.10   Reaction pathway for the hydrogenolysis of cellulose, hemicellulose and inulin to ethylene glycol and 1,2-propylene glycol. Adapted from refs. 107 and 109.

catalyzed by metal was considered as the fastest step. The methoxycyclohexanol could be hydrolyzed to methanol and 1,2-cyclohexanediol, which was dehydrated to cyclohexanone by acid catalysis. Subsequently, cyclohexanone was hydrogenated to cyclohexanol, which is dehydrated and hydrogenated, forming cyclohexane.

Wang and Rinaldi investigated the hydrogenolysis of lignin-derived model compounds and organosolv lignin in the presence of Raney Ni.[131,132] They found that the secondary alcohols (*e.g.* 2-propanol and 2-butanol) could be used as a hydrogen source for the hydrogenation and hydrogenolysis of phenol and diphenyl ether, respectively. The solvent properties are of high importance in the hydrogenolysis of lignin. Indeed, they found that Raney nickel was very active for the hydrogenolysis and hydrogenation in nonbasic solvents. In contrast, when basic solvents is used (*e.g.* primary alcohols, THF), the catalyst became less active but much more selective for hydrogenolysis, which led to the generation of phenols as the main products.

Recently, Fukuoka and coworkers[133] found that the hydrogenolysis of phenol compounds to cycloalkanes could also be realized by Pt/C under acid-free hydrothermal condition. The *in situ* generated H_3O^+ species in water could aid the dehydration of cyclohexanol intermediates. Based on these results, a general pathway for the conversion of phenolic lignin-models into alkanes is proposed in Scheme 2.11.

The reaction mechanism for the ether cleavage of some representative native lignin linkages in phenolic dimers, *e.g.* α-O-4 (8%), β-O-4 (55%), 4-O-5 (5%), by the concurrent use of Pd/C and H-ZSM-5 was also investigated.[127] The catalyst mixture of Pd/C and H-ZSM-5 leads to phenol and ethylbenzene as the primary products at a low β-O-4 dimer conversion. Without using of Pd/C catalyst, the conversion of α-O-4 dimer yields 63% recombination products (*e.g.* 2-benzylphenol) and 37% phenol. In the presence of H-ZSM-5 (at 473 K for 2 h), benzyl phenyl ether was totally converted into smaller cycloalkanes. Based on these results, the reaction pathways for the conversion of β-O-4 and α-O-4 model compounds were proposed to begin with the ether hydrogenolysis to arenes and substituted phenols on the Pd catalyst. Subsequently, cycloalkanes were produced by hydrogenation and

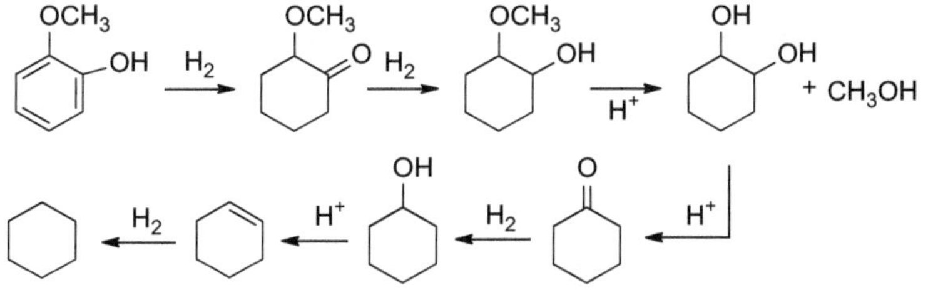

Scheme 2.11 Reaction pathways for the hydrogenolysis of guaiacol by the concurrent use of Pd/C and H_3PO_4 catalysts.
Adapted from refs. 131 and 132.

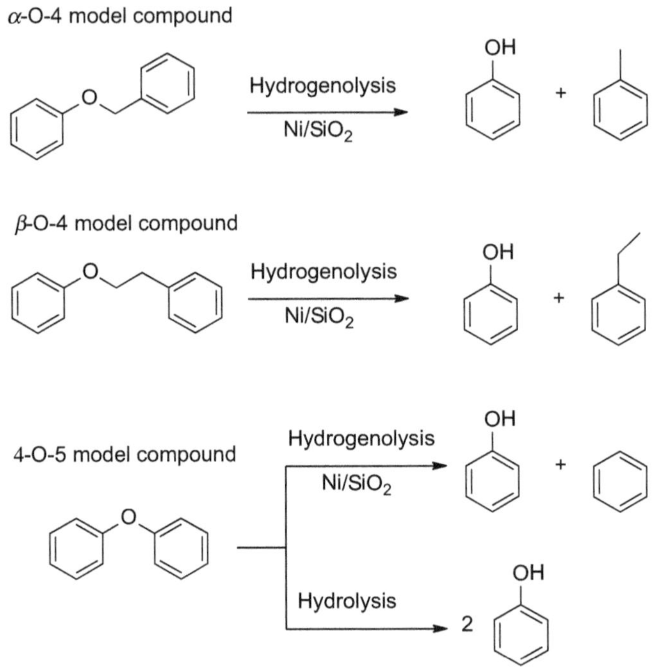

α-O-4 model compound

β-O-4 model compound

4-O-5 model compound

Scheme 2.12   Reaction pathways for the cleavage of ether linkages in lignin. Adapted from ref. 134.

dehydration catalyzed by a metal/acid bifunctional catalyst. Moreover, it was found that diphenyl ether (DPE) was quantitatively converted into cyclohexane at temperatures of 473 K and H_2 pressures of 5 MPa by the concurrent use of Pd/C and H-ZSM-5. In contrast, H-ZSM-5 was inactive for the ether cleavage in DPE under the same conditions. This result suggested that the synergism of Pd and H-ZSM-5 is very important for the cleavage of the diaryl ether bond, forming phenol that undergoes hydrogenation to cyclohexane. Similar reaction pathways for the α-O-4, β-O-4, 4-O-5 cleavage (see Scheme 2.12) were also studied using a Ni/SiO$_2$ catalyst.[134]

2.9   Hydrodeoxygenation of Carbohydrates to Biofuels

The conversion of C_5 and C_6 sugars into alkanes has drawn considerable attention.[135–147] In the early work by Dumesic and coworkers,[78,148] it was found that sorbitol can be hydrodeoxygenated to alkanes over a series of noble metal (Pt- or Pd-) loaded solid acid catalysts. Nb-based catalysts demonstrated superior activity in comparison to a silica-alumina based catalysts. Huber and coworkers[149] screened a series of solid acid loaded Pt catalysts in the hydrodeoxygenation (HDO) of sorbitol. Among the investigated candidates, Pt/zirconium phosphate (Pt/ZrP) demonstrated the best catalytic performance. In the presence of this catalyst, a product mixture with high research octane numbers were obtained at a high carbon yield (73%) by the aqueous phase HDO of C_6 and C_5 sugar alcohol. Analogously, gasoline-like products with an octane number about 96.5 were achieved at a carbon yield of 57% by a two-step HDO process of aqueous carbohydrate solution from maple wood.[150]

To understand the reaction mechanism for the HDO of carbohydrates, some experiments were carried out by identification of the intermediates generated in the HDO of sorbitol over the Pt/SiO$_2$-Al$_2$O$_3$ catalyst.[79] As shown in Scheme 2.13, the overall reaction pathway from sorbitol to light alkanes is mainly composed of four key reactions: hydrogenation, dehydration, retro-aldol condensation and decarbonylation. At the very beginning of the reaction, the sorbitol is dehydrated and produces 1,4-sorbitan and isosorbide. In turn, the isosorbide generated from the dehydration of sorbitol undergoes hydrogenolysis and dehydration/hydrogenation rendering 1,2,6-hexanetriol. This intermediate is then dehydrated and hydrogenated to hexanol or converted into lighter alcohols and polyols by decarbonylation reactions. The hexanols can be converted either into CO and *n*-pentane by dehydrogenation/decarbonylation, or into *n*-hexane by dehydration/hydrogenation. The levels of aldehydes detected in the product mixtures were very low. Lercher and coworkers explained this phenomenon by the disproportionation of aldehydes into alcohol and acid products.[151] Most of the C_4 and C_5 oxygenates detected in products contain less than two oxygen atom(s), indicating that these oxygenates are mainly produced by decarbonylation reactions.

Similar to C_6 oxygenates, the C_4 and C_5 oxygenates can also be converted into *n*-butane and *n*-pentane by dehydration/hydrogenation reactions or into lighter oxygenates and alkanes by decarbonylation reactions. The C_2 and C_3 alcohols were generated primarily by the retro-aldol condensation of sorbitol. These polyols were further converted into alcohols or alkanes by dehydration and hydrogenation. Moreover, the C_2 and C_3 lighter oxygenates can also be produced by the decarbonylation reactions. Methanol and methane in the products may be generated from the glycerol retro-aldol condensation and dehydrogenation/decarbonylation of glycol or ethanol.

From the aforementioned studies, the HDO of sorbitol was proposed to involve three main kinds of reactions:

1. *Carbon–carbon bond cleavage* takes place on metal surfaces. At the initial reaction stage, when the oxygen content in the feed molecule is relatively high. C–C bond is primarily cleaved by retro-aldol condensation. The formation of carbonyl group is necessary for retro-aldol condensation. For primary alcohol or polyol, the cleavage of the C–C bond can also occur by the decarbonylation after dehydrogenation of the adsorbed species to aldehydes.

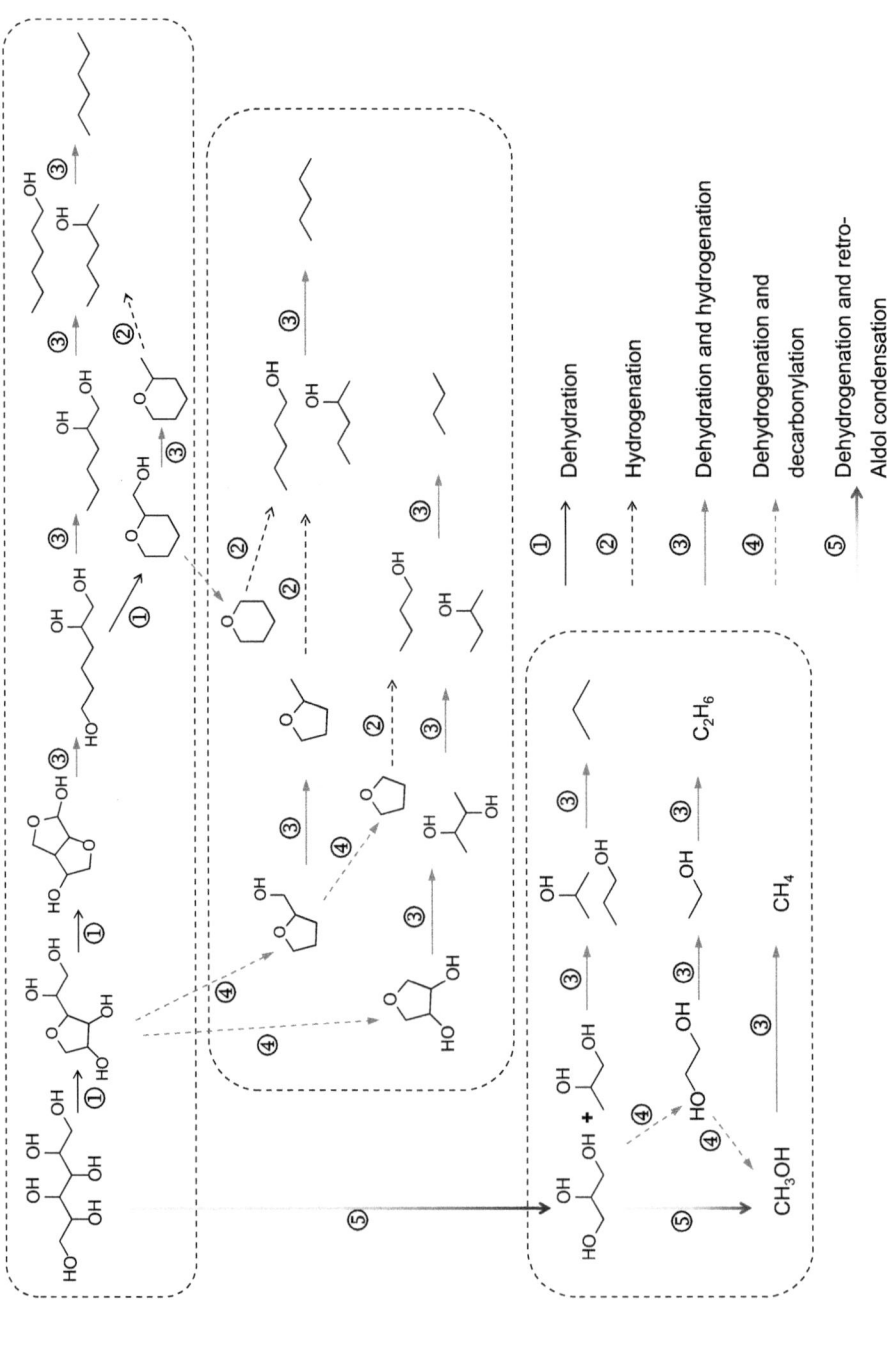

Scheme 2.13 Major reaction pathways and chemical reactions for the HDO of sorbitol in the presence of a Pt/SiO$_2$-Al$_2$O$_3$ catalyst. Adapted from ref. 79.

2. *Carbon–oxygen bond cleavage* mainly takes place on Brønsted acid sites by dehydration. Three kinds of functional groups (C=O bond, C=C bond and ether linkage) are formed during the dehydration. The ether linkages are formed at the initial stage of the reaction with the generation of structurally more stable cyclic ethers. As a result, the C–C bond cleavage is suppressed. This was confirmed by the distribution of products when isosorbide was used as the feedstock. C=O bonds are generated from the dehydration of diols with adjacent hydroxyl groups. C=C bonds are formed by the dehydration of alcohols. The dehydration of a secondary alcohol group is evidently faster than that of a primary one, which can be rationalized by the stability of carbocation.

3. *Hydrogenation* takes place on metal surfaces. According to the distribution of the intermediated in liquid product, the rate of hydrogenation decreases in the order of C=C > C=O > C–O–C bonds. Because the carbonyl groups are usually involved in C–C bond cleavage, the metal catalysts with higher activity for the hydrogenation of C=O and C=C bond is preferred in order to suppress the C–C cleavage reactions. Considering that dehydrogenation is a reversible reaction, a high H_2 pressure is also favorable for the suppression of C–C bond-cleavage reactions.

2.10 Conclusion

Over the past decade, significant progress has been made in lending insight into the reaction mechanisms in biomass conversion. XRD, TEM, TPD, TPR, IR, DFT and DRIFT have been demonstrated to be powerful tools in achieving this goal. Although there is still debate about some of the proposed mechanisms, remarkable consensus has been achieved on the reaction pathways and the functionalities of catalysts involved in hydrogenation and hydrogenolysis. However, challenges have yet to be overcome. Because of the intrinsic feature of the transition metals, metal species at various oxidation states might occur in the reactions, and the exact active species are still not clear enough. Rare examples of structurally defined intermediates have been verified experimentally in order to support the proposed pathways. Furthermore, the extension of the hydrogenation and hydrogenolysis scope as well as the improvement of efficiency and practicality for biorefining is still at the primary stage. In spite of these drawbacks, the advantages of hydrogenation and hydrogenolysis for biorefining have been realized gradually with exciting developments underway.

References

1. P. Gallezot and D. Richard, *Catal. Rev.*, 1998, **40**, 81–126.
2. M. S. Ide, B. Hao, M. Neurock and R. J. Davis, *ACS Catal.*, 2012, **2**, 671–683.
3. H. Wang, Y. Shu, M. Zheng and T. Zhang, *Catal. Lett.*, 2008, **124**, 219–225.

4. X. Yang, A. Wang, X. Wang, T. Zhang, K. Han and J. Li, *J. Phys. Chem. C*, 2009, **113**, 20918–20926.
5. M. Niu, Y. Wang, W. Li, J. Jiang and Z. Jin, *Chin. J. Catal.*, 2013, **34**, 674–678.
6. M. Lashdaf, A. O. I. Krause, M. Lindblad, M. Tiitta and T. Venäläinen, *Appl. Catal. A: Gen.*, 2003, **241**, 65–75.
7. F. Delbecq and P. Sautet, *J. Catal.*, 1995, **152**, 217–236.
8. U. K. Singh and M. A. Vannice, *J. Catal.*, 2001, **199**, 73–84.
9. P. Kluson and L. Cerveny, *Appl. Catal. A: Gen.*, 1995, **128**, 13–31.
10. F.-A. Khan and G. Süss-Fink, *Eur. J. Inorg. Chem.*, 2012, **2012**, 727–732.
11. P. Beccat, J. C. Bertolini, Y. Gauthier, J. Massardier and P. Ruiz, *J. Catal.*, 1990, **126**, 451–456.
12. P. Sautet, *Top. Catal.*, 2000, **13**, 213–219.
13. A. Giroir-Fendler, D. Richard and P. Gallezot, *Catal. Lett.*, 1990, **5**, 175–181.
14. L. Zhu, G.-S. Hu, J.-Q. Lu, G.-Q. Xie, P. Chen and M.-F. Luo, *Catal. Commun.*, 2012, **21**, 5–8.
15. M. De bruyn, S. Coman, R. Bota, V. I. Parvulescu, D. E. De Vos and P. A. Jacobs, *Angew. Chem. Int. Ed.*, 2003, **42**, 5333–5336.
16. T. B. L. W. Marinelli, S. Nabuurs and V. Ponec, *J. Catal.*, 1995, **151**, 431–438.
17. Rodiansono, S. Khairi, T. Hara, N. Ichikuni and S. Shimazu, *Catal. Sci. Technol.*, 2012, **2**, 2139–2145.
18. Rodiansono, T. Hara, N. Ichikuni and S. Shimazu, *Chem. Lett.*, 2012, **41**, 769–771.
19. J. L. Margitfalvi, G. Vankó, I. Borbáth, A. Tompos and A. Vértes, *J. Catal.*, 2000, **190**, 474–477.
20. S. Nishiyama, T. Kubota, K. Kimura, S. Tsuruya and M. Masai, *J. Mol. Catal. A: Chem.*, 1997, **120**, L17–L22.
21. M. G. Musolino, C. Busacca, F. Mauriello and R. Pietropaolo, *Appl. Catal. A Gen.*, 2010, **379**, 77–86.
22. A. Sepúlveda-Escribano, F. Coloma and F. Rodrıguez-Reinoso, *J. Catal.*, 1998, **178**, 649–657.
23. J. Silvestre-Albero, F. Rodrıguez-Reinoso and A. Sepúlveda-Escribano, *J. Catal.*, 2002, **210**, 127–136.
24. A. Dandekar and M. A. Vannice, *J. Catal.*, 1999, **183**, 344–354.
25. M. Consonni, D. Jokic, D. Yu Murzin and R. Touroude, *J. Catal.*, 1999, **188**, 165–175.
26. E. Gebauer-Henke, J. Grams, E. Szubiakiewicz, J. Farbotko, R. Touroude and J. Rynkowski, *J. Catal.*, 2007, **250**, 195–208.
27. K. Liberková and R. Touroude, *J. Mol. Catal. A: Chem.*, 2002, **180**, 221–230.
28. E. V. Ramos-Fernández, B. Samaranch, P. Ramírez de la Piscina, N. Homs, J. L. G. Fierro, F. Rodríguez-Reinoso and A. Sepúlveda-Escribano, *Appl. Catal. A: Gen.*, 2008, **349**, 165–169.
29. M. Tamura, K. Tokonami, Y. Nakagawa and K. Tomishige, *Chem. Commun.*, 2013, **49**, 7034–7036.

30. T. Eggeman and D. Verser, *Appl. Biochem. Biotechnol.*, 2005, **122**, 605–618.
31. A. Primo, P. Concepcion and A. Corma, *Chem. Commun.*, 2011, **47**, 3613–3615.
32. B. K. Ly, D. P. Minh, C. Pinel, M. Besson, B. Tapin, F. Epron and C. Especel, *Top. Catal.*, 2012, **55**, 466–473.
33. H. G. Manyar, C. Paun, R. Pilus, D. W. Rooney, J. M. Thompson and C. Hardacre, *Chem. Commun.*, 2010, **46**, 6279–6281.
34. Y. Takeda, Y. Nakagawa and K. Tomishige, *Catal. Sci. Technol.*, 2012, **2**, 2221–2223.
35. Z. Wang, G. Li, X. Liu, Y. Huang, A. Wang, W. Chu, X. Wang and N. Li, *Catal. Commun.*, 2014, **43**, 38–41.
36. M. Li, G. Li, N. Li, A. Wang, W. Dong, X. Wang and Y. Cong, *Chem. Commun.*, 2014, **50**, 1414–1416.
37. D. C. Elliott, *Energy Fuel*, 2007, **21**, 1792–1815.
38. B. X. Peng, X. G. Yuan, C. Zhao and J. A. Lercher, *J. Am. Chem. Soc.*, 2012, **134**, 9400–9405.
39. B. X. Peng, Y. Yao, C. Zhao and J. A. Lercher, *Angew. Chem. Int. Ed.*, 2012, **51**, 2072–2075.
40. G. W. Huber, P. O'Connor and A. Corma, *Appl. Catal. A: Gen.*, 2007, **329**, 120–129.
41. H. Olcay, L. Xu, Y. Xu and G. W. Huber, *ChemCatChem*, 2010, **2**, 1420–1424.
42. L. Chen, Y. Zhu, H. Zheng, C. Zhang, B. Zhang and Y. Li, *J. Mol. Catal. A: Chem.*, 2011, **351**, 217–227.
43. D. P. Minh, M. Besson, C. Pinel, P. Fuertes and C. Petitjean, *Top. Catal.*, 2010, **53**, 1270–1273.
44. L. Corbel-Demailly, B.-K. Ly, D.-P. Minh, B. Tapin, C. Especel, F. Epron, A. Cabiac, E. Guillon, M. Besson and C. Pinel, *ChemSusChem*, 2013, **6**, 2388–2395.
45. D. T. Johnson and K. A. Taconi, *Environ. Prog.*, 2007, **26**, 338–348.
46. Y. Nakagawa and K. Tomishige, *Catal. Sci. Technol.*, 2011, **1**, 179–190.
47. S. Hirasawa, Y. Nakagawa and K. Tomishige, *Catal. Sci. Technol.*, 2012, **2**, 1150–1152.
48. J. Hu, X. Liu, Y. Fan, S. Xie, Y. Pei, M. Qiao, K. Fan, X. Zhang and B. Zong, *Chin. J. Catal.*, 2013, **34**, 1020–1026.
49. J. Hu, X. Liu, B. Wang, Y. Pei, M. Qiao and K. Fan, *Chin. J. Catal.*, 2012, **33**, 1266–1275.
50. S. Pan, L. Zheng, R. Nie, S. Xia, P. Chen and Z. Hou, *Chin. J. Catal.*, 2012, **33**, 1772–1777.
51. Y. Nakagawa and K. Tomishige, *Catal. Surv. Asia*, 2011, **15**, 111–116.
52. Y. Kusunoki, T. Miyazawa, K. Kunimori and K. Tomishige, *Catal. Commun.*, 2005, **6**, 645–649.
53. T. Miyazawa, Y. Kusunoki, K. Kunimori and K. Tomishige, *J. Catal.*, 2006, **240**, 213–221.
54. T. Miyazawa, S. Koso, K. Kunimori and K. Tomishige, *Appl. Catal. A: Gen.*, 2007, **318**, 244–251.

55. J. Feng, J. Wang, Y. Zhou, H. Fu, H. Chen and X. Li, *Chem. Lett.*, 2007, **36**, 1274–1275.

56. Y. Shinmi, S. Koso, T. Kubota, Y. Nakagawa and K. Tomishige, *Appl. Catal. B: Environ.*, 2010, **94**, 318–326.

57. J. Guan, X. Chen, G. Peng, X. Wang, Q. Cao, Z. Lan and X. Mu, *Chin. J. Catal.*, 2013, **34**, 1656–1666.

58. Y. Nakagawa, X. Ning, Y. Amada and K. Tomishige, *Appl. Catal. A: Gen.*, 2012, **433–434**, 128–134.

59. S. Koso, H. Watanabe, K. Okumura, Y. Nakagawa and K. Tomishige, *Appl. Catal. B: Environ.*, 2012, **111**, 27–37.

60. Y. Amada, Y. Shinmi, S. Koso, T. Kubota, Y. Nakagawa and K. Tomishige, *Appl. Catal. B: Environ.*, 2011, **105**, 117–127.

61. Y. Nakagawa, Y. Shinmi, S. Koso and K. Tomishige, *J. Catal.*, 2010, **272**, 191–194.

62. S. Koso, Y. Nakagawa and K. Tomishige, *J. Catal.*, 2011, **280**, 221–229.

63. Y. Amada, H. Watanabe, M. Tamura, Y. Nakagawa, K. Okumura and K. Tomishige, *J. Phys. Chem. C*, 2012, **116**, 23503–23514.

64. Y. Zhang, X.-C. Zhao, Y. Wang, L. Zhou, J. Zhang, J. Wang, A. Wang and T. Zhang, *J. Mater. Chem. A*, 2013, **1**, 3724–3732.

65. L. Liu, Y. Zhang, A. Wang and T. Zhang, *Chin. J. Catal.*, 2012, **33**, 1257–1261.

66. Y. Nakagawa and K. Tomishige, *Catal. Today*, 2012, **195**, 136–143.

67. F. Yang, Q. Liu, X. Bai and Y. Du, *Bioresour. Technol.*, 2011, **102**, 3424–3429.

68. H. Zhao, J. E. Holladay, H. Brown and Z. C. Zhang, *Science*, 2007, **316**, 1597–1600.

69. M. Chia, Y. J. Pagán-Torres, D. Hibbitts, Q. Tan, H. N. Pham, A. K. Datye, M. Neurock, R. J. Davis and J. A. Dumesic, *J. Am. Chem. Soc.*, 2011, **133**, 12675–12689.

70. L. E. Schniepp and H. H. Geller, *J. Am. Chem. Soc.*, 1946, **68**, 1646–1648.

71. S. Koso, I. Furikado, A. Shimao, T. Miyazawa, K. Kunimori and K. Tomishige, *Chem. Commun.*, 2009, **45**, 2035–2037.

72. S. Koso, N. Ueda, Y. Shinmi, K. Okumura, T. Kizuka and K. Tomishige, *J. Catal.*, 2009, **267**, 89–92.

73. W. Xu, H. Wang, X. Liu, J. Ren, Y. Wang and G. Lu, *Chem. Commun.*, 2011, **47**, 3924–3926.

74. T. Buntara, S. Noel, P. H. Phua, I. Melián-Cabrera, J. G. de Vries and H. J. Heeres, *Angew. Chem. Int. Ed.*, 2011, **50**, 7083–7087.

75. K. Y. Chen, S. Koso, T. Kubota, Y. Nakagawa and K. Tomishige, *ChemCatChem*, 2010, **2**, 547–555.

76. E. Söderling, K. K. Mäkinen, C. Y. Chen, H. R. Pape, Jr., W. Loesche and P. L. Mäkinen, *Caries Res.*, 1989, **23**, 378–384.

77. R. H. Manning, W. M. Edgar and E. A. Agalamanyi, *Caries Res.*, 1992, **26**, 104–109.

78. G. W. Huber, R. D. Cortright and J. A. Dumesic, *Angew. Chem. Int. Ed.*, 2004, **43**, 1549–1551.

79. N. Li and G. W. Huber, *J. Catal.*, 2010, **270**, 48–59.
80. N. Li, G. A. Tompsett and G. W. Huber, *ChemSusChem*, 2010, **3**, 1154–1157.
81. A. W. Heinen, J. A. Peters and H. van Bekkum, *Carbohydr. Res.*, 2001, **330**, 381–390.
82. W. P. Deng, Y. L. Wang, Q. H. Zhang and Y. Wang, *Catal. Surv. Asia*, 2012, **16**, 91–105.
83. R. Rinaldi and F. Schüth, *ChemSusChem*, 2009, **2**, 1096–1107.
84. W. P. Deng, Q. H. Zhang and Y. Wang, *Dalton Trans.*, 2012, **41**, 9817–9831.
85. A. Fukuoka and P. L. Dhepe, *Angew. Chem. Int. Ed.*, 2006, **45**, 5161–5163.
86. R. Palkovits, K. Tajvidi, J. Procelewska, R. Rinaldi and A. Ruppert, *Green Chem.*, 2010, **12**, 972–978.
87. J. Geboers, S. Van de Vyver, K. Carpentier, P. Jacobs and B. Sels, *Green Chem.*, 2011, **13**, 2167–2174.
88. J. Geboers, S. Van de Vyver, K. Carpentier, K. de Blochouse, P. Jacobs and B. Sels, *Chem. Commun.*, 2010, **46**, 3577–3579.
89. W. P. Deng, M. Liu, X. S. Tan, Q. H. Zhang and Y. Wang, *J. Catal.*, 2010, **271**, 22–32.
90. M. Liu, W. Deng, Q. Zhang, Y. Wang and Y. Wang, *Chem. Commun.*, 2011, **47**, 9717–9719.
91. L.-N. Ding, A.-Q. Wang, M.-Y. Zheng and T. Zhang, *ChemSusChem*, 2010, **3**, 818–821.
92. P. F. Yang, H. Kobayashi, K. Hara and A. Fukuoka, *ChemSusChem*, 2012, **5**, 920–926.
93. H. Kobayashi, H. Matsuhashi, T. Komanoya, K. Hara and A. Fukuoka, *Chem. Commun.*, 2011, **47**, 2366–2368.
94. G. S. Yi and Y. G. Zhang, *ChemSusChem*, 2012, **5**, 1383–1387.
95. N. Meine, R. Rinaldi and F. Schuth, *ChemSusChem*, 2012, **5**, 1449–1454.
96. J. Hilgert, N. Meine, R. Rinaldi and F. Schuth, *Energy Environ. Sci.*, 2013, **6**, 92–96.
97. H. Kobayashi, Y. Ito, T. Komanoya, Y. Hosaka, P. L. Dhepe, K. Kasai, K. Hara and A. Fukuoka, *Green Chem.*, 2011, **13**, 326–333.
98. H. Kobayashi, T. Komanoya, K. Hara and A. Fukuoka, *ChemSusChem*, 2010, **3**, 440–443.
99. L. Zhao, J. H. Zhou, Z. J. Sui and X. G. Zhou, *Chem. Eng. Sci.*, 2010, **65**, 30–35.
100. J. Sun and H. Liu, *Green Chem.*, 2011, **13**, 135–142.
101. N. Ji, T. Zhang, M. Zheng, A. Wang, H. Wang, X. Wang and J. G. Chen, *Angew. Chem. Int. Ed.*, 2008, **47**, 8510–8513.
102. A. Wang and T. Zhang, *Acc. Chem. Res.*, 2013, **46**, 1377–1386.
103. Y. Zhang, A. Wang and T. Zhang, *Chem. Commun.*, 2010, **46**, 862–864.
104. G. H. Zhao, M. Y. Zheng, A. Q. Wang and T. Zhang, *Chin. J. Catal.*, 2010, **31**, 928–932.
105. L. Zhou, J. Pang, A. Wang and T. Zhang, *Chin. J. Catal.*, 2013, **34**, 2041–2046.

106. J. Pang, M. Zheng, A. Wang and T. Zhang, *Ind. Eng. Chem. Res.*, 2011, **50**, 6601–6608.

107. C. Z. Li, M. Y. Zheng, A. Q. Wang and T. Zhang, *Energy Environ. Sci.*, 2012, **5**, 6383–6390.

108. J. Pang, M. Zheng, A. Wang, R. Sun, H. Wang, Y. Jiang and T. Zhang, *AIChE J.*, 2014, **60**, 2254–2262.

109. L. Zhou, A. Wang, C. Li, M. Zheng and T. Zhang, *ChemSusChem*, 2012, **5**, 932–938.

110. X. Wang, L. Meng, F. Wu, Y. Jiang, L. Wang and X. Mu, *Green Chem.*, 2012, **14**, 758–765.

111. X. Wang, F. Wu, S. Yao, Y. Jiang, J. Guan and X. Mu, *Chem. Lett.*, 2012, **41**, 476–478.

112. Z. Tai, J. Zhang, A. Wang, M. Zheng and T. Zhang, *Chem. Commun.*, 2012, **48**, 7052–7054.

113. M.-Y. Zheng, A.-Q. Wang, N. Ji, J.-F. Pang, X.-D. Wang and T. Zhang, *ChemSusChem*, 2010, **3**, 63–66.

114. Y. Liu, C. Luo and H. C. Liu, *Angew. Chem. Int. Ed.*, 2012, **51**, 3249–3253.

115. Z. Tai, J. Zhang, A. Wang, J. Pang, M. Zheng and T. Zhang, *ChemSusChem*, 2013, **6**, 652–658.

116. G. Zhao, M. Zheng, J. Zhang, A. Wang and T. Zhang, *Ind. Eng. Chem. Res.*, 2013, **52**, 9566–9572.

117. S. Suganuma, K. Nakajima, M. Kitano, D. Yamaguchi, H. Kato, S. Hayashi and M. Hara, *J. Am. Chem. Soc.*, 2008, **130**, 12787–12793.

118. J. F. Pang, A. Q. Wang, M. Y. Zheng and T. Zhang, *Chem. Commun.*, 2010, **46**, 6935–6937.

119. X. Zhao, J. Wang, C. Chen, Y. Huang, A. Wang and T. Zhang, *Chem. Commun.*, 2014, **50**, 3439–3442.

120. J. E. Holladay, J. F. White, J. J. Bozell and D. Johnson, *Top Value-Added Chemicals from Biomass, Volume II—Results of Screening for Potential Candidates from Biorefinery Lignin*, http://www1.eere.energy.gov/bioenergy/pdfs/pnnl-16983.pdf.

121. Q. Song, J. Cai, J. Zhang, W. Yu, F. Wang and J. Xu, *Chin. J. Catal.*, 2013, **34**, 651–658.

122. Q. Song, F. Wang, J. Cai, Y. Wang, J. Zhang, W. Yu and J. Xu, *Energy Environ. Sci.*, 2013, **6**, 994–1007.

123. H. Wang, M. Tucker and Y. Ji, *J. Appl. Chem.*, 2013, **2013**, 1–9.

124. P. Gao, C. Li, H. Wang, X. Wang and A. Wang, *Chin. J. Catal.*, 2013, **34**, 1811–1815.

125. J. Zakzeski, P. C. A. Bruijnincx, A. L. Jongerius and B. M. Weckhuysen, *Chem. Rev.*, 2010, **110**, 3552–3599.

126. N. Yan, C. Zhao, P. J. Dyson, C. Wang, L. T. Liu and Y. Kou, *ChemSusChem*, 2008, **1**, 626–629.

127. C. Zhao and J. A. Lercher, *ChemCatChem*, 2012, **4**, 64–68.

128. C. Zhao, S. Kasakov, J. Y. He and J. A. Lercher, *J. Catal.*, 2012, **296**, 12–23.

129. C. Zhao, Y. Kou, A. A. Lemonidou, X. B. Li and J. A. Lercher, *Angew. Chem. Int. Ed.*, 2009, **48**, 3987–3990.
130. C. Zhao, J. Y. He, A. A. Lemonidou, X. B. Li and J. A. Lercher, *J. Catal.*, 2011, **280**, 8–16.
131. X. Wang and R. Rinaldi, *ChemSusChem*, 2012, **5**, 1455–1466.
132. X. Wang and R. Rinaldi, *Energy Environ. Sci.*, 2012, **5**, 8244–8260.
133. H. Ohta, H. Kobayashi, K. Hara and A. Fukuoka, *Chem. Commun.*, 2011, **47**, 12209–12211.
134. J. Y. He, C. Zhao and J. A. Lercher, *J. Am. Chem. Soc.*, 2012, **134**, 20768–20775.
135. G. W. Huber, J. N. Chheda, C. J. Barrett and J. A. Dumesic, *Science*, 2005, **308**, 1446–1450.
136. J. Q. Bond, D. M. Alonso, D. Wang, R. M. West and J. A. Dumesic, *Science*, 2010, **327**, 1110–1114.
137. A. Corma, O. de la Torre, M. Renz and N. Villandier, *Angew. Chem. Int. Ed.*, 2011, **50**, 2375–2378.
138. A. Corma, O. de la Torre and M. Renz, *ChemSusChem*, 2011, **4**, 1574–1577.
139. W. Xu, Q. Xia, Y. Zhang, Y. Guo, Y. Wang and G. Lu, *ChemSusChem*, 2011, 1758–1761.
140. R. Xing, W. Qi and G. W. Huber, *Energy Environ. Sci.*, 2011, **4**, 2193–2205.
141. G. Li, N. Li, Z. Wang, C. Li, A. Wang, X. Wang, Y. Cong and T. Zhang, *ChemSusChem*, 2012, **5**, 1958–1966.
142. A. Corma, O. de la Torre and M. Renz, *Energy Environ. Sci.*, 2012, **5**, 6328–6344.
143. J. Yang, N. Li, G. Li, W. Wang, A. Wang, X. Wang, Y. Cong and T. Zhang, *Chem. Commun.*, 2014, **50**, 2572–2574.
144. G. Li, N. Li, J. Yang, A. Wang, X. Wang, Y. Cong and T. Zhang, *Bioresour. Technol.*, 2013, **134**, 66–72.
145. J. Yang, N. Li, G. Li, W. Wang, A. Wang, X. Wang, Y. Cong and T. Zhang, *ChemSusChem*, 2013, **6**, 1149–1152.
146. G. Li, N. Li, S. Li, A. Wang, Y. Cong, X. Wang and T. Zhang, *Chem. Commun.*, 2013, **49**, 5727–5729.
147. G. Li, N. Li, J. Yang, L. Li, A. Wang, X. Wang, Y. Cong and T. Zhang, *Green Chem.*, 2014, **16**, 594–599.
148. R. M. West, M. H. Tucker, D. J. Braden and J. A. Dumesic, *Catal. Commun.*, 2009, **10**, 1743–1746.
149. N. Li, G. A. Tompsett and G. W. Huber, *ChemSusChem*, 2010, **3**, 1154–1157.
150. N. Li, G. A. Tompsett, T. Y. Zhang, J. A. Shi, C. E. Wyman and G. W. Huber, *Green Chem.*, 2011, **13**, 91–101.
151. A. Wawrzetz, B. Peng, A. Hrabar, A. Jentys, A. A. Lemonidou and J. A. Lercher, *J. Catal.*, 2010, **269**, 411–420.

CHAPTER 3

Noble-Metal Catalysts for Conversion of Lignocellulose under Hydrogen Pressure

HIROKAZU KOBAYASHI,[a] HIDETOSHI OHTA[b] AND ATSUSHI FUKUOKA*[a]

[a] Catalysis Research Center, Hokkaido University, Sapporo 001-0021, Japan; [b] Department of Materials Science and Biotechnology, Graduate School of Science and Engineering, Ehime University, Matsuyama 790-8577, Japan *Email: fukuoka@cat.hokudai.ac.jp

3.1 Introduction

Lignocellulose (*e.g.* wood, crop residues, *etc.*) consists of cellulose (40–50%), hemicellulose (20–40%) and lignin (20–30%).[1] Cellulose is a water-insoluble polymer composed of glucose units linked by 1,4-β-glycosidic bonds (Figure 3.1(a)),[2] while hemicellulose contains pentoses and hexoses connected through several types of glycosidic bonds (Figure 3.1(b)).[3] The components of hemicelluloses are varied depending on the plant. Lignin is a three-dimensional aromatic polymer generated by the radical polymerization of *p*-coumaryl alcohol, coniferyl alcohol and sinapyl alcohol (Figure 3.1(c)).[4] Since cellulose, hemicellulose and lignin are highly oxygenated polymers, they should first undergo depolymerization and then reduction and/or hydrodeoxygenation (HDO), allowing for their utilization as feedstock for the production of chemicals and fuels. Severe reaction conditions (*e.g.* hot-compressed water) are often needed for the

RSC Energy and Environment Series No. 13
Catalytic Hydrogenation for Biomass Valorization
Edited by Roberto Rinaldi

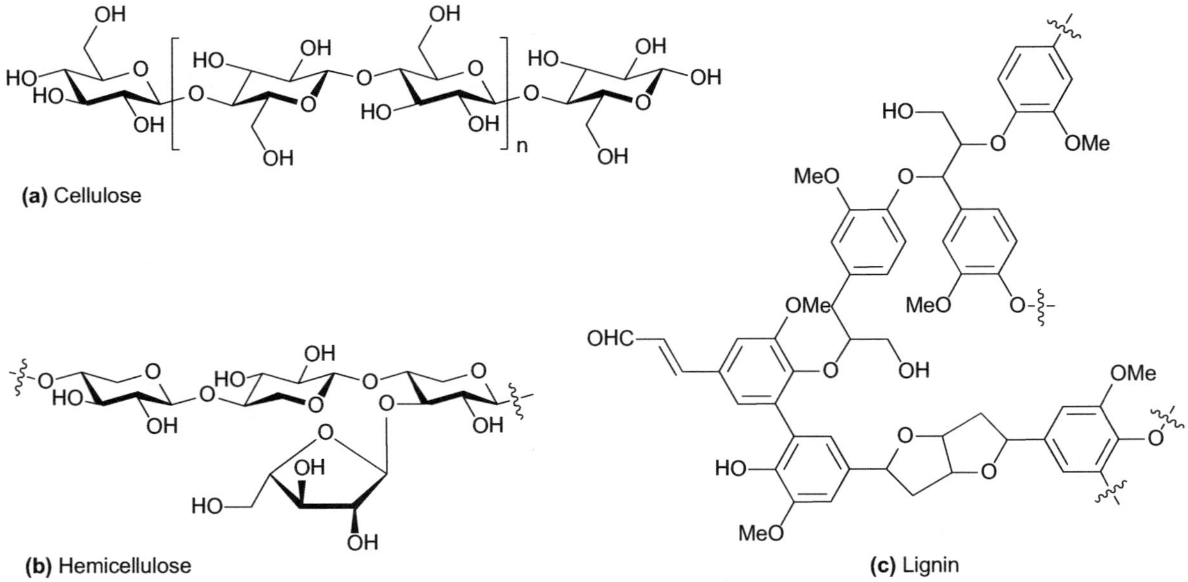

Figure 3.1 Structure of cellulose and typical examples of proposed structures of hemicellulose and lignin.

decomposition of the robust polymers in addition to the HDO of the molecular species formed. These features show that precious-metal catalysts are favorable choices for the processing of plant biomass because they are remarkably active for the reduction reactions and more stable than base metals under harsh conditions. In this chapter, we address the advances in biomass conversion through noble-metal catalysts. Since, analytical techniques are extremely important to understand the action of heterogeneous catalysis on biomass conversion, we also present important points regarding the use of high-performance liquid chromatography and gas chromatography in this chemistry.

3.2 Hydrolytic Hydrogenation of Cellulose and Hemicellulose

3.2.1 Catalysis of Noble Metals for Hydrolytic Hydrogenation

Cellulose undergoes hydrolysis, releasing glucose into the reaction medium. This reaction provides an entry point for the production of sorbitol through the subsequent hydrogenation of glucose (Scheme 3.1). Sorbitol serves as a platform chemical for the production of plastics, surfactants, plasticizers and medicines (Scheme 3.2). A derivative of sorbitol, poly-(ethylene-*co*-isosorbide terephthalate) (PEIT; glass-transition temperature $T_g = $ up to 470 K) is a heat-resistant plastic alternative to PET ($T_g = 353$ K).[5] Poly(isosorbide carbonate) (PIC) is a more environmentally benign plastic than polycarbonate (PC) synthesized from bisphenol-A, which is suspected of acting as an endocrine disruptor. Overall, these examples show that the biomass-derived chemicals are not only mere replacements but in part have more attractive properties than oil-derived chemicals. Since the annual productions of PET and PC are 50 and 3.5 million tons, respectively, PEIT and PIC can be major targets for biorefineries. The one-pot synthesis of sorbitol from cellulose has another advantage in terms of the selective production; sorbitol is more stable than glucose under the severe reaction conditions required for the depolymerization of cellulose.

In the 1950s, the hydrolytic hydrogenation of cellulose to sorbitol and sorbitan was first accomplished by using mineral acids and supported Ru catalysts under H_2 pressure of 7 MPa (Table 3.1, entry 1).[6] However, the products in the hydrolytic hydrogenation using soluble acids must be

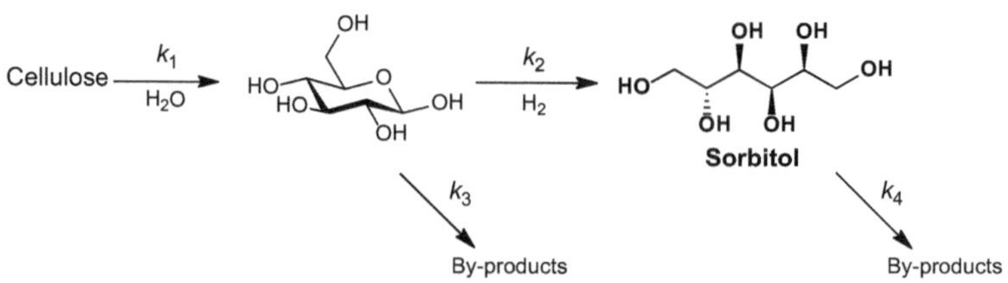

Scheme 3.1 Conversion of cellulose into sorbitol.

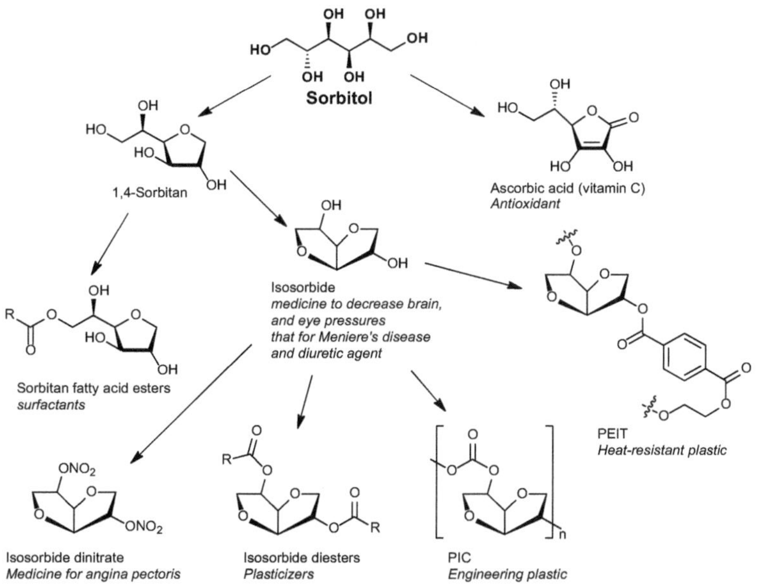

Scheme 3.2 Value-added chemicals obtained from sorbitol.

neutralized and then purified for further use. Development of a hydrolytic hydrogenation process using only solid catalysts has remained a challenge for half a century, due to the poor contact established between the solid cellulose and the solid catalyst. In contrast, a patent from 1989 claimed that the one-pot conversion of starch to sorbitol was possible using Ru/USY. In this catalyst, USY acts as a solid acid suitable to the hydrolysis of water-soluble starch. In turn, Ru catalyzes the reduction of glucose to sorbitol.[7]

The first conversion of cellulose to sugar alcohols by solid catalysts without adding a molecular acid catalyst was reported in 2006.[8] Microcrystalline cellulose was converted into sorbitol (25% yield) and mannitol (6%) by Pt/Al_2O_3 catalyst under a H_2 pressure of 5 MPa at 463 K for 24 h (entry 2). For the production of the sugar alcohols, the turnover number (TON) relative to bulk Pt atoms was 34. The solid catalyst was easily separated from the products by filtration. Supported Ru catalysts were also effective for this reaction, whereas Rh, Pd and Ir catalysts gave only small amounts of sugar alcohols regardless of their hydrogenation abilities.

Scheme 3.1 shows a simplified pathway for the hydrolytic hydrogenation of cellulose. Pseudo-first-order rate constants are given with respect to hydrolysis of cellulose to glucose (k_1), hydrogenation of glucose to sorbitol (k_2), side reactions of glucose (k_3), and decomposition of sorbitol (k_4). The rate constant k_2 should be larger than k_3 to gain acceptable yields of sorbitol, as the maximum yield of sorbitol is limited to $k_2/(k_2+k_3)$ even if $k_4=0$. Furthermore, k_1, which is usually much lower than k_2, must be higher than k_4; otherwise, the degradation of sorbitol is not negligible. Assuming that k_2 is far larger than k_1, k_3 and k_4, the yield of sorbitol is expressed as given by

Table 3.1 Hydrolytic hydrogenation of cellulose by noble-metal catalysts in water.

Entry	Catalysts	Substrate	T/K	Reaction time/h	$P(H_2)$/MPa	Conv./%	Yield /% Sorbitol	Mannitol	Total	Ref.
1	Ru/C, H_2SO_4	Sulfite-cellulose	433	2	7	–	–	–	82[a]	6
2	2.5 wt% Pt/Al_2O_3	MC[b]	463	24	5	–	25	6	31	8
3	2 wt% Pt/BP2000	BC[c]	463	24	5	82	49	9	58	11
4	4 wt% Ni-4 wt% Ir/MC	BC[c]	518	0.5	6	100	52	10	62	17
5	4 wt% Ru/AC	MC[b]	518	0.5	6	86	30	10	39	14
6	1 wt% Ru/CNT	H_3PO_4-treated cellulose	458	24	5	–	69	4	73	20
7	5 wt% Ru/C, $H_{0.5}Cs_{3.5}SiW_{12}O_{40}$[a]	BC[c]	443	48	5	100	–	–	70	21a
8	1 wt% Ru/$Cs_3PW_{12}O_{40}$	BC[c]	433	24	2	–	43	2	45	21b
9	5 wt% Ru/C, $H_4SiW_{12}O_{40}$ (pH 1.9)	BC[c]	463	1	9.5	100	–	–	85	22a
10	5 wt% Ru/C, $H_4SiW_{12}O_{40}$ (pH 1.3)	α-Cellulose[d]	433	7	5	99	–	–	53	22b
11	Pt/$H_4SiW_{12}O_{40}$ ($H_0 = ca.$ −2)	MC[b]	333	24	0.7	–	54	–	–	23
12	0.2 wt% Ru/HUSY, HCl (pH 2.3)	BC[c]	463	24	5	100	–	–	66	25
13	5 wt% Ru/C, H_2SO_4 (pH 1)	Oligomers[e]	423	1	5	–	86	5	91	26b
14	2 wt% Ru/AC	BC[c]	463	18	0.8	83	30	8	38	27
15	2 wt% Ru/BP2000	Cellobiose	393	16	0	98	23	35[f]	58[g]	28

[a]Monoanhydrides of sorbitol.
[b]Microcrystalline cellulose.
[c]Ball-milled cellulose.
[d]Containing 10% hemicellulose, thus the yields of the products were apparently slightly lower.
[e]Cellulose was preliminary hydrolyzed by H_2SO_4 under mechanocatalytic conditions.
[f]Gluconic acid.
[g]Sum of sorbitol and gluconic acid.

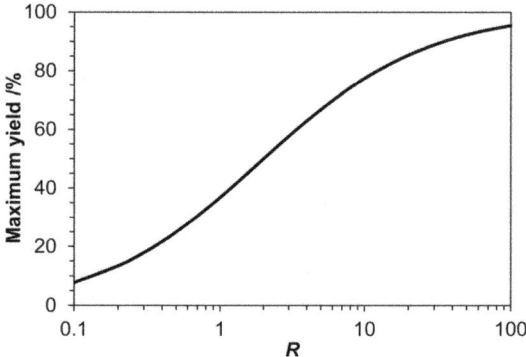

Figure 3.2   Theoretical maximum yield of sorbitol. $R = k_1/k_4$.

eqn (3.1). The derivative of eqn (3.1) provides the maximum yield of sorbitol, as described by eqn (3.2). This equation indicates that the maximum yield of sorbitol increases with raising the ratio of k_1 to k_4, denoted as R. The value of R should be greater than 12 to achieve 80% yield of sorbitol (Figure 3.2). Metal catalysts affect not only k_2 but also k_1, k_3 and k_4, and this is why the yields of sorbitol are significantly changed by the different kinds of metal catalysts.

$$\text{Yield of sorbitol} = \begin{cases} k_1 t e^{-k_1 t} & \text{if } k_1 = k_4, \\ \dfrac{k_1}{k_4 - k_1}\left(e^{-k_1 t} - e^{-k_4 t}\right) & \text{if } k_1 \neq k_4, \end{cases} \qquad (3.1)$$

where t is reaction time.

$$\text{Maximum yield of sorbitol} = \begin{cases} e^{-1} & \text{if } R = 1, \\ \dfrac{1}{1-R}\left(R^{\frac{1}{1-R}} - R^{\frac{2-R}{1-R}}\right) & \text{if } R \neq 1, \end{cases} \qquad (3.2)$$

where $R = k_1/k_4$.

The metal catalysts active for the hydrolytic hydrogenation of cellulose are basically applicable to that of hemicellulose. Ru/AC (AC: activated carbon) produced arabitol from beet fiber, containing arabinan-type hemicellulose,[9] and Pt/MCM-41, xylitol from bleached birch kraft pulp comprising xylan-type hemicellulose.[10] However, a different trend could be seen in the conversion of cellulose and hemicellulose, that is, a Ru/γ-Al$_2$O$_3$ catalyst active in the cellulose conversion did not produce arabitol from beet fiber. This phenomenon might be due to the lower chemical stability of arabinose compared with glucose. However, other factors could also account for the aforementioned observation, namely a distinguished adsorption profile of arabinose on the catalysts as well as impurities (*e.g.* sulfur, nitrogen and minerals) in the hemicellulose biomass.

Regarding the catalyst supports, we should take into account their stability as many supports are unstable in hot-compressed water. In effect, carbons are the first choices under these conditions. Additionally, TiO$_2$ and ZrO$_2$ are relatively stable. Although γ-Al$_2$O$_3$ is frequently used as a high surface area support, this material undergoes hydration at high temperatures (*e.g.* 463 K) rendering boehmite [AlO(OH)], which has no acidity and shows a low specific surface area.[11,12] A specific example is the gradual decrease in activity of Pt/γ-Al$_2$O$_3$ and Ru/γ-Al$_2$O$_3$ in their reuse on the hydrolytic hydrogenation of cellulose at 463 K. For comparison, Pt/BP2000 (BP2000: a carbon black, Cabot) and Ru/AC were stable at least for three times of reuse.[11,14] Moreover, SiO$_2$ is slightly soluble in hot-compressed water.[13] Under this condition, the zeolite skeleton is prone to collapse due to extensive dealuminization and silica solubilization.

3.2.2   Optimization of Catalysts

Since the hydrolysis of cellulose is the rate-determining step, the hydrolytic functions of the catalyst are of utmost importance (see Chapter 5 for further mechanistic detail). Pt/BP2000 accelerated the hydrolysis of cellulose 2.5 times against that without catalysts, affording 58% yield of sorbitol and mannitol from amorphous cellulose at 463 K under 5 MPa of H$_2$ (entry 3).[11] The origin of the hydrolytic function of this catalyst has not yet been clarified. Importantly, slower hydrolysis rates in hot-compressed water without the catalyst occurs by the direct attack of water molecules.[15] Ru/CMK-3 (CMK-3: a mesoporous carbon) is also active for the hydrolysis of cellulose to glucose.[16] Hence, the high valence of Ru(IV) in this catalyst creates the acidic properties and CMK-3 containing oxygenated functional groups helps in the hydrolysis of the substrate. In contrast, a Ru/AC catalyst having zero-valent Ru was inactive for the conversion of cellulose to glucose, but gave sorbitol and mannitol in 39% yield from crystalline cellulose at 518 K under 6 MPa of H$_2$ as the hydrolysis took place by hot-compressed water (entry 5).[14] Ir-Ni/MC (MC: a mesoporous carbon) was useful for the hydrolytic hydrogenation of cellulose (entry 4) in which MC had a hydrolytic activity similar to CMK-3.[17] It is noteworthy that the addition of Ir to Ni/MC remarkably increases the durability and hydrogenation activity, showing the superior properties of the noble metal.

Oxygenated carbons are indeed useful catalysts for the hydrolysis of cellulose,[18] as they facilitate the adsorption of glucose or oligossacharides, keeping the substrate close to the active sites.[19] Moreover, the catalyst support was treated with an acid to accelerate the hydrolysis step. HNO$_3$-oxidized CNT (CNT: a carbon nanotube) was used as a support of Ru, and this catalyst provided 73% yield of sorbitol and mannitol at 458 K under 5 MPa of H$_2$ (entry 6).[20] In the best examples, Ru catalysts with Cs-substituted heteropolyacids led to 70% yield of sorbitol and mannitol at 433–443 K (entries 7 and 8).[21] Although the heteropolyacids are not stable in hot-compressed water, the hydrothermal pretreatment improves their durability.

Another strategy is the refinement of the usability of homogeneous acids. Cellulose was transformed to sorbitol and mannitol up to 85% in only 1 h using H$_4$SiW$_{12}$O$_{40}$ and Ru/C catalysts under aqueous conditions (entries

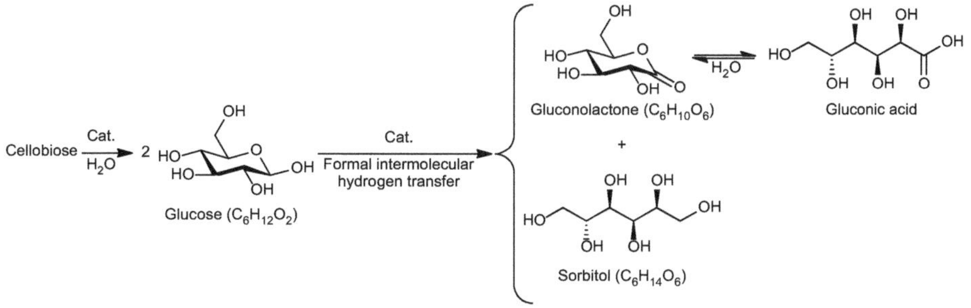

Scheme 3.3   Hydrolytic disproportionation of cellobiose.

9 and 10).[22] A concentrated $H_4SiW_{12}O_{40}$ aqueous solution (0.7 M) with Pt nanoparticles gave 54% yield of sorbitol under mild conditions [333 K, $P(H_2) = 0.7$ MPa] (entry 11).[23] This reaction proceeds at low temperature due to the strong acidity of the heteropolyacids in addition to the hydrogen-bonding-accepting ability of the negatively charged heteropolyanions, which decreases the crystallinity of cellulose. Recrystallization and extraction methods are available to recover the heteropolyacids.[24] Moreover, a 66% yield of sorbitol and mannitol was obtained by promoting the hydrolysis step with diluted HCl (0.0177 wt%, pH 2.3) (entry 12).[25] Some types of stainless steel used for general industrial reactors are fairly resistant to such low concentrations of acid (see Chapter 13). Recently, cellulose was hydrolyzed by H_2SO_4 under mechanocatalytic conditions using planetary ball milling, and the hydrolytic hydrogenation of the produced oligomers were conducted in dilute H_2SO_4 solution (pH 1, entry 13).[26] This two-step reaction provided 91% yield of sorbitol and mannitol.

One of the disadvantages of the hydrolytic hydrogenation is the need of high pressures of H_2 (≥ 2 MPa). A Ru/AC catalyst prepared by the conventional impregnation method using $RuCl_3$ was active for this reaction under H_2 pressures of 0.8 MPa at 463 K (entry 14).[27] Physicochemical measurements imply that the high dispersion of Ru (particle diameters 1–2 nm) would be conducive to decreasing the pressure of H_2. Interestingly, this catalyst can also utilize 2-propanol as a source of hydrogen in the hydrolytic transfer hydrogenation. Pt nanoparticles worked with $H_4SiW_{12}O_{40}$ for the hydrolytic hydrogenation of cellulose to sorbitol (54% yield) under 0.7 MPa of H_2 at 333 K (entry 11).[23] Low-pressure reactions with acceptable yields are becoming more feasible. In addition, the hydrolytic disproportionation of cellobiose to sorbitol and gluconic acid was achieved without a H_2 pressure (Scheme 3.3) (entry 15).[28]

3.2.3   Analysis of Products

In this section, we address the analysis of products from the hydrolytic hydrogenation of cellulose, which is thoughtfully performed, since many similar compounds can form in the reactions. First, the definitions of conversion, carbon-based yield and selectivity are given by eqns (3.3)–(3.5), respectively.

$$\text{Conversion} = \frac{\text{(the amount of cellulose used)} - \text{(remaining cellulose)}}{\text{(the amount of cellulose used)}}$$

$$(3.3)$$

$$\text{Yield} = \frac{\text{(mole of carbon in the product)}}{\text{(mole of carbon in cellulose used)}} \qquad (3.4)$$

$$\text{Selectivity} = \frac{\text{(yield of the product)}}{\text{(conversion)}} \qquad (3.5)$$

Regarding conversion of cellulose, to determine the amount of "unreacted" cellulose, the dried solid residue after the reaction is often weighed. This method is convenient only if the formation of solid byproducts (humins) and the dissolution of solid catalysts are negligible. Otherwise, analysis of cellulose after dissolution through total organic carbon (TOC) or other methods are necessary. TOC is also useful to measure the water content in pure cellulose before reaction. It is noteworthy that the yields should not be based on the conversion with distribution of products, but on their detected amount, because the former method allows the overestimation of the yield by ignoring undetected products. Typically, using the absolute calibration method in the HPLC analysis with an autosampler, the error of the determined amounts of products is about 1%. The product selectivity should not be calculated among identified or some selected compounds without designations, but determined as the ratio of the yield of a product to the conversion.

Figure 3.3(a) shows an HPLC chromatogram (refractive-index detector) of the reaction catalyzed by Ru/AC.[27] The reaction mixture was analyzed using a Shodex Sugar SH-1011 column (ø 8 × 300 mm with a guard column) with water as the mobile phase (0.5 mL min^{-1}) at 323 K. This column is H$^+$-type (crosslinked sulfonated styrene divinylbenzene copolymer) and products are separated by size exclusion in addition to their affinity with the stationary phase. In this column, cellobiose and cellobitol elute together at 13.3–13.7 min, followed by glucose at 15.2 min. Sorbitol and mannitol appear at the same retention time of 16.6 min, and sorbitan and pentitols produced by the C–C bond cleavage of C$_6$ sugar compounds make a shoulder peak at 17.5 min. Hence, this column just shows the total yield of sorbitol and mannitol with a small error. In contrast, ethylene glycol (23.7 min) and propylene glycol (25.0 min) can be well determined. In addition, inorganic compounds are typically eluted within 20 min, which lies well within the range of the retention times found for sugar

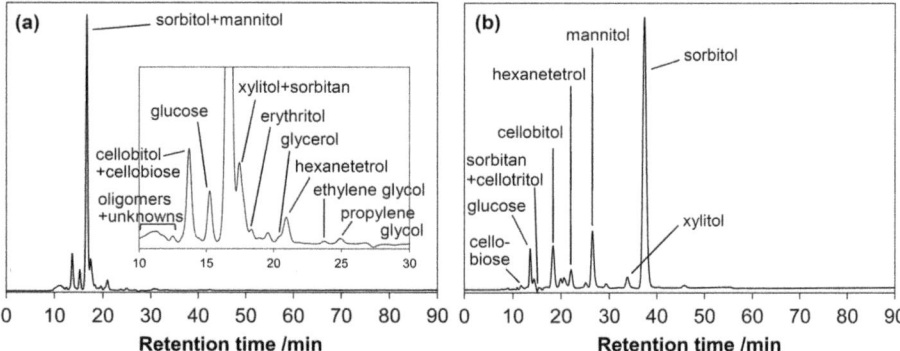

Figure 3.3   Analysis of products for the hydrolytic hydrogenation of cellulose by Ru/AC catalyst[27] by HPLC with (a) a Sugar SH-1011 column and (b) a RPM-Monosaccharide Pb++ column.

products. Thus, one should pay special attention to dissolved catalyst species or impurities.

Apart from the H^+-type columns, there are other column types commercially available, namely Na^+, Ca^{2+} and Pb^{2+} columns, which are also based on crosslinked sulfonated styrene divinylbenzene copolymer. These columns separate products by the formation complex between M^{n+} and polyols. As a result, the retention time found for sugar alcohols increases in the following order: H^+, $Na^+ < Ca^{2+} < Pb^{2+}$. Phenomenex Rezex RPM-Monosaccharide Pb++ (ø 7.8×300 mm with a guard column) is typically used for the analysis of sugar alcohols with water as the mobile phase (0.6 mL min^{-1}) at 343 K (Figure 3.3(b)). Sugar alcohols are specifically retained by this column compared to that in H^+ form. Sorbitol and mannitol give resolved peaks at 37.4 min and 26.5 min, respectively, and thus they can be quantified. In addition, a small peak of xylitol is found at 33.7 min. However, smaller polyhydric alcohols, such as ethylene glycol (19.7 min), glycerol (20.0 min), erythritol (20.7 min) and propylene glycol (20.9 min), overlap in the Pb^{2+}-type column.

Figure 3.4 represents the analysis for the hydrolytic hydrogenation of hemicellulose (arabinoxylan) containing a small amount of C_6 sugars. The SH-1011 column provides a single peak for arabitol and xylitol at 17.9 min, whereas the RPM-Monosaccharide Pb++ column resolves the two peaks (arabitol at 26.4 min and xylitol at 33.7 min). However, arabitol overlaps with mannitol or xylitol. This fact is of importance in the analysis of products containing both C_6 and C_5 sugar alcohols derived from complex substrates (*e.g.* α-cellulose, lignocellulosic materials).

Figure 3.5 depicts a chromatogram (GC, thermal conductivity detector) of gaseous products in the hydrolytic hydrogenation of cellulose, for which an activated carbon column (ø 3×2 m) was used with He as the carrier gas (30 mL min^{-1}) at 403 K. This column can separate air, CO, CO_2 and light hydrocarbons, allowing for their quantification.

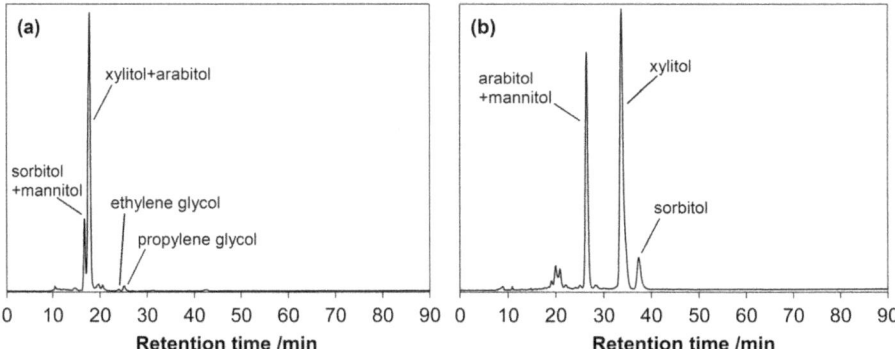

Figure 3.4 Analysis of products for the hydrolytic hydrogenation of arabinoxylan by Pt/BP2000 catalyst[11] by HPLC with a (a) Sugar SH-1011 column, and (b) RPM-Monosaccharide Pb++ column.

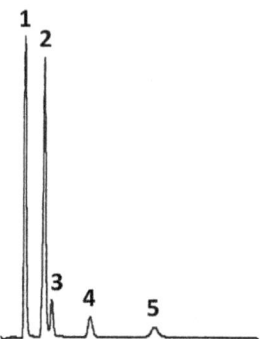

Figure 3.5 Gas-phase analysis by GC-TCD with an activated carbon column. 1: H_2, 0.85 min; 2: air (contaminated during sampling), 1.55 min; 3: CO, 1.82 min; 4: CH_4, 3.16 min, 5: CO_2, 5.4 min. Reliable quantification of H_2 in He is impossible due to their similar thermal conductivities.

3.3 Hydrodeoxygenation of Lignin and Related Compounds

3.3.1 Hydrogenolysis of Lignin

Development of processes for conversion of lignin into chemicals or fuels is of great importance to effectively utilize the whole lignocellulosic biomass. Catalytic hydrogenolysis and subsequent HDO of lignin generally produces a mixture of gases, solids and liquid products (*e.g.* phenols, catechols, aromatics, cycloalkanes, cycloalkanols and cycloalkanones). Current major targets are aliphatic/aromatic hydrocarbons for fuel applications, aromatic commodity chemicals (*e.g.* phenol, benzene, toluene and xylene) in addition to some value-added chemicals (*e.g.* vanillin).[29,30] Catalytic valorization of lignin has been investigated mainly by using cost-effective base-metal catalysts, such as $CoMo/Al_2O_3$ and $NiMo/Al_2O_3$,[29] but they are still insufficient with regard to activity and durability (as discussed in detail in Chapter 8).[31] In recent contributions, Raney Ni and other nickel catalysts were found effective in the conversion of lignin and its model compounds either through conventional reductions or hydrogen-transfer reactions.[32] Nonetheless, noble-metal catalysts continue to attract interest due to their high performance in the hydrogenolysis of lignin.

Catalytic hydrogenolysis of organosolv lignins, which are partially depolymerized by acidic or nonacidic alcoholysis and therefore partially soluble lignins (*e.g.* THF, dioxanes, ethanol/water 1:1 v/v solutions, and other polar solvent), has been investigated in the presence of supported noble-metal catalysts under H_2 pressure (Table 3.2).

Rh/C catalyzed the hydrogenolysis of organosolv softwood (spruce)[33] and hardwood (aspen)[34] lignins in dioxane-H_2O at 468 K (entries 1 and 2). The lignins were mainly converted into monomeric phenol (*i.e.* guaiacol and syringol) derivatives, at 20–40 wt% yields. $Ru/\gamma-Al_2O_3$[35] and Pd/C[36] were highly effective for the conversion of organosolv lignins at 573–653 K, leading to liquid products with decreasing oxygen content (entries 3 and 4). The yields of liquid products were higher than those of $CoMo/\gamma-Al_2O_3$ and $NiMo/\gamma-Al_2O_3$ catalysts, suggesting superior performance of noble-metal catalysts compared with base metals. Pt/C facilitated the hydrogenolysis of organosolv birch lignin in acidic H_2O/dioxane at 473 K to yield four phenolic monomers and some dimers in a selective manner (entry 5).[37] The reaction products obtained were completely hydrodeoxygenated in water by the combination of Pd/C catalyst and H_3PO_4, affording 42 wt% C_8–C_9 alkanes, 10 wt% C_{14}–C_{18} alkanes and 11 wt% methanol (entry 6). As shown here, noble-metal catalysts are highly active for the hydrogenolysis of organosolv lignins even in aqueous media in which conventional $CoMo/\gamma-Al_2O_3$ and $NiMo/\gamma-Al_2O_3$ catalysts are deactivated.[31]

The above-mentioned processes operate under high pressures of hydrogen. Novel approaches to convert lignin into liquid products have been investigated by using *in situ* generated hydrogen from a bioderived feedstock (Table 3.3). Depolymerization and HDO of organosolv switchgrass lignin was achieved with Pt/C catalyst and formic acid as a hydrogen source, giving liquid products with decreasing molecular weight and oxygen contents (entry 1).[38] Cu/PMO (PMO: porous metal oxide), which can utilize methanol as a hydrogen source, catalyzes the conversion of organosolv poplar lignin in supercritical methanol to yield a complex mixture composed of monomeric cyclohexyl derivatives with reduced oxygen content (entry 2).[39]

3.3.2 Hydrodeoxygenation of Lignin-Related Phenols

Hydrodeoxygenation (HDO) is the most efficient method to convert lignin-related phenols into upgraded biofuels and aromatic commodity chemicals.[29] The reaction of phenols usually involves two reaction pathways (Scheme 3.4). The first is the direct C–O bond hydrogenolysis of phenols into

Table 3.2 Hydrogenolysis of organosolv lignin under pressurized H_2.

Entry	Catalyst	T /K	P(H$_2$) /MPa[a]	Solvent	Lignin[b]	Product[c]	Ref.
1	Rh/C	468	3.4	Dioxane-H$_2$O	Spruce lignin	Monomeric phenols[h]	33
2	Rh/C	468	3.4	Dioxane-H$_2$O	Aspen lignin	Monomeric phenols[i]	34
3	Ru/γ-Al$_3$O$_3$	573	2	EtOH	Organosolv lignin (Aldrich)[f]	Monomeric phenols[j]	35
4[d]	Pd/C	653	10	–[e]	Organocell lignin[g]	Monomeric phenols and cyclohexanones[k]	36
5	Pt/C + H$_3$PO$_4$	473	4	Dioxane-H$_2$O	Birch lignin	Monomeric and dimeric phenols[l]	37
6	Pd/C + H$_3$PO$_4$	523	4	H$_2$O	The mixture obtained in entry 5	C$_8$–C$_9$ and C$_{14}$–C$_{18}$ alkanes[m]	37

[a]Initial H$_2$ pressure at 298 K.
[b]Organosolv lignins were isolated from the corresponding milled woods by acidic or nonacidic ethanolysis.
[c]Identified major products.
[d]The reaction was performed with a mixture of lignin, catalyst and balance material.
[e]Without solvent.
[f]Purchased from Aldrich.
[g]Obtained from Organocell GmbH Munich, FRG.
[h]4-Propylguaiacol and 4-propylsyringol were formed as major products in at least 20% yield.
[i]1-Propylguaiacol, 4-propylsyringol, 3-guaiacyl-1-propanol and 3-syringyl-1-propanol were formed in at least 40% yield.
[j]The yield of liquid products is 92% based on lignin by TOC analysis.
[k]The yield of liquid products is 81 wt% based on lignin.
[l]1-Propylguaiacol, 4-propylsyringol, 3-guaiacyl-1-propanol and 3-syringyl-1-propanol and some dimers were formed.
[m]42 wt% C$_8$–C$_9$ alkanes, 10 wt% C$_{14}$–C$_{18}$ alkanes and 11 wt% MeOH.

Table 3.3 Hydrogenolysis of organosolv lignin by *in situ* generated H_2.

Entry	Catalyst	T/K	H_2 Source	Solvent	Lignin[a]	Product[b]	Ref.
1	Pt/C	623	HCO_2H	EtOH	Switchgrass lignin	Monomeric phenols[e]	38
2	Cu/PMO[c]	573	–[d]	MeOH	Poplar lignin	Monomeric cyclohexyl derivatives	39

[a]Organosolv lignins isolated from the corresponding milled woods by acidic alcoholysis.
[b]Identified major products.
[c]PMO = porous metal oxide.
[d]H_2 is generated from the solvent.
[e]Alkylated phenol and guaiacol derivatives.

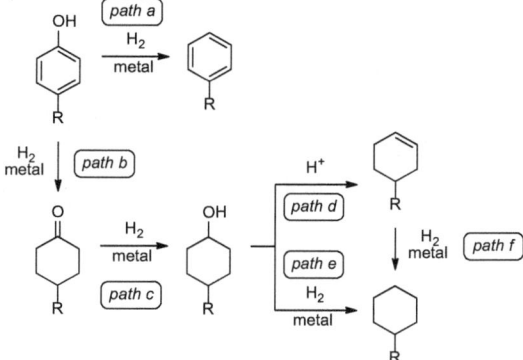

Scheme 3.4 Reaction pathways for the hydrodeoxygenation of phenols.

benzenes (path a). The second is the sequential conversion of phenols into cyclohexanes *via* cyclohexanol intermediates (paths b–f).[40] The selectivity of the products can be tuned by catalyst design and reaction conditions. In addition, the HDO catalyst requires high efficiency for deoxygenation, high tolerance to water and resistance to coke deposition. Several groups have recently investigated the effect of noble metals, supports, additives and solvents on the catalytic activity of supported metal catalysts.

The HDO of 4-propylphenol (4-PrPhOH) into propylcyclohexane (PrCyH) was investigated in water at 553 K by using various noble-metal catalysts (Pt, Rh, Ru and Pd) supported on carbon without any additives (Table 3.4, entries 1–4).[41] Among the catalysts, Pt/C showed the highest activity and selectivity to PrCyH. The catalyst converted important lignin-related phenols, guaiacol and syringol derivatives, with good yield of cycloalkanes.[41] Also, the catalyst support has a significant influence on the HDO activity. Pt/CeO$_2$, Pt/TiO$_2$ and Pt/ZrO$_2$ effected a decrease in the selectivity to PrCyH (entries 5–7). Pt/γ-Al$_2$O$_3$ showed no activity due to the structural change of γ-Al$_2$O$_3$ into boehmite [AlO(OH)] under reaction conditions (entry 8). Altogether, these

Table 3.4 Hydrodeoxygenation of phenols over supported metal catalysts.[a–c]

Entry	Catalyst[d]	Substrate	Solvent	T/K	Conv./%	Major products (Selectivity/%)	Ref.
1	Pt/C[e]	4-PrPhOH[m]	H_2O	553	100	PrCyH[o] (97)	41
2	Rh/C[e]	4-PrPhOH[m]	H_2O	553	100	PrCyH[o] (83)	41
3	Ru/C[e]	4-PrPhOH[m]	H_2O	553	100	PrCyH[o] (55) PrCyH[o] (26),	41
4	Pd/C[e]	4-PrPhOH[m]	H_2O	553	84	4-PrCyOH[p] (68)	41
5	Pt/CeO$_2$[f]	4-PrPhOH[m]	H_2O	553	100	PrCyH[o] (83)	41
6	Pt/TiO$_2$[g]	4-PrPhOH[m]	H_2O	553	100	PrCyH[o] (77)	41
7	Pt/ZrO$_2$[h]	4-PrPhOH[m]	H_2O	553	100	PrCyH[o] (66)	41
8	Pt/γ-Al$_2$O$_3$[i]	4-PrPhOH[m]	H_2O	553	<1	−	41
9	Pd/C[j] + H$_3$PO$_4$	4-PrPhOH[m]	H_2O	523	100	PrCyH[o] (96)	42
10	Pd/C[j] + HZSM-5[k]	4-PrPhOH[m]	H_2O	473	100	PrCyH[o] (98)	43
11	Pt/C[l]	4-MePhOH[n]	H_2O	573	>95	MePhH[q] (ca 45), MeCyH[r] (ca 50)	44
12	Pt/C[l]	4-MePhOH[n]	Heptane	573	>95	4-MeCyOH[s] (91)	44

[a]Initial H$_2$ pressure at RT: 4 MPa (entries 1–8), 5 MPa (entries 9 and 10).
[b]Under pressurized H$_2$, total pressure at 573 K = 8.3 MPa (entries 11 and 12).
[c]Reaction time: 1 h (entries 1–8, 11 and 12), 0.5 h (entry 9), 2 h (entry 10).
[d]Supported metal catalysts were prepared by an impregnation method.
[e]Activated carbon, SX-Ultra (Norit).
[f]JRC-CEO-2.
[g]P-25, Degussa.
[h]JRC-ZRO-2.
[i]JRC-ALO-2.
[j]Purchased from Aldrich.
[k]Si/Al = 45, Brønsted acid site density = 0.278 mmol g^{-1}.
[l]Purchased from Fisher Scientific.
[m]4-Propylphenol.
[n]4-methylphenol.
[o]Propylcyclohexane.
[p]4-Propylcyclohexanol.
[q]Toluene.
[r]Methylcyclohexane.
[s]4-Methylcyclohexanol.

results clearly indicate that a carbon support is one of the most favorable choice for the aqueous-phase HDO of phenols.

The reaction listed in entry 1 refers to a Pt-catalyzed hydrogenation of 4-PrPhOH to 4-propylcyclohexanol (4-PrCyOH), followed by the direct C–O bond hydrogenolysis of 4-PrCyOH into 4-PrCyH (Scheme 3.4: paths b, c and e). Recently, it has been reported that the addition of acid catalysts (e.g. H$_3$PO$_4$[42] and H-ZSM-5[43]) increases the efficiency of hydrodeoxygenation (entries 9 and 10). Notably, the combination of Pd/C with H-ZSM-5 enabled the reaction at temperatures as low as 473 K. This reaction takes place via Pd-catalyzed hydrogenation and acid-catalyzed dehydration (paths b, c, d and f) in which H-ZSM-5 is proposed to promote the rate-determining dehydration step. The solvent effects (see Chapter 4 for a detailed discussion) on the HDO

Table 3.5   Catalytic activity of modified Pt catalysts.[a]

Entry	Catalyst	Conv./%	Selectivity/%					Ref.
			MePhH[k]	3-MeCyOH[l]	MeCyH[m]	Others[n]		
1	Pt/γ-Al$_2$O$_3$[b]	38	67	17	14	2		45
2	Pt/F-γ-Al$_2$O$_3$[c,d]	63	82	1	16	1		45
3	Pt/K-γ-Al$_2$O$_3$[e,f]	26	78	16	4	2		45
4	Pt/SiO$_2$[g]	55	78	17	4	1		45
5	Pt/K-SiO$_2$[h,f]	14	5	85	6	4		45
6	Pt-Co/γ-Al$_2$O$_3$[i]	57	45	1	52	2		46
7	Pt-Ni/γ-Al$_2$O$_3$[j]	63	51	3	43	3		46

[a]533 K, partical H$_2$ pressure = 0.5 atm, 63 g(cat)/mol(4-MePhOH).
[b]1.7 wt% Pt.
[c]1.5 wt% Pt and 1.5 wt% F.
[d]The support was treated with NH$_4$F.
[e]1.8 wt% Pt, 0.3 wt% K.
[f]The support was treated with K$_2$CO$_3$ and KHCO$_3$.
[g]1.6 wt% Pt.
[h]1.6 wt% Pt, 0.3 wt% K.
[i]1.3 wt% Pt, 4.3 wt% Co.
[j]1.5 wt% Pt, 4.9 wt% Ni.
[k]MePhH = toluene.
[l]3-MeCyOH = 3-methylcyclohexanol.
[m]MeCyH = methylcyclohexane.
[n]Methylcyclohexene, phenol, xylenol, *etc.*

of 4-methylphenol (4-MePhOH) by Pt/C catalyst was investigated.[44] The reaction in water was nearly completed at 573 K rendering toluene (MePhH) and methylcyclohexane (MeCyH) with about 45% and 50% selectivity, respectively (entry 11). In contrast, the reaction in *n*-heptane produced 4-methylcyclohexanol (4-MeCyOH) in 91% selectivity (entry 12). It is suggested that the solvents affect the adsorption mode of 4-MePhOH onto the catalyst surface, giving different product distributions in each solvent.

The effect of acid/base pretreatment on the catalyst supports was investigated in the gas-phase HDO of 4-methylphenol (4-MePhOH) at 533 K in the presence of Pt/γ-Al$_2$O$_3$ and Pt/SiO$_2$ (Table 3.5).[45] First, unmodified Pt/ γ-Al$_2$O$_3$ and Pt/SiO$_2$ catalysts resulted in toluene (MePhH) as a primary product (entries 1 and 4). Acid treatment (NH$_4$F) of γ-Al$_2$O$_3$ support was more effective than base treatment (K$_2$CO$_3$ and KHCO$_3$) in order to increase the reaction rate and the selectivity of MePhH (entries 2 *vs.* 3). On the one hand, this result suggests an enhancement of the acid character in addition to an alteration of the binding strength of reactive species on the γ-Al$_2$O$_3$ surface. On the other hand, base treatment of the SiO$_2$ support markedly changed the product distribution. Pt/K-SiO$_2$ led to 3-methylcyclohexanol (3-MeCyOH) in 85% selectivity at low conversion (14%, entries 4 *vs.* 5). Potassium cations are able to poison the Brønsted acid sites on SiO$_2$ surface. In addition, these cations can block catalytically active sites on the Pt surface.

Bimetallic catalysts have been developed to improve catalytic activity of noble metals. The addition of Co and Ni to Pt/γ-Al$_2$O$_3$ improved the catalytic activities and modified the product distribution (entries 6 and 7,

Table 3.5).[46] Similar effects exerted by bimetallic catalysts on the HDO of phenols were reviewed elsewhere.[47]

The usual HDO process requires high reaction temperatures exceeding 573 K.[29] Active catalytic systems are required to make the reaction more efficient and a less energy-consuming process. Recently, a highly active catalytic system based on noble-metal nanoparticles (Rh, Ru and Pt NPs) and an ionic liquid bearing Brønsted acid functionality was reported (Scheme 3.5).[48] The HDO of 4-ethylphenol could be performed at a mild temperature of 403 K to give ethylcyclohexane in high yield. However, the catalyst was moderately effective for the conversion of guaiacol derivatives.

Lignin-derived phenols are important feedstock for the production of hydrocarbon fuels.[49] However, most of the hydrocarbons have low molecular weights (C_6–C_9) and moderate fuel values. Sequential hydroalkylation-HDO of phenols can provide hydrocarbons with appropriate molecular weights (C_{12}–C_{18}) for fuel application. Efficient aqueous-phase conversion of phenols into bicycloalkanes was demonstrated in the presence of a Pd/H-BEA catalyst at 523 K (Scheme 3.6).[50] The selectivity for hydroalkylation of phenols depends on the concentration and molar ratio of phenols to *in situ* formed cycloalkanols.

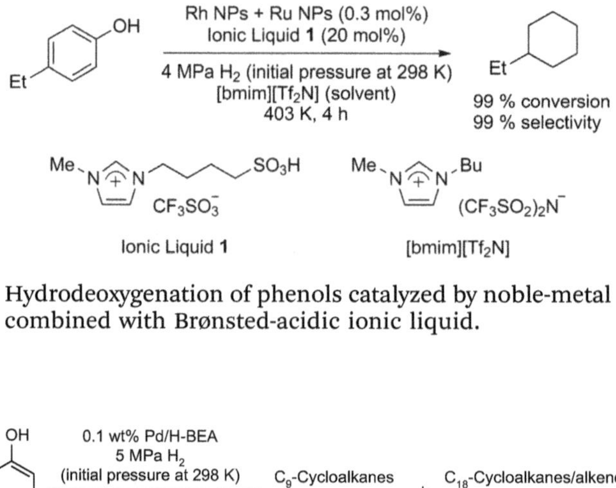

Scheme 3.5 Hydrodeoxygenation of phenols catalyzed by noble-metal nanoparticles combined with Brønsted-acidic ionic liquid.

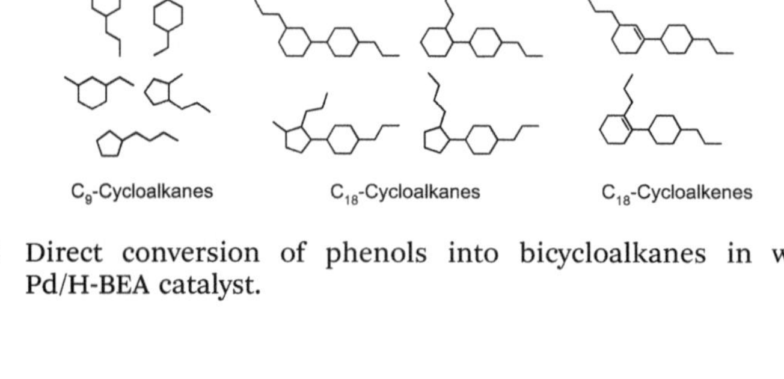

Scheme 3.6 Direct conversion of phenols into bicycloalkanes in water using Pd/H-BEA catalyst.

Table 3.6 Selective hydrodeoxygenation of vanillin in water.

Entry	Catalyst	Solvent	Temp./ K	Conv./ %	Selectivity/% A [f]	B [g]	C [h]	Ref.
1[a]	Pd/SWNT-SiO$_2$[c]	H$_2$O-decalin[e]	373	86	53	47	−	51
2[a]	Pd/SWNT-SiO$_2$[c]	H$_2$O-decalin[e]	473	>99	7	93	−	51
3[a]	Pd/SWNT-SiO$_2$[a]	H$_2$O-decalin[e]	523	>99	−	7	93	51
4[b]	Pd/CN$_{0.132}$[d]	H$_2$O	363	>99	−	>99	−	52

[a]0.3 MPa H$_2$ (initial pressure at RT).
[b]0.1 MPa H$_2$ (balloon).
[c]Single-walled carbon nanotube (SWCN)-silica nanohybrid.
[d]Mesoporous N-doped carbon, N/C atom ratio = 0.132.
[e]H$_2$O:decalin = 1:1 vol/vol.
[f]4-Hydroxymethylguaiacol.
[g]4-Methylguaiacol.
[h]Guaiacol.

Selective aqueous-phase HDO of vanillin was investigated (Table 3.6). In water–decalin biphasic medium, Pd/SWNT-SiO$_2$ (SWNT: single-walled carbon nanotube) catalyst, in which the Pd nanoclusters were mainly located on the SiO$_2$ surface, stabilizes the water–oil emulsions and converts vanilin into 4-hydroxymethylguaiacol (A), 4-methylguaiacol (B) and guaiacol (C) at the liquid/liquid interface.[51] The product selectivity can be controlled by the reaction temperature, since the product partition can be thus tuned (entries 1–3). Furthermore, Pd/CN$_{0.132}$ (CN$_{0.132}$: N-doped mesoporous carbon) promoted the HDO of vanillin in water, giving B in >99% yield as the sole product (entry 4).[52]

3.3.3   Analysis of Products

In this section, we address the analysis of products obtained from the catalytic conversion of lignin and related phenols. The definitions of conversion, yield and selectivity are summarized in Section 3.2.3 (eqns (3.3)–(3.5)). The absolute calibration method can be used in order to determine the amount of gaseous products, such as CO, by a GC-TCD with a packed column. However, an internal standard is necessary to obtain good accuracy for a flow reaction system in which the rate of outlet flow is not well determined (*e.g.*, a small amount of argon is mixed in a He carrier gas at the top of the catalyst bed). Similarly, the use of an internal or an external standard is highly recommended for the quantification of liquid products by a GC equipped with a capillary column and a flame ionization detector (FID). For instance, 2-isopropylphenol can be added to the reaction mixture after the reaction (as an external standard). The organic products are extracted with ethyl acetate to separate the aqueous layer before analysis. Figure 3.6 depicts a GC-FID chromatogram and mass spectra for product analysis in the HDO of 4-propylphenol in water with Pt/C catalyst (Table 3.4, entry 1). The analysis was performed on a capillary column (HR-1, ø 0.25 i.d. × 50 m) using He as the carrier gas (gauge pressure: 150 kPa). The column was heated from 313 to 523 K with a ramping rate of 10 K min^{-1}.

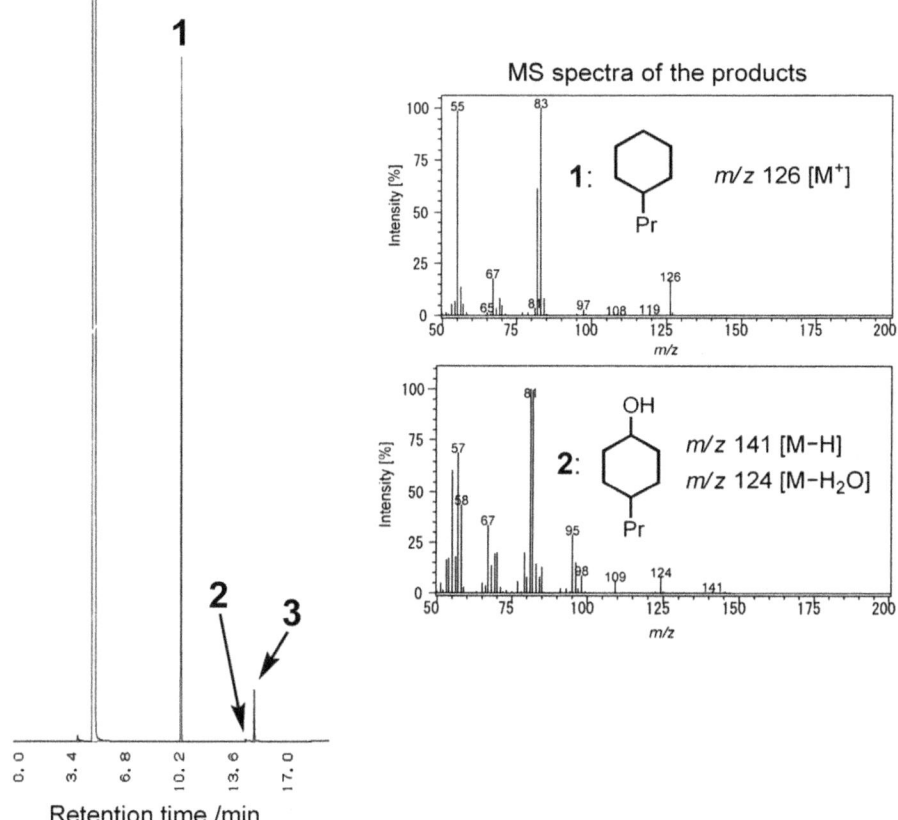

Figure 3.6 GC-FID analysis with a capillary column (HR-1). **1:** propylcyclohexane, 10.69 min; **2:** 4-propylcyclohexanol, 14.80 min; **3:** 2-isopropylphenol (internal standard), 15.28 min.

Elemental analysis (CHN) is very useful in order to evaluate the degree of HDO, which is often expressed as H/C and O/C molar ratios, and to calculate higher heating values (HHV) of the obtained liquid products (pyrolysis oils) by using Dulong's formula.[53] Gel permeation chromatography (GPC) is also useful to monitor the molecular weight distribution of lignin and products.

3.4 Summary

Supported Pt and Ru catalysts transform cellulose to sorbitol and mannitol in high yields, due to the high hydrogenation activity and limited decomposition of the sugar alcohols. In combination with carbon supports, the noble-metal catalysts can be reused several times because of the high water tolerance. Carbon-supported noble-metal catalysts are also useful for the HDO of lignin and related phenolic compounds even in the presence of water. A variety of active catalytic systems enable HDO under mild conditions.

Noble metals are expensive since their low occurrence on the Earth crust; however, it is worth choosing precious-metal catalysts when they give higher catalytic performances and better selectivity than base metals. Indeed, many platinum group catalysts have been used in industrial chemical processes even for the production of bulk chemicals (*e.g.*, Pt–Re catalysts for reforming gasoline). Furthermore, the valuable elements can be recovered from the deactivated solid catalysts, thus indicating the good applicability of noble-metal catalysts in catalytic hydrogenation for biomass valorization.

References

1. Y. Sun and J. Cheng, *Bioresour. Technol.*, 2002, **83**, 1.
2. P. Zugenmaier, in *Crystalline Cellulose and Derivatives: Characterization and Structures. Springer Series in Wood Science* ed. T. E. Timell and R. Wimmer, Springer-Verlag, Berlin, Heidelberg, 2007, pp. 101.
3. H. V. Scheller and P. Ulvskov, *Annu. Rev. Plant Biol.*, 2010, **61**, 263.
4. L. B. Davin, M. Jourdes, A. M. Patten, K.-W. Kim, D. G. Vassão and N. G. Lewis, *Nature Prod. Rep.*, 2008, **25**, 1015.
5. F. Fenouillot, A. Rousseau, G. Colomines, R. Saint-Loup and J.-P. Pascault, *Prog. Polym. Sci.*, 2010, **35**, 578–622; M. Rose and R. Palkovits, *ChemSusChem*, 2012, **5**, 167.
6. A. A. Balandin, N. A. Vasyunina, G. S. Barysheva and S. V. Chepigo, *Bull. Acad. Sci. USSR*, 1957, **6**, 403.
7. P. Jacobs and H. Hinnekens, *EP Patent*, 0 329 923, 1989.
8. A. Fukuoka and P. L. Dhepe, *Angew. Chem. Int. Ed.*, 2006, **45**, 5161.
9. S. K. Guha, H. Kobayashi, K. Hara, H. Kikuchi, T. Aritsuka and A. Fukuoka, *Catal. Commun.*, 2011, **12**, 980.
10. M. Käldström, N. Kumar and D. Y. Murzin, *Catal. Today*, 2011, **167**, 91; M. Käldström, N. Kumar, M. Tenho, M. V. Mokeev, Y. E. Moskalenko and D. Y. Murzin, *ACS Catal.*, 2012, **2**, 1381.
11. H. Kobayashi, Y. Ito, T. Komanoya, Y. Hosaka, P. L. Dhepe, K. Kasai, K. Hara and A. Fukuoka, *Green Chem.*, 2011, **13**, 326.
12. R. M. Ravenelle, J. R. Copeland, W.-G. Kim, J. C. Crittenden and C. Sievers, *ACS Catal*, 2011, **1**, 552.
13. G. B. Alexander, W. M. Heston and R. K. Iler, *J. Phys. Chem.*, 1954, **58**, 453.
14. C. Luo, S. Wang and H. Liu, *Angew. Chem. Int. Ed.*, 2007, **46**, 7636.
15. O. Bobleter, *Prog. Polym. Sci.*, 1994, **19**, 797.
16. H. Kobayashi, T. Komanoya, K. Hara and A. Fukuoka, *ChemSusChem*, 2010, **3**, 440; T. Komanoya, H. Kobayashi, K. Hara, W.-J. Chun and A. Fukuoka, *Appl. Catal. A: Gen.*, 2011, **407**, 188.
17. J. Pang, A. Wang, M. Zheng, Y. Zhang, Y. Huang, X. Chen and T. Zhang, *Green Chem.*, 2012, **14**, 614.
18. H. Kobayashi, M. Yabushita, T. Komanoya, K. Hara, I. Fujita and A. Fukuoka, *ACS Catal.*, 2013, 3, 581.

19. (a) O. M. Gazit and A. Katz, *Langmuir*, 2012, **28**, 431–437; (b) O. M. Gazit and A. Katz, *J. Am. Chem. Soc.*, 2013, **135**, 4398; (c) P.-W. Chung, A. Charmot, O. M. Gazit and A. Katz, *Langmuir*, 2012, **28**, 15222; (d) M. Yabushita, H. Kobayashi, J. Hasegawa, K. Hara and A. Fukuoka, *ChemSusChem*, 2014, **7**, 1443.

20. W. Deng, X. Tan, W. Fang, Q. Zhang and Y. Wang, *Catal. Lett.*, 2009, **133**, 167.

21. (a) J. Geboers, S. Van de Vyver, K. Carpentier, P. Jacobs and B. Sels, *Green Chem.*, 2011, **13**, 2167; (b) M. Liu, W. Deng, Q. Zhang, Y. Wang and Y. Wang, *Chem. Commun.*, 2011, **47**, 9717.

22. (a) J. Geboers, S. Van de Vyver, K. Carpentier, K. de Blochouse, P. Jacobs and B. Sels, *Chem. Commun.*, 2010, **46**, 3577; (b) R. Palkovits, K. Tajvidi, A. M. Ruppert and J. Procelewska, *Chem. Commun.*, 2011, **47**, 576.

23. Y. Ogasawara, S. Itagaki, K. Yamaguchi and N. Mizuno, *ChemSusChem*, 2011, **4**, 519.

24. W. Deng, Q. Zhang and Y. Wang, *Dalton Trans.*, 2012, **41**, 9817–9831; E. N. Dorokhova and I. P. Alimarin, *Russ. Chem. Rev.*, 1979, **48**, 502.

25. J. Geboers, S. Van de Vyver, K. Carpentier, P. Jacobs and B. Sels, *Chem. Commun.*, 2011, **47**, 5590.

26. (a) N. Meine, R. Rinaldi and F. Schüth, *ChemSusChem*, 2012, **5**, 1449; (b) J. Hilgert, N. Meine, R. Rinaldi and F. Schüth, *Energy Environ. Sci.*, 2013, **6**, 92; (c) A. Shrotri, L. K. Lambert, A. Tanksale and J. Beltramini, *Green Chem.*, 2013, **15**, 2761.

27. H. Kobayashi, H. Matsuhashi, T. Komanoya, K. Hara and A. Fukuoka, *Chem. Commun.*, 2011, **47**, 2366.

28. T. Komanoya, H. Kobayashi, K. Hara, W.-J. Chun and A. Fukuoka, *J. Energy Chem.*, 2013, **22**, 290.

29. J. Zakzeski, P. C. A. Bruijnincx, A. L. Jongerius and B. M. Weckhuysen, *Chem. Rev.*, 2010, **110**, 3552.

30. T. Yoshikawa, D. Na-Ranong, T. Tago and T. Masuda, *J. Jpn. Petrol. Inst.*, 2010, **53**, 178.

31. E. Furimsky and F. E. Massoth, *Catal. Today*, 1999, **52**, 381.

32. (a) V. Molinari, C. Giordano, M. Antonietti and D. Esposito, *J. Am. Chem. Soc.*, 2014, **136**, 1758; (b) A. G. Sergeev, J. D. Webb and J. F. Hartwig, *J. Am. Chem. Soc.*, 2012, **134**, 20226; (c) M. R. Sturgeon, M. H. O'Brien, P. N. Ciesielski, R. Katahira, J. S. Kruger, S. C. Chmely, J. Hamlin, K. Lawrence, G. B. Hunsinger and T. D. Foust, *Green Chem.*, 2014, **16**, 824; (d) X. Wang and R. Rinaldi, *ChemSusChem*, 2012, **5**, 1455; (e) J. He, C. Zhao and J. A. Lercher, *J. Catal.*, 2014, **309**, 362; (f) X. Wang and R. Rinaldi, *Energy Environ. Sci.*, 2012, **5**, 8244; (g) X. Wang and R. Rinaldi, *Angew. Chem. Int. Ed.*, 2013, **52**, 11499.

33. J. M. Pepper and Y. W. Lee, *Can. J. Chem.*, 1969, **47**, 723.

34. J. M. Pepper and R. W. Fleming, *Can. J. Chem.*, 1978, **56**, 896.

35. P. T. Patil, U. Armbruster, M. Richter and A. Martin, *Energy Fuels*, 2011, **25**, 4713.

36. D. Meier, R. Ante and O. Faix, *Biores. Techol.*, 1992, **40**, 171.

37. N. Yan, C. Zhao, P. J. Dyson, C. Wang, L.-T. Liu and Y. Kou, *ChemSusChem*, 2008, **1**, 626.
38. W. Xu, S. J. Miller, P. K. Agrawal and C. W. Jones, *ChemSusChem*, 2012, **5**, 667.
39. K. Barta, T. D. Matson, M. L. Fettig, S. L. Scott, A. V. Iretskii and P. C. Ford, *Green Chem.*, 2010, **12**, 1640.
40. F. E. Massoth, P. Politzer, M. C. Concha, J. S. Murray, J. Jakowski and J. Simons, *J. Phys. Chem. B*, 2006, **110**, 14283.
41. H. Ohta, H. Kobayashi, K. Hara and A. Fukuoka, *Chem. Commun.*, 2011, **47**, 12209.
42. C. Zhao, Y. Kou, A. A. Lemonidou, X. Li and J. A. Lercher, *Angew. Chem. Int. Ed.*, 2009, **48**, 3987.
43. C. Zhao and J. A. Lercher, *ChemCatChem*, 2012, **4**, 64.
44. H. Wan, R. V. Chaudhari and B. Subramaniam, *Top Catal.*, 2012, **55**, 129.
45. A. J. Foster, P. T. M. Do and R. F. Lobo, *Top Catal.*, 2012, **55**, 118.
46. P. T. M. Do, A. J. Foster, J. Chen and R. F. Lobo, *Green Chem.*, 2012, **14**, 1388.
47. D. M. Alonso, S. G. Wettstein and J. A. Dumesic, *Chem. Soc. Rev.*, 2012, **41**, 8075.
48. N. Yan, Y. Yuan, R. Dykeman, Y. Kou and P. J. Dyson, *Angew. Chem. Int. Ed.*, 2010, **49**, 5549.
49. G. W. Huber, S. Iborra and A. Corma, *Chem. Rev.*, 2006, **106**, 4044.
50. C. Zhao, D. M. Camaioni and J. A. Lercher, *J. Catal.*, 2012, **288**, 92.
51. S. Crossley, J. Faria, M. Shen and D. E. Resasco, *Science*, 2010, **327**, 68.
52. X. Xu, Y. Li, Y. Gong, P. Zhang, H. Li and Y. Wang, *J. Am. Chem. Soc.*, 2012, **134**, 16987.
53. B. Scholze and D. Meier, *J. Anal. Appl. Pyrol.*, 2001, **60**, 41.

Solvents and Solvent Effects in Biomass Conversion

ROBERTO RINALDI

Max-Planck-Institut für Kohlenforschung, Kaiser-Wilhelm-Platz 1, D-45470 Mülheim an der Ruhr, Germany
Email: rinaldi@kofo.mpg.de

4.1 Introduction

Hydrogenation and hydrogenolysis of molecules derived from plant biomass are reactions often performed in liquid phase, since the highly functionalized biomass molecules (*e.g.* cellulose, hemicellulose, sugars, and lignin) cannot be vaporized without undergoing thermal decomposition. Hence, a solvent usually serves as a medium in which both a substrate and a solid catalyst are suspended. Nonetheless, the contact between the solid substrate and the solid catalyst is still ineffective. Therefore, substrate solubilization—either by dissolution or by solvolysis/thermolysis—is required. Importantly, solvent effects on heterogeneous catalysts play crucial roles in controlling both catalytic activity and selectivity of a chemical transformation. In this chapter, solvents and solvent effects in hydrogenation and hydrogenolysis of biomass-derived molecules are examined. This chapter is structured as follows. In the next section, key aspects of plant biomass and hydrogen solubilization will be addressed. Thereafter, solvent properties commonly used to rationalize solvent effects in heterogeneous catalysis will be presented. Lastly, the solvent role in competitive adsorption, catalyst activity and selectivity will be discussed in detail for selected case studies.

RSC Energy and Environment Series No. 13
Catalytic Hydrogenation for Biomass Valorization
Edited by Roberto Rinaldi
© The Royal Society of Chemistry 2015
Published by the Royal Society of Chemistry, www.rsc.org

4.2 Plant Biomass and Hydrogen Solubilization

4.2.1 Cellulose Solubilization

For catalytic conversion of cellulose, there are two types of processes leading to the biopolymer solubilization. The first is the "physical" dissolution, that is, the polymer is swollen, and then dissolved in a solvent. Herein, the dissolution is referred to as a nonreactive process (*i.e.* cellulose does not undergo derivatization or depolymerization). The second process is the chemical solubilization. In this case, the polymer is either depolymerized, rendering small oligosaccharides soluble in the reaction medium, or the polymer is derivatized, leading to a soluble macromolecule under the reaction conditions. In the next sections, these two types of solubilization will be addressed.

4.2.1.1 Dissolution

The supramolecular structure of cellulose is perhaps one of the most difficult hurdles facing heterogeneously catalyzed processes starting with cellulosic fibers. In effect, the dissolution of cellulose activates the biopolymer through disassembling its supramolecular structure. For instance, cellulose, a recalcitrant polymer in the solid state, becomes a reactive macromolecule in solution. Under homogeneous conditions (*i.e.* cellulose is solubilized in the reaction medium), cellulose can undergo hydrolysis at temperatures as low as 353 K. Nonetheless, under heterogeneous conditions (*i.e.* cellulose is not dissolved in the reaction medium), the reaction happens to a large extent only at temperatures of 473 K or higher.[1] Accordingly, understanding the physical chemistry aspects of cellulose dissolution is certainly important for making the best use of catalysis for cellulose conversion.[2,3]

The dissolution of cellulose begins with the diffusion of solvent molecules through the supramolecular structure. Hence, the biopolymer swells and becomes highly solvated, forming a gel. In this state, cellulose is still unable to interact efficiently with the catalyst surface. The macromolecules become well dispersed in the medium only upon the collapse of the gel structure. It is important to keep in mind that a cellulose solution is a colloidal system because of the macromolecule dimensions (*e.g.* a straight chain of cellulose comprising 200 AGUs measures *ca.* 100 nm in length).[2] For catalysis, the "swollen state" of cellulose facilitates homogeneous processes (*e.g.* solvolysis or depolymerization) that generate oligomers, which can be then transformed by a solid catalyst.

To understand which properties a solvent should possess in order to dissolve or swell cellulose, the thermodynamics of the dissolution process is briefly addressed here. For those readers who wish to obtain a more detailed discussion, refer to Ref. 4. Logically, the dissolution of cellulose is a spontaneous process, at a given temperature and pressure, only if the free energy of solution, $\Delta G (= \Delta H - T\Delta S)$, assumes a negative value. Considering that the entropy of solution, ΔS, is always a positive value because of the increased conformational mobility of the polymeric chains in solution, the term "$- T\Delta S$" always contributes to $\Delta G < 0$. In this manner, the sign of ΔG is determined by the enthalpy of solution, which is approximately equal to the heat of mixing, ΔH_{mix}, given by:

$$\Delta H_{mix} = V_{mix}(\delta_1 - \delta_2)^2 \phi_1 \phi_2 \tag{4.1}$$

$$\delta = \left(\frac{E}{V}\right)^{1/2} = \left(\frac{\Delta H_{vap} - RT}{V}\right)^{1/2} \tag{4.2}$$

where, V_{mix} is the mixture volume, ϕ_1 and ϕ_2 are the volume fractions of the two components, and δ is the solubility parameter (or also called Hildebrand parameter). For eqn (4.2), ΔE and V stand for the energy of vaporization and the molar volume of the component, respectively.

Analyzing eqn (4.1) shows that $\delta_1 \approx \delta_2$ is required in order to obtain an enthalpy of mixing near zero, and thus, $\Delta G < 0$. As defined by eqn (4.2), the δ parameter is directly related to the term (E/V), that is, the *cohesive energy density*. The physical meaning of this term is the energy required to remove a molecule from its nearest neighbor, forming a cavity in the solvent for insertion of solute molecules. In this manner, only solvents with cohesive energy density similar to cellulose are able to dissolve the biopolymer. This fact translates into the requirement of solvents with an extremely high boiling, as $E = \Delta H_{vap} - RT$ (where ΔH_{vap} is the latent heat of vaporization, R is the gas constant, and T is the absolute temperature in Kelvin). In practice, this requirement poses problems for solvent reclamation through distillation.

To date, there has been little agreement on δ values of microcrystalline cellulose. The value determined by a direct method, based on inverse gas chromatography, is 39.9 MPa$^{1/2}$.[5] However, an indirect mechanical measurement revealed a much lower δ value of 25.7 MPa$^{1/2}$.[6] Furthermore, the δ parameter calculated from group molar attraction constants results in a value of 30.2 MPa$^{1/2}$.[6] In turn, Hansen and Bjorkman considered that amorphous cellulose would have a δ value similar to that of Dextran C, 38.6 MPa$^{1/2}$ (Table 4.1, entry 37).[7] Finally, the solubility of cellulose in 1-butyl-3-methylimidazolium chloride, [BMIM]Cl, suggests a δ value of about 35.0 MPa$^{1/2}$.

Considering 39.9 MPa$^{1/2}$ as the δ value of cellulose, the analysis of Table 4.1 reveals that only the molecular solvent, formamide (Table 4.1, entry 26), and the ionic liquid, [BMIM]Cl (Table 4.1, entry 26) would have solubility parameters similar to that of cellulose. Formamide does not dissolve cellulose, whereas [BMIM]Cl does. The false prediction exposes an important point for the solubility prediction by Hildebrand theory. The cohesive energy, δ^2, is indeed the sum of three distinct cohesive energies resulting from (1) nonpolar, atomic (dispersion) interactions, δ_D^2, (2) permanent dipole–permanent dipole molecular interactions, δ_P^2, and

Table 4.1 Solubility parameters of selected solvents and cellulose. Compiled from Refs. 87 (molecular solvents), 88 and 89 (ionic liquids), and 5–7 (cellulose).

Entry	Molecular solvents[87]	δ $(MPa)^{1/2}$	δ_P	δ_D	δ_H
1	*n*-Hexane	14.9	14.9	0	0
2	Diethyl ether	15.8	14.5	2.9	5.1
3	Ethyl acetate	18.1	15.8	5.3	7.2
4	Toluene	18.2	18.0	1.4	2.0
5	Methyl ethyl ketone	19.0	16.0	9.0	5.1
6	Tetrahydrofuran	19.4	16.8	5.7	8
7	Cyclohexanone	19.6	17.8	6.3	5.1
8	Acetone	20.0	15.5	10.4	7.0
9	1,4-Dioxane	20.5	19.0	1.8	7.4
10	Carbon disulfide	20.5	20.5	0	0.6
11	Acetic acid	21.4	14.5	8.0	13.5
12	Pyridine	21.8	19.0	8.8	5.9
13	*N*-Methyl-2-pyrrolidone	22.9	18.0	12.3	7.2
14	2-Propanol	23.5	15.8	6.1	16.4
15	*N*,*N*-Dimethylformamide	24.8	17.4	13.7	11.3
16	Formic acid	24.9	14.3	11.9	16.6
17	Ethanol	26.5	15.8	8.8	19.4
18	Dimethylsulfoxide	26.7	18.4	16.4	10.2
19	NMNO	26.9	19.0	16.1	10.2
20	Triethyleneglycol	27.5	16.0	12.5	18.6
21	Methanol	29.6	15.1	12.3	22.3
22	Diethylene glycol	29.9	16.2	14.7	20.5
23	Propylene glycol	30.2	16.8	9.4	23.3
24	Ethanolamine	31.5	17.2	15.6	21.3
25	Dipropylene glycol	31.7	16.0	20.3	18.4
26	Ethylene glycol	32.9	17.0	11.0	26.0
27	Glycerol	36.1	17.4	12.1	29.3
28	Formamide	36.6	17.2	26.2	19.0
29	Water	47.8	15.6	16.0	42.3
Ionic liquids[88,89]					
30	[BMIM]PF$_6$	29.3	21.0	17.2	10.9
31	[OMIM]PF$_6$	27.8	20.0	16.5	10.0
32	[BMIM]BF$_4$	31.5	23.0	19.0	10.0
33	[BMIM]Cl	35.0	19.1	20.7	20.7
Substrates					
34	Cellulose[5]	39.9	19.4	12.7	31.3
35	Cellulose[6]	25.7	–	–	–
36	Cellulose[6]	30.2	15.8	6.8	24.8
37	Dextran C[7]	38.6	24.3	19.9	22.5

(3) hydrogen-bonding interactions, δ_H^2, as proposed by Hansen and indicated in:

$$\delta^2 = \delta_D^2 + \delta_P^2 + \delta_H^2 \qquad (4.3)$$

According to Hansen,[7] dissolution or swelling takes place when a solvent has δ_D, δ_P and δ_H similar to those of the polymer. It is noteworthy that

Hildebrand theory and HSP are *thermodynamic approaches* to predict solubility. It may happen that the prediction indicates that the material is soluble, but in reality, the material is only swollen or even insoluble. Two main reasons may account for the "incorrect" prediction. The first is a process predicted to be spontaneous may never happen due to kinetic reasons. For instance, solvent molecules, which are too large to diffuse through the polymer supramolecular structure, make the swelling process extremely slow. The second reason is both Hildebrand theory and HSP approach consider a solution as an ideal mixture. Thus, specific interactions between a solvent and a polymer are not considered in these analyses.

4.2.1.2   Chemical Solubilization

By chemical solubilization, cellulose is *in situ* converted into oligosaccharides or glucose (*e.g.* through hydrolysis) or other intermediates (*e.g.* through solvolysis or thermolysis) that are soluble in the reaction medium. Therefore, not cellulose itself, but sugars or other intermediates are the species participating in the heterogeneously catalyzed reactions. Quantitative predictions of the pathways for the acid-catalyzed hydrolysis of cellobiose are now available by DFT studies at the BB1K/6-31++G(d,p) level[8] and Car–Parrinello MD simulations combined with metadynamics.[9] By analogy with cellobiose, cellulose hydrolysis should be strongly affected by the proton affinity of the reaction medium, because the proton transfer from H_3O^+ to the glycosidic O site is hindered by the strong proton affinity of water. Both DFT calculations and Car–Parrinello simulations predict that the protonation of the glycosidic bond accounts for 90% of the apparent activation energy required for the hydrolysis of cellobiose, that is, about 117 out of *ca.* 130 kJ mol^{-1}.[8]

Due to cellulose recalcitrance, the solvolysis processes (which also include hydrolysis) require high-severity conditions (*e.g.* sub/supercritical water) to take place at appreciable reaction rates.[1,2] Under these conditions, both the reaction performance and the stability of the solid catalyst are unnecessarily deteriorated. Therefore, *ex situ* depolymerization of cellulose seems to be a more attractive option for the valorization cellulose sugars by hydrogenation and hydrogenolysis.

Due to the high costs associated with solvent recycling and wastewater treatment,[2] a better process option would be to perform the depolymerization of cellulose under solvent-free conditions. Progress in this new research frontier has been achieved by innovative approaches to depolymerize cellulose by solid-state reactions performed either with nonthermal atmospheric plasma[10] or by mechanocatalysis.[11–17] Regarding mechanocatalysis, we demonstrated the simple impregnation of substrate fibers with a low loading of a strong acid (*e.g.* HCl, H_2SO_4)—prior to the ball-milling treatment—holds the key for highly efficient depolymerization of lignocellulose (*i.e.* full conversion of lignocellulose into water-soluble products at milling durations as short as 2 h).[16]

Importantly, since the acid content (H_2SO_4 or HCl) is not consumed during the mechanocatalytic depolymerization,[16] the dissolution of the processed substrate (*e.g.* "water-soluble wood"), at concentrations as high as 10 wt%, produces an acidic solution (pH 1). At 413 K for 1 h, the water-soluble oligosaccharides undergoes hydrolysis, rendering very high yields of glucose (> 90% relative to glucans) and xylose (> 95% relative to xylans). Under these conditions, lignin precipitates, and thus, can be easily separated from the solution of sugars by filtration. Recently, we demonstrated the "water-soluble cellulose" to be a suitable feedstock for efficient production of sugar alcohols.[13] Under low-severity conditions (*i.e.* 423 K for 1 h), 94% yield of hexitols was obtained from the microcrystalline cellulose depolymerized by the mechanocatalytic approach, in the presence of pre-activated Ru/C. It is noteworthy that such a high yield was achieved in an overall time of only 3 h, that is, *24–36 times faster* than the best examples so far reported for (insoluble) milled cellulose.[18–21]

4.2.2　Lignin Solubilization

Unlike cellulose, which is a linear homopolymer of glucose, lignin is a network aromatic polymer composed of phenylpropenyl units (coumaryl, coniferyl and syringyl alcohols), which are randomly connected through carbon–carbon and carbon–oxygen (ether) bonds (Scheme 4.1).[22] The linkage distribution in the biopolymer structure depends not only upon the type of wood but also upon the part of the plant. The most predominant ether linkage is β-O-4 (softwood: 39 to 48%, hardwood: 32 to 37%) followed by α-O-4 (softwood: 11 to 16%, hardwood: 28 to 32%).[23] Bond-dissociation enthalpy (BDE) values reveal the following order of bond strength α-O-4 < β-O-4 < 4-O-5. In turn, the carbon–carbon bonds α-1 and 5'-5 are more stable than β-O-4 by about 100 to 200 kJ mol^{-1}, respectively (Scheme 4.1).[24,25]

During chemical pulping of wood, α-O-4 and β-O-4 ether linkages are cleaved, while native carbon–carbon linkages are largely preserved, and many other carbon-carbon linkages are formed by condensation reactions.[26] In contrast, through processes based on solvent extraction (*e.g.* Organosolv,[27] and Organocat[28]), α-O-4 linkages are predominately cleaved, whereas there is evidence indicating β-O-4 and carbon–carbon linkages to be largely preserved in the isolated organosolv lignins.[26,27] Therefore, extent of chemical modification of lignin obtained from chemical pulping (*e.g.* Kraft and sulfite processes) is larger compared with that from solvent extraction (*e.g.* Organosolv). Moreover, the apparent molecular weight (M_w) of Kraft lignin or sulfonated lignin is often higher than that of Organosolv lignins,[23] because condensed units are much more prone to be produced by oxidative coupling and condensation reactions under the harsh conditions of the Kraft and sulfite pulping.[27] Accordingly, lignin degradation upon its extraction dramatically changes the identity of its chemical structure. Therefore, the structural features of technical lignins, and hence, the

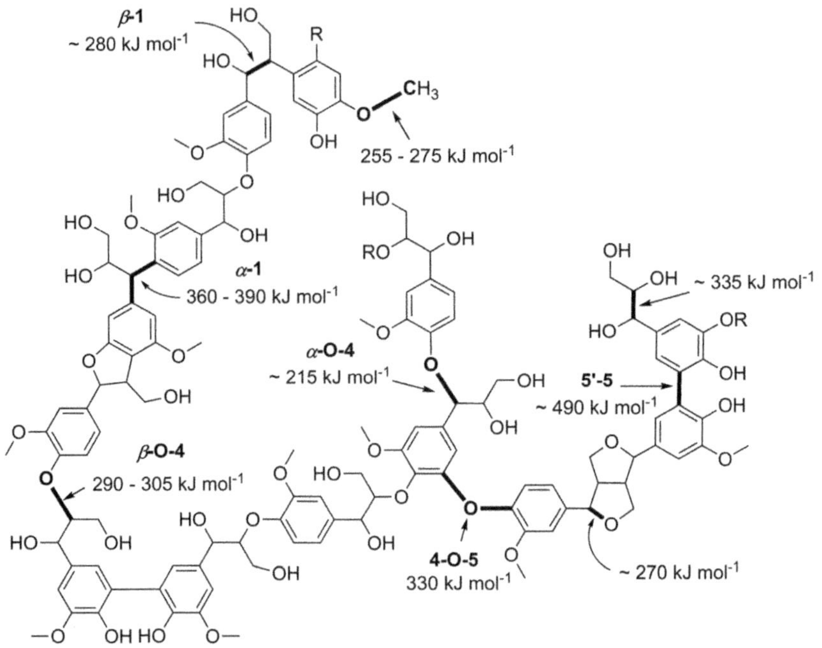

Scheme 4.1 Example of a proposed lignin structure depicting some of the primary lignin subunits and their estimated BDE values (adapted from Ref. 22). The BDE values were obtained from Ref. 25.

solubility properties of these polymers are mostly defined by the delignification method.

4.2.2.1 Dissolution

In 1952, Schuerch identified two solvent properties as being key for solubilization of technical lignins. The first is a δ parameter of *ca.* 22 MPa$^{1/2}$.[29] Interestingly, this value is lower than that estimated for the lignin monomers (coumaryl alcohol, 28.9 MPa$^{1/2}$; coniferyl alcohol, 27.6 MPa$^{1/2}$; syringyl alcohol, 29.0 MPa$^{1/2}$).[30,31] The second property is a good hydrogen-bonding capacity.[29] The hydrogen-bond capacity of a solvent was determined using CH$_3$OD as a probe molecule. In 1940, Gordy and Stanford reported the use of CH$_3$OD as a probe molecule to assess the hydrogen-bonding capacity of a medium.[32] The spectrum of liquid CH$_3$OD shows a broad O–D vibrational band at 2494 cm^{-1} (or 4.01 µm). However, this band is sharp and appears at 2681 cm^{-1} (or 3.73 µm) for a 0.1 mol·L^{-1} CH$_3$OD solution in benzene.[32] Using this value as a reference, Schuerch found that lignin dissolution takes place in the solvents ($\delta \approx$ 22 MPa$^{1/2}$) in which the O–D vibrational band shows Δv values of 97 cm^{-1} or higher.[29] Table 4.2 summarizes Schuerch's observations on the solubility of technical lignins in several solvents.

Table 4.2 Solubility of isolated lignins in single solvents according to Schuerch. Adapted from Ref. 29.

Solvent	δ^a (MPa$^{1/2}$)	Δv^b (cm^{-1})	"Indulin" kraft pine[c]	"Meadol" soda hardwood[c]	Ethanol maple[c]	Native spruce[c]	Native aspen[c]
Hexane	14.9	L	Ins	Ins	9	Ins	Ins
Diethyl ether	15.8	130	Ins	Ins	6	Ins	Ins
CCl$_4$	17.5	L	Ins	Ins	Ins		Ins
Xylene	17.9	L	Ins	Ins			Ins
Butyl Cellosolve	18.2	H	Par	Par			
Diethylene glycol	18.6	H	Sol	Sol^-			
Ethyl acetate	18.1	84	Sli^-	Sli^-			
Benzene	18.7	L	Ins	Ins	7		
Methyl ethyl ketone	19.0	77	Par^-	Par^-			
Chloroform	19.0	–	Ins	Sli^+	4		
Chlorobenzene	19.4	14	Ins	Ins			
Carbon disulfide	20.4	L	Ins	Ins	Ins	Ins	Ins
Dioxane	20.4	97	Sol	Sol	3	Sol	Sol
Nitrobenzene	20.4	28	Ins	Sli			
Acetone	20.4	97	Par	Par^-	2	Sol	Sol
Methyl formate	20.7	ca. 84	Par^-	Par^-			
Acrylonitrile	21.4	ca. 56	Sli	Sli			
1-Nitropropane	21.8	L	Ins	Ins			
Pyridine	21.8	181	Sol	Sol	1	Sol	Sol
Methyl Cellosolve	22.0	H	Sol	Sol	Sol	Sol	Sol
Nitroethane	22.6	L	Ins	Ins			
1-Butanol	23.3	H	Ins	Ins			
2-Propanol	23.5	H	Sli	Ins			
Acetonitrile	24.3	63	Sli	Sli^+			
Nitromethane	25.7	L	Sli^-	Sli			
Ethanol	25.9	H	Sli^+	Sli	5		
Ethylene glycol	29.0	206	Sol	Sol	Sol	Sol	Sol
Methanol	29.2	187	Par^-	Par^-		Sol	Sol
Glycerol	33.7	H	Ins	Ins			
Water	47.7	H	Ins	Ins	8	Ins	Ins

$^a\delta$ values from Ref. 29 are in good agreement with those from Ref. 87 (Table 4.1).
^{b}Values determined by Gordy and Stanford[32] except those labeled "ca." which are estimates from Gordy's data for analogous compounds.[29] Where close analogues were not possible, H for high and L for low is used.[29]
^{c}Ins = insoluble, Sli = slightly soluble, Sol = soluble, Par = partially soluble, Sol = soluble, numerals are in order of decreasing solubility for a given lignin.[29]

The addition of water to organic solvents is useful in order to adjust the solvent system properties (δ parameter and hydrogen-bonding capacity), improving the solubilization of lignin. Wang *et al.*[31] demonstrated that the fraction of water in the 1,4-butanediol/water system can be used to adjust the δ parameter of the solvent system, and eventually improve the hydrogen-bonding capacity of the solvent system. As a result, they found sugarcane bagasse lignin to be soluble in 1,4-butanediol/water (80 : 20, v/v). Under this condition, the value for the δ parameter was 30.7 MPa$^{1/2}$, which was close to the δ value of the enzymatic hydrolysis/mild acidolysis lignin (28 MPa$^{1/2}$).

Curvelo and coworkers[33] studied organosolv pulping of *Pinus caribaea* in nine organic solvents (in 9 : 1 mixtures with 2 mol L^{-1} HCl aqueous solutions) at 398 K for 6 h. They found the lignin yields to be dependent on the δ value. The plot of lignin yield against δ value showed a "Bell shape" curve with maximum at 23 MPa$^{1/2}$. This value is in agreement with the δ value proposed by Schuerch.[29,34]

Technical lignins are soluble in alkaline aqueous solutions. Evstigneev[35] studied the factors affecting the solubility of several technical lignins recovered both by chemical pulping (*i.e.* soda, soda-anthraquinone, sulfate, and sulfate-anthraquinone) and by mild treatments (*i.e.* Björkman method, Pepper method, Freundberg method, and 1,4-dioxane extraction). The lignin solubility in aqueous alkali solutions is determined by the number of phenol groups (OH_{phen}) per phenylpropane units (PPU) in the macromolecule. A minimal content of 31 OH_{phen}/100 PPU was found to be required for lignin solubilization in aqueous alkali solutions.

4.2.2.2   *Chemical Solubilization*

Lignin subunits are predominantly connected by thermally labile ether linkages (α-O-4 and β-O-4). This feature can be exploited in the design of catalytic systems for lignin conversion. We demonstrated that lignin conversion could be performed even in solvents in which the substrate is initially insoluble (*e.g.* alkanes).[22] Nonetheless, temperatures higher than 523 K are required for the polymer thermolysis or solvolysis, which generates soluble fragments that are then transported into the catalyst pore, and converted on the catalyst surface.[22,36] As indicated in Scheme 4.2, solvents exert effects both upon the (noncatalytic) thermolysis of lignin and hydrogenolysis and hydrogenation of the lignin fragments. Another aspect to consider is the role of the catalyst in the conversion of lignin fragments. Should a catalyst not allow for the fast conversion of aldehyde and ketone species, formed upon thermolysis of lignin, these fragments would most likely condense. This reaction decreases the yield of liquids products upon formation of recalcitrant, carbon–carbon crosslinked chars.

In 1931, Kleinert and Tayenthal[37] proposed the cooking of wood in ethanol–water (1 : 1, v/v) solution as a method to extract lignin from the lignocellulosic matrix. Clearly, the efficiency of this process is intricately related to the solvent properties;[33] however, the process nature is now known to

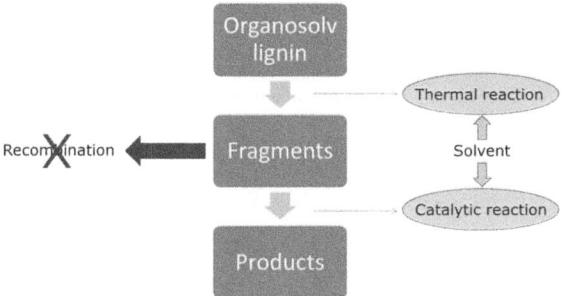

Scheme 4.2 Steps of lignin conversion.

be autocatalytic.[26,27] When wood is thermally treated in alcohol-water solutions, hemicellulose undergoes deacetylation at temperatures as low as 453 K.[26] As a result, acetic acid is *in situ* generated, catalyzing the partial hydrolysis of hemicellulose, and most importantly, catalyzing solvolytic processes on α-O-4 and, to a lesser extent, on β-O-4 linkages (Scheme 4.3).[26] The solvolytic processes are responsible for the release of reactive lignin fragments into the extraction liquor, as displayed in Scheme 4.3. Should these fragments have a long residence time in the plant tissue or liquor, the fragments recombine generating a less reactive, condensed lignin stream of high M_w.[38]

We demonstrated lignin to be solvolytically released from the plant cell wall as oligomers of low molecular weight ($200 < M_w < 1000$ Da).[38] Such oligomers offer a great opportunity for efficient upgrading of the lignin stream already at the extraction step.[38] We showed the reactive fragments to be easily converted into small molecules by hydrogen-transfer reactions catalyzed by inexpensive Raney® Ni. In this process, the reactive functionalities (*e.g.* aldehydes and ketones), occurring in lignin oligomers, are reduced to alcohols or hydrocarbons,[22] thus minimizing the likelihood of lignin fragments to condense. Consequently, the lignin stream is obtained as a viscous oil instead of a polymeric solid. Moreover, pulps (*i.e.* cellulose and hemicellulose stream) obtained by this method have a low lignin content, and thus, they are amenable to enzymatic hydrolysis.[38]

Figure 4.1 compares a two-dimensional gas chromatography image (2D GC×GC) of organosolv lignin and lignin stream obtained from catalytic biorefining in the presence of Raney® Ni in 2-PrOH/H$_2$O (7:3, v/v) at 453 K for a duration of 3 h. As organosolv lignin is a nonvolatile polymer, Figure 4.1(a) shows only some weak spots, which are associated with traces of low M_w compounds formed upon lignin extraction. In contrast, Figure 4.1(b) shows phenols to be the predominant species present in the depolymerized lignin oil stream, aside from a secondary fraction of cyclohexanols, 1,2-cyclohexanediols, and other diols (derived from the hydrogenolysis of hemicellulose sugars, or side reactions involving 2-PrOH). By thermogravimetric analysis, the fraction of volatile components should correspond to approximately 55 wt% of the entire oil sample, as estimated by the weight loss determined at 573 K (GC injector temperature).[38]

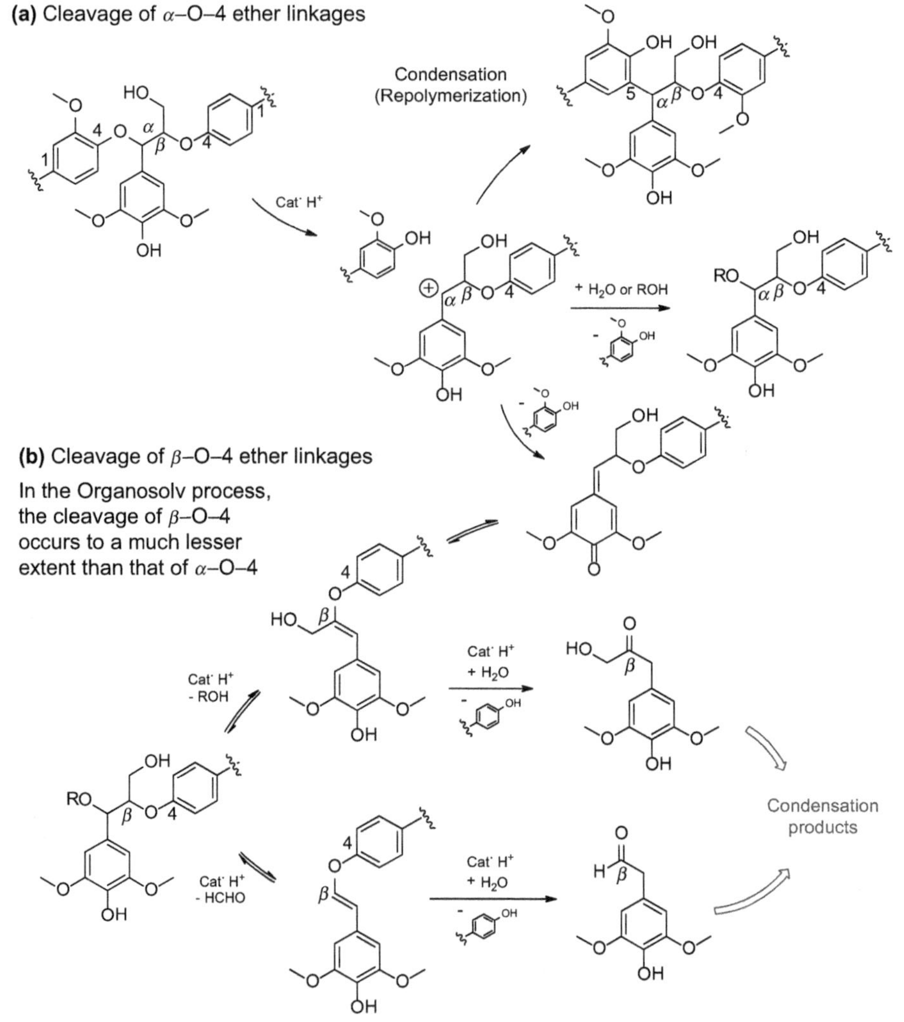

(a) Cleavage of α–O–4 ether linkages

(b) Cleavage of β–O–4 ether linkages

In the Organosolv process, the cleavage of β–O–4 occurs to a much lesser extent than that of α–O–4

Scheme 4.3 Solvolysis of lignin linkages by the Organosolv process. (a) Reaction network proposed for the solvolysis of α-O-4 (dominant in the Organosolv process), and (b) β-O-4 ether linkages, respectively.[26]

4.2.3 Hydrogen Solubility

An efficient transport of H_2 from the gas phase through the liquid phase, and then onto the catalyst surface is key for catalytic hydrogenations. In this context, hydrogen solubility in the reaction medium is of fundamental importance for the investigation of solvent effects on catalytic hydrogenations (see also Sections 12.2.1.1 and 13.4.5). Molecular hydrogen is more soluble in nonpolar solvents than in the polar counterparts, as shown in Figure 4.2. Logically, hydrogen solubility depends on the reaction conditions (*i.e.* temperature and pressure) in addition to the composition of the reaction medium.

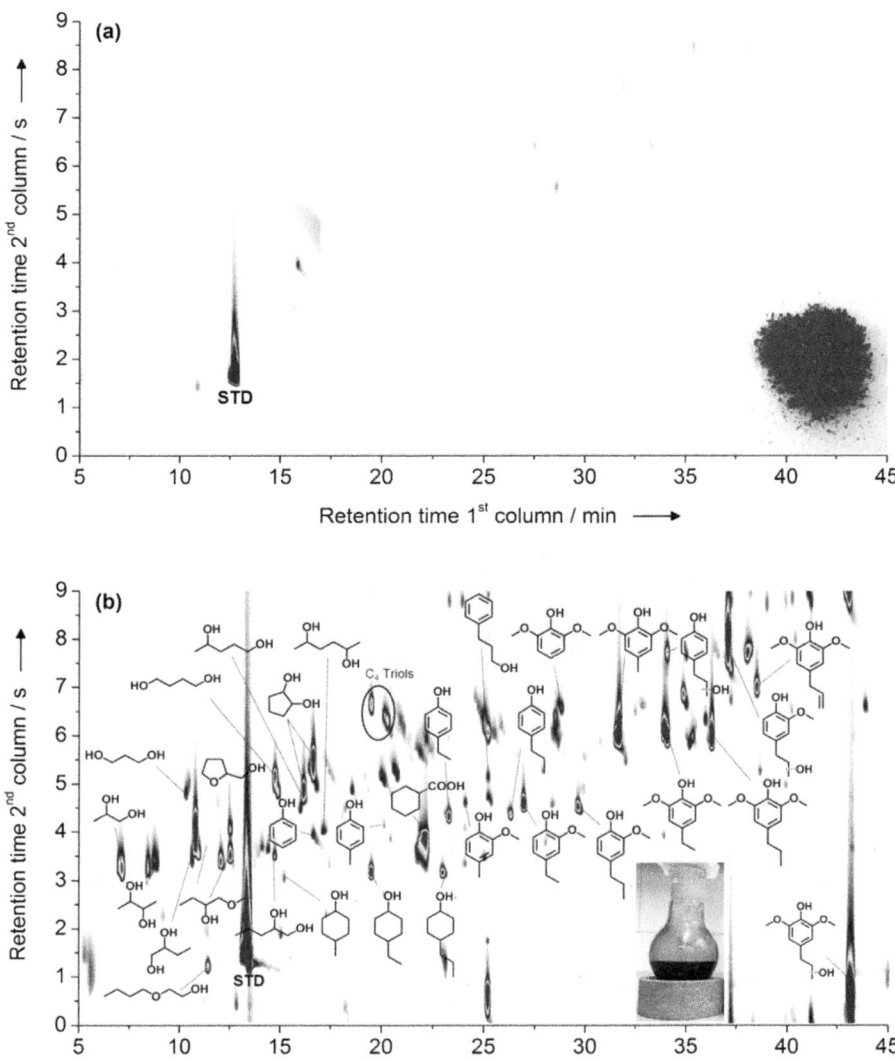

Figure 4.1 GC×GC images of (a) Poplar organosolv lignin and (b) lignin stream (lignin oil) obtained from the catalytic biorefining of poplar wood in the presence of Raney® Ni at 453 K for 3 h (2-PrOH/H2O, 7 : 3, v/v). The inset shows the visual appearance of organosolv lignin and lignin oil.[38]

Unfortunately, the hydrogen solubility in water—a typical solvent used in carbohydrate conversion—is lower than that in methanol by approximately an order of magnitude. However, sugars and sugar alcohols were found to slightly improve the hydrogen solubility in water.[39] In contrast, inorganic or organic salts dramatically reduce the hydrogen solubility in a medium.[40,41] Table 4.3 lists the literature on the H_2-solubility data by solvent type.

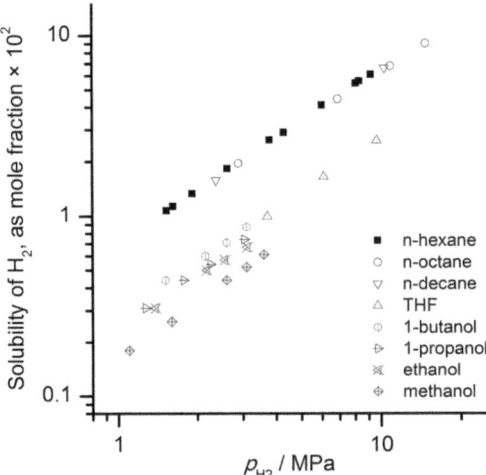

Figure 4.2 Solubility of hydrogen in some organic solvents.
Compiled from Refs. 42 and 43.

Table 4.3 Literature on hydrogen solubility in several
solvent classes.

Solubility in	Ref.
Alcohols	42, 44–49
Alkanes	43, 49–52
Alkenes	51
Arenes	51, 53
Aqueous solutions of sugars and sugar alcohols	39
Aqueous solutions of inorganic salts	40, 41
Amides	43
Amines	54
CO_2-expanded liquids	55
Diols	56
Esters	42
Ethers	43, 57
Ketones	43, 51
Ionic liquids	58–60, 90
Nitro compounds	48
Other solvents	43, 61
Water	40, 62

4.3 Solvent Properties

Correlations between solvents properties and reaction rates have been extensively studied in order to rationalize solvent effects. Recently, Adjiman and coworkers[63] introduced an elegant *in silico* method for prediction of the best solvent candidates for organic reactions. Unfortunately, in heterogeneous catalysis, the influence of the reaction medium in catalytic hydrogenation is often neglected, and therefore, still poorly understood. In this field, solvent properties are also useful in order to propose hypotheses on the nature of the interactions of solvent molecules occurring at the liquid/solid interface, where the catalytic reactions happen. The most considered properties for the analysis of solvent effects are: (1) solvent polarity or Reichardt parameter $(E_T{}^N)$, (2) Kamlet–Taft parameters, *i.e.* H-bonding properties (α, β) and polarity/polarizibility (π^*), (3) acceptor number (AN), and (4) donor number (DN).

4.3.1 Polarity $(E_T{}^N)$

Solvent polarity qualitatively expresses the overall capacity of a solvent for solvating dissolved species, which can be charged or neutral, nonpolar or dipolar. This term is, however, difficult to quantitatively express or define.[71] Some approaches for a quantitative definition of polarity involve the consideration of physical solvent properties (*e.g.* dielectric constant, dipole moment, or refractive index) as a measure of solvent polarity.[64] However, these approaches treat the solvent as a nonstructured isotopic continuum, not composed of individual solvent molecules with their own solvent–solvent interactions, and most critically, they do not take into account specific solute–solvent interactions (*i.e.* hydrogen bonding, and electron-pair donor [EPD]/electron-pair acceptor [EPA] complexes).[64]

To overcome these limitations, the use of probe molecules (*i.e.* solvatochromic compounds) was introduced to report the overall characteristics of the chemical environment surrounding a solute.[64] Solvatochromic compounds are solutes that change their colors due to a change in solvent polarity.[65] For this purpose, a pyridinium-*N*-phenolate betaine dye is often used.[64] $E_T(30)$ values are based on 4-(2,4,6-triphenylpyridinium)-2,6-diphenylphenoxide dye as a probe molecule. According to eqn (4.1), $E_T(30)$ values are defined as the molar electronic transition energies (E_T) of the dissolved dye, given in kcal mol^{-1}, at 298 K and normal pressure (0.1 MPa):[64]

$$E_T(30) = hc\tilde{\nu}_{max}N_A = 28591/\lambda_{max} \text{ (nm)} \qquad (4.4)$$

where, λ_{max} is the wavelength of the maximum of the longest wavelength, intramolecular charge-transfer π–π^* adsorption band of the dye.

E_T values can be given as normalized, dimensionless $E_T{}^N$ values, as defined by:

$$E_T{}^N = \frac{E_T(\text{solvent}) - E_T(\text{TMS})}{E_T(\text{water}) - E_T(\text{TMS})} \qquad (4.5)$$

Hence, normalized E_T values to give $E_T{}^N(\text{water}) = 1$ and $E_T{}^N(\text{TMS}) = 0$, where TMS stands for tetramethylsilane, a nonpolar reference solvent. Comprehensive lists of $E_T{}^N$ are given in Refs. 64 and 66. The solvent polarity is broadly used as a correlation parameter in the analysis of solvent effects on chemo- and regioselectivity of chemical transformations catalyzed by solids.[67,68]

4.3.2 Kamlet–Taft Parameters (α, β and π^*)

A more comprehensive description of solvent polarity is given by the so-called Kamlet–Taft parameters that separately measure the hydrogen-bond donor (α), hydrogen bond acceptor (β), and dipolarity/polarizability (π^*) abilities of a solvent. Similarly to E_T^N, the parameters α and β are solvatochromic properties determined by the energies of the longest wavelength adsorption peaks of certain carefully selected probe solutes in the solvents in question.[69] Unlike E_T^N, α and β are based on solvatochromic analysis performed on two different probe solutes in separate experiments.[70] The parameter α, which describes the ability of a solvent to donate a hydrogen bond, is normalized to give α(non-HBD solvent) = 0 and α(methanol) = 1.0. The parameter β, which describes the ability of a solvent to accept a hydrogen bond, is normalized to give β(cyclohexane) = 0 and β(HMPT) = 1.0. Details on the empirical determination of these parameters are given in ref. 64. The solvatochromic parameter π^* is a measure of the exoergic effects of solute-solvent, dipole–dipole, and dipole–induced dipole interactions. Hence, it describes the ability of a solvent to stabilize a neighboring charge or dipole through nonspecific dielectric interactions.[64] The π^* scale is normalized to give π^*(cyclohexane) = 0 and π^*(dimethyl sulfoxide) = 1. A comprehensive list of Kamlet–Taft parameters for several solvent classes is given in Refs. 64, 69, and 70. It is noteworthy that α, β, and π^* are dimensionless parameters.

Marcus[69] showed several binary correlations of solvent parameters. In this context, the parameter α is closely associated with the acceptor number (AN), a quantitative measure of Lewis acidity. Likewise, he also showed that the parameter β strongly correlates with the donor number (DN), a quantitative measure of Lewis basicity. In turn, the parameter π^*, which describes a combination of polarity and polarizability of a solvent, correlates neither with acidity nor basicity. Overall, should a catalytic property (conversion, selectivity, *etc.*) show a good correlation with α or β, this correlation might not be directly associated with solvation phenomena, but with Lewis acid–base interactions between solvent and catalyst surface.[22]

4.3.3 Acceptor Number (AN)

Acceptor number (AN) is a measure of the Lewis acidity of a solvent. It is obtained from the chemical shift in the ^{31}P NMR spectrum of triethylphosphine oxide, a Lewis-base probe, dissolved in the respective solvents.[72] The ^{31}P NMR shift depends strongly on the acid–base interaction between acidic solvent and triethylphosphine oxide. This correlation enables the use of ^{31}P NMR shifts to establish a scale of the solvent Lewis acidity. The shift is scaled as AN(hexane) = 0 and AN(SbCl$_5$) = 100. These values are *dimensionless*. AN values for several solvents are given in Refs. 69 and 73.

4.3.4 Donor Number (DN)

Unlike the aforementioned solvent properties, which are determined by spectroscopic methods (UV-Vis and ^{31}P NMR), the donor number is determined by calorimetry. The donor number value is defined as $DN = -\Delta_r H$ obtained from the reaction of a Lewis base and a standard Lewis acid (SbCl$_5$) to form a 1:1 complex in a dilute solution of 1,2-dichloroethane at 298 K, as given by:

$$SbCl_5 + Solv: \leftrightarrows Solv\text{-}SbCl_5 \qquad (4.6)$$

DN values for several solvents are given in Refs. 69 and 73. Although these values are presented sometimes without a unit (*e.g.* in Ref. 69), the values are often given in kcal mol^{-1}, but sometimes, values in kJ mol^{-1} are also reported. Therefore, DN values should be considered with caution. It is noteworthy that the attempt to provide a "balance" of acid–base properties of a solvent by assuming the difference "DN–AN" as a correlation parameter constitutes an inadequate approach for the analysis of solvent effects on a chemical reaction. This is because DN is a calorimetric value (kJ–mol^{-1}), whereas AN (dimensionless value) is given relative to chemical shift in the ^{31}P NMR spectrum of triethylphosphine oxide. Although chemists are used to a reciprocity of acid–base properties (*e.g.* pH + pOH = 14, for dilute aqueous solutions at 298 K and 0.1 MPa), AN and DN values are not reciprocal scales of Lewis acidity and basicity. Unfortunately, this point is often unclear in the literature, and still causes much confusion among freshmen and experienced researchers.

4.4 Selected Examples of Solvent Effects in Biomass Hydrogenation

The correlation of reaction rates and product distributions with solvent polarity or dielectric constant has been one of the main tools towards rationalization of solvent effects on heterogeneous catalysts. However, it remains unclear how the solvent properties, presented in the previous section, are able to steer the activity and selectivity of metal catalysts. On the one hand, there is no reason to assume that the weak specific (H-bonding) and/or nonspecific (van der Waals) interactions involving solvent and substrate would greatly influence chemical transformations, which take place on the catalyst surface through (very) strong, well-established interactions between substrate and active sites/support. On the other hand, the mechanism of adsorption of a substrate on the catalytic surface may considerably change depending on the solvent properties. In addition, the solvent itself may compete for the catalytic sites. In this section, some selected examples will be discussed in order to show the significance of such effects for the design of new catalytic systems for biomass valorization through hydrogenation and hydrogenolysis.

4.4.1 Preferential Adsorption

Perhaps one of the most comprehensive accounts on the influence of solvent effects in hydrogenation was given by Singh and Vannice in 2001.[68] Overall, there is clear evidence that a polar solvent enhances adsorption and reaction of a nonpolar reactant, while a nonpolar solvent enhances the adsorption and reaction of a polar reactant.[68] Unfortunately, to the best of our knowledge, there is no study on the solvent effects on the preferential adsorption of biomass-derived molecules. For instance, in almost all the studies, hydrogenation of sugars is performed in water because sugars are simply insoluble in alcohols (*e.g.* ethanol) or other common organic solvents. Nonetheless, Mikkola *et al.*[91] studied the hydrogenation of xylose in water and aqueous ethanol solutions. They found the organic solvent component to exert an enhancing effect on the hydrogenation of sugars. The authors proposed the improved hydrogen solubility as the main factor accounting for the "rate-accelerating" effect of alcohols. However, the decrease in polarity of the medium upon addition of ethanol could also be another factor favoring the adsorption of xylose, and thus, its conversion into xylitol.

Unlike sugars, which are almost insoluble in most of the common organic solvents used for performing hydrogenations, lignin and its model compounds are soluble in a multitude of organic solvents, thus enabling the study of solvent effects on their catalytic conversions.[74–76] In this context, we studied hydrogenolysis of diphenyl ether in fourteen solvents using Raney® Ni as a catalyst (Figure 4.3).[22]

Figure 4.3 shows the reactions conducted in common solvents for lignin conversion (*e.g.* methanol, ethanol, tetrahydrofuran and 1,4-dioxane) achieved the lowest conversions in the presence of Raney® Ni. Among the alcohols studied, the reaction conversion decreases in the following order: 1,1,1,6,6,6-hexafluoropropanol (Hex-F-2-PrOH, 100%), 2-propanol (2-PrOH, 73%), 2-butanol (55%), ethanol (33%), 1-butanol (32%), *t*-butanol (22%), and methanol (12%). In turn, among the cyclic ethers, the reaction shows the best conversion in 2-methyltetrahydrofuran (2-methyl-THF, 61%), whereas poor conversions are obtained in 1,4-dioxane (18%) and tetrahydrofuran (THF, 30%). Markedly, quantitative conversion of the substrate is reached, when the reaction is conducted in Hex-F-2-PrOH, decaline, methylcyclohexane (MCH), or *n*-heptane.[22]

The product distribution pattern found for the hydrogenolysis of diphenyl ether (Figure 4.3) was also observed for the hydrogenolysis of lignin. We found that subjecting organosolv lignin to Raney® Ni in methanol under an initial H_2 pressure of 7 MPa (r.t.) at 573 K for 8 h is an effective strategy to produce a complex mixture of phenols from lignin. Under the same conditions, reaction in methylcyclohexane resulted in cyclic alcohols and cyclic alkanes, whereas reaction in 2-propanol led to cyclic alcohols, cyclic ketones, and unsaturated products.[22]

The correlations of solvent properties and conversion indicated the Lewis basicity of the solvent, as expressed by DN, to be an important parameter for

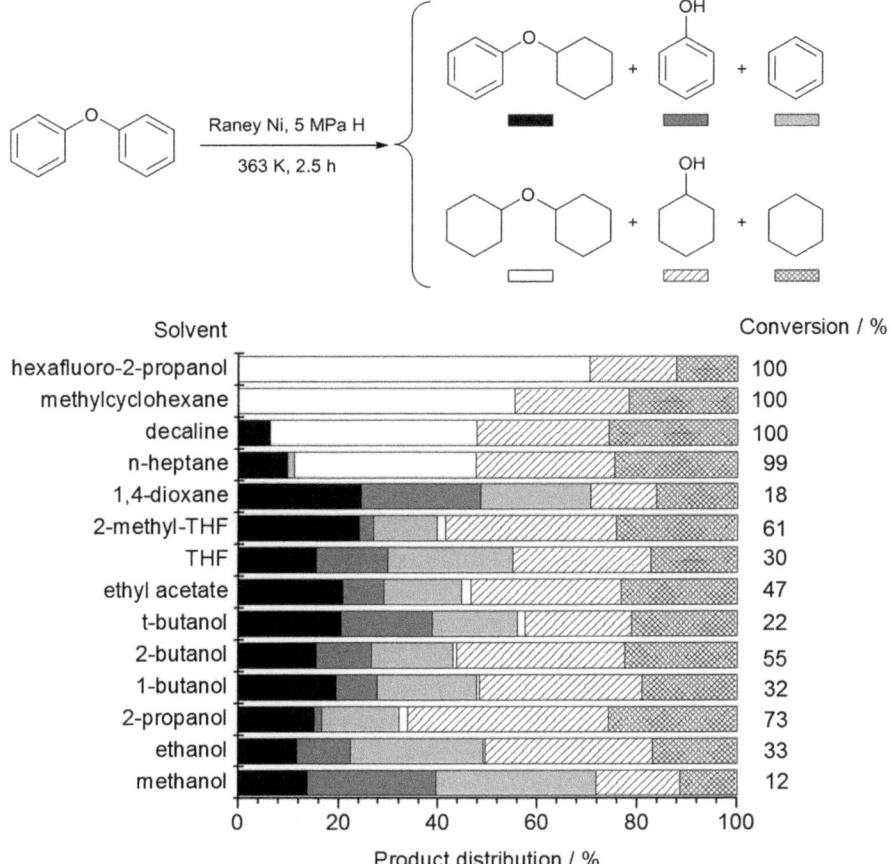

Figure 4.3 Results from the conversion of diphenyl ether with Raney® Ni in several solvents.[22]

the reaction performance. In Figure 4.4, it is possible to identify a correlation between DN of the solvents and substrate conversions. As indicated by the dashed line, an exponential decay model fits the conversions—achieved in methylcyclohexane (MCH), decaline, *n*-heptane, Hex-F-2-PrOH, 2-Me-THF, ethyl acetate, tetrahydrofuran, 1-butanol, ethanol and *t*-butanol—very well with the respective DN of the solvents (adjusted-R-Square 0.97).[22]

The results from the mathematical model suggests competitive adsorption of the solvent to be steering the activity of Raney® Ni.[22] As diphenyl ether shows a DN value of 21.1 kJ mol^{-1}, the solvents—MCH, decaline, *n*-heptane, and Hex-F-2-PrOH (DN = 0)—should not compete directly with diphenyl ether for surface sites. In contrast, solvents—showing a higher DN value than diphenyl ether—should compete with the substrate for catalytic surface. Nonetheless, the mathematical model could not describe the low conversions achieved in methanol and 1,4-dioxane, and the high conversions achieved in 2-propanol and 2-butanol. Studies on the adsorption of

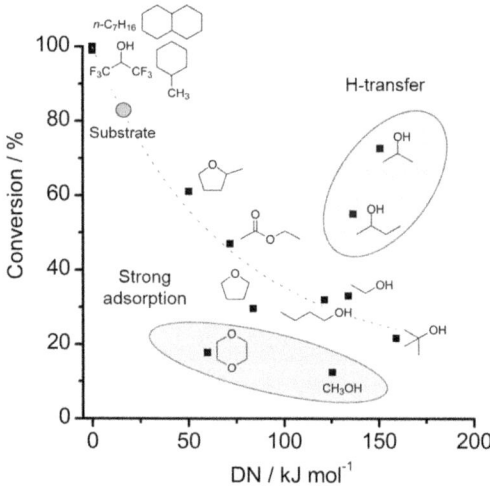

Figure 4.4   Conversion of diphenyl ether *versus* the Lewis basicity of the solvents as described by the donor number (DN).
Adapted from Ref. 22.

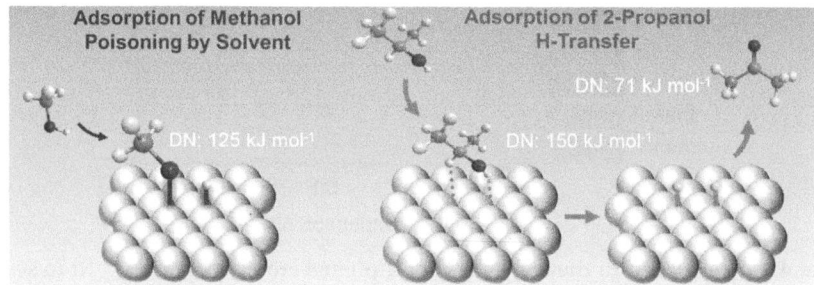

Figure 4.5   Qualitative view of the interactions of a primary (a, methanol) and secondary alcohol (b, 2-propanol) with a Ni surface.
Adapted from Ref. 81.

methanol on Ni(111) and Ni(100) surfaces showed that the process involves the cleavage of the O–H bond, generating a surface methoxyl group and a H adatom (Figure 4.5).[77] Likewise, ethanol adsorbs on Ni surfaces leading to a surface ethoxyl group and a H adatom.[78] Al-Mawlali and Saleh[77,78] showed that Ni films take up 1.9 times more methanol than ethanol in the adsorption process. Hence, it is reasonable to assume that the extensively surface coverage by methoxyl groups is perhaps one of the reasons for the inhibition, and in some cases, the deactivation of Raney® Ni catalysts in reactions conducted in methanol (or even ethanol).[79] With regard to 1,4-dioxane, this molecule can use both *O*-sites to simultaneously interact with the Ni surface, establishing a strong interaction with the catalyst surface, extensively blocking the active sites.

Table 4.4 Influence of the solvents in the reactions rates and activation energies found for the individual steps of phenol hydrodeoxygenation in water, methanol and hexadecane at temperature of 473 K under a H_2 pressure of 4 MPa. Adapted from ref. 84.

Solvent	Phenol hydrogenation over Pd/C		Cyclohexanone hydrogenation over Pd/C		Cyclohexanol dehydration over HZSM-5		Cyclohexene hydrogenation over Pd/C	
	$TOF_1{}^a$	E_a (kJ mol^{-1})	$TOF_2{}^a$	E_a (kJ mol^{-1})	$TOF_3{}^b$	E_a (kJ mol^{-1})	$TOF_4{}^a$	E_a (kJ mol^{-1})
Water	4.2×10^3	70	2.1×10^4	37	1.6×10^3	115	6.0×10^6	25
Methanol	1.3×10^3	50	3.5×10^3	28	1.8×10^3	124	5.6×10^6	12
Hexadecane	7.8×10^2	79	5.1×10^3	48	1.1×10^3	124	7.7×10^6	24

[a]TOF_1, TOF_2 and TOF_4 are given as mol mol$^{-1}{}_{Pd\ surf}$ h^{-1}.
[b]TOF_3 is given as mol mol$^{-1}{}_{BAS}$ h^{-1}, where BAS stands for Brønsted acid sites. The kinetic parameters were determined at 473 K and 4 MPa H_2 at a stirring speed of 700 rpm.

Remarkably, better conversions than those predicted from the DN values were achieved in 2-propanol and in 2-butanol. Again, the nature of the interactions of secondary alcohols with Ni surface should be taken into account. Raval and coworkers[80] reported the interaction of 2-propanol with a Ni(111) surface to cleave both the O–H and the α-C–H bonds. As a result, acetone is released into the gas phase and two H atoms remain adsorbed on the Ni surface. The H atoms can be used in hydrogenation or hydrogenolysis (Figure 4.5).[38,81–83] In liquid-phase systems, the equilibrium 2-PrOH ⇌ acetone + 2 H_{ad} at the liquid/solid interface is conducive to the conversion of diphenyl ether, as acetone (DN = 70 kJ mol^{-1}) is less basic than 2-propanol (DN = 150 kJ mol^{-1}), and therefore, should alleviate problems associated with the competitive adsorption of the solvent.

Recently, Lercher and coworkers[84] examined the solvent effects on the individual steps of phenol hydrodeoxygenation to cyclohexane with Pd/C and HZSM-5 as catalysts. Table 4.4 summarizes the influence of the solvents in the reactions rates and activation energies found for the individual steps of phenol hydrodeoxygenation in water, methanol and hexadecane at temperature of 473 K under a H_2 pressure of 4 MPa. The activation energies decreased in the order E_a(phenol hydrogenation) > E_a(cyclohexanone hydrogenation) > E_a(cyclohexene hydrogenation). In turn, the reaction rates increased in the inverse order, *i.e.* TOF_1(phenol hydrogenation) < TOF_2(cyclohexanone hydrogenation) < TOF_4(cyclohexene hydrogenation). The apparent activation energies for the three hydrogenations were lower in methanol than in water or hexadecane. The rates and activation energies for dehydration catalyzed by HZSM-5 were similar in the three solvents.[84]

4.4.2 Reactive Solvent

When performed in alcohols, hydrogenation of ketones and aldehydes can take place through a hemiacetal or acetal intermediate, which can

Table 4.5 Hydrogenation products of 4-methylcyclohexanone in ethanol in the presence of noble-metal catalysts and H_2.[a] Adapted from Ref. 86.

		Composition of the reaction mixture (mol%)						
Catalyst	Conversion (%)			OH	OH		Ketal	Ketone
Pd	100	86.9	9.7	2.3	1.1	0	0	0
Pt	100	9.2	3.9	40.2	34.5	12.2	0	0
Rh[b]	64.5	4.0	0.8	48.9	9.9	9.9	12.3	16.4
Ru	100	0	0	69.0	31.0	31.0	0	0

[a]Reaction conditions: 4-methylcyclohexanone (0.2 mL) was hydrogenated using 40 mg of hydroxide (Pd, Rh, Ru) or oxide (Pt) catalysts in ethanol (10 mL) at 298 K under a H_2 pressure of 0.1 MPa. The catalysts were prereduced in water, washed with water and then with ethanol prior to use.
[b]The authors reported an unknown compound formed in this reaction that has the same retention time as 1-ethoxy-4-methylcyclohexene by GC analysis.

subsequently be reduced to the corresponding ether. Van Bekkum and coworkers[85] studied the conversion of 5-hydroxymethylfurfural (HMF) with Pd/C in alcoholic solution. In 1-propanol, the formation of an intermediate, 2-alkoxy-5-hydroxymethylfuran, at yields exceeding 85% at 0.5 h, was obtained. Similar results were achieved in 2-propanol.

The formation of ethers is typically obtained by Pd-catalyzed hydrogenation of C=O using alcohols as solvents. Nishimura *et al.*[86] showed the selective formation of ethers in the hydrogenation of 4-methylcyclohexanone using a Pd catalyst. With other noble-metal catalysts, high selectivity to cyclohexanols is observed. Notably, acetalization seems to be fully suppressed in the presence of Ru catalysts. In effect, the reaction exclusively produces cyclohexanols (Table 4.5).

Some solvents (*e.g.*, primary and secondary alcohols) can act as an H-source for hydrogenations even under low severity conditions (*e.g.*, 323–423 K). In this context, Raney® Ni is able to catalyze the reduction of many organic functionalities by transfer hydrogenation using 2-propanol as an H-donor.[38,81,83] Nonetheless, transfer hydrogenations catalyzed by Raney® Ni have not found extensive uses in organic synthesis due to their lack of chemoselectivity, which often leads to defunctionalization of complex molecules. This "disadvantage" attracted our attention because the upgrade of highly functionalized phenols, such as those found for the lignin substructures or pyrolysis oils, requires catalysts capable of simultaneously reducing several functionalities under low-severity conditions. Recently, we demonstrated the transfer hydrodeoxygenation of the phenolic fraction of pyrolysis oils, rendering a mixture of cyclic alcohols and diols, to be feasible under unprecedented low-severity conditions (*e.g.* 433 K and autogeneous pressure) using 2-propanol and Raney® Ni.[83]

4.5   Concluding Remarks

Solvents and solvent effects are of extreme importance for the development of catalytic systems for valorization of biomass through hydrogenation and hydrogenolysis. Solvents are not limited to serve as media for the solubilization of substrate and intermediates. In fact, solvents exert key effects upon solid catalysts. These effects are related to competitive adsorption of the solvent, and most importantly, to reactions of the solvent on the catalyst surface. These reactions can lead to formation of stable surface species, which may block the surface. Therefore, solvent screening is always recommended when starting the study of a new catalytic system for a particular liquid-phase hydrogenation/hydrogenolysis. An "outstanding" catalyst performance (*i.e.* activity and selectivity) may be a result of solvent effects instead of catalyst properties or preparation. In spite of this, solvent effects on heterogeneous catalysts still constitute a neglected field in catalysis research. Therefore, there is a serious lack of real understanding of the solvent interactions at the liquid/solid interfaces. Such interfaces are not easy to analyze spectroscopically. Accordingly, most of the knowledge gained so far is a result from indirect correlations between catalytic data and (solvatochromic) solvent properties. Such correlations offer good "hints" about eventual interactions occurring on the catalyst surface. However, they do not replace more consistent conclusions drawn by surface-science studies.

Acknowledgments

R.R. acknowledges the financial support by Alexander von Humboldt Foundation (Sofja Kovalevskaja Award 2010). This work was performed as part of the Cluster of Excellence "Tailor-Made Fuels from Biomass", funded by the German Federal and State governments to promote science and research at German universities.

References

1. R. Rinaldi and F. Schüth, *ChemSusChem*, 2009, **2**, 1096–1107.
2. R. Rinaldi and F. Schüth, *Energy Environ. Sci.*, 2009, **2**, 610–626.
3. S. VandeVyver, J. Geboers, P. A. Jacobs and B. F. Sels, *ChemCatChem*, 2011, **3**, 82–94.
4. R. Rinaldi and J. Reece, in *Catalysis for the Conversion of Biomass and its Derivatives*, eds. M. Behrens and A. K. Datye, Max Planck Research Library for History and Development of Knowledge, Berlin, Open Access Editon, 2013.
5. N. Huu-Phuoc, H. Nam-Tran, M. Buchmann and U. W. Kesselring, *Int. J. Pharm.*, 1987, **34**, 217–223.
6. R. J. Roberts and R. C. Rowe, *Int. J. Pharm.*, 1993, **99**, 157–164.
7. C. M. Hansen and A. Bjorkman, *Holzforschung*, 1998, **52**, 335–344.

8. C. Loerbroks, R. Rinaldi and W. Thiel, *Chem. Eur. J.*, 2013, **19**, 16282–16294.
9. X. Liang, A. Montoya and B. S. Haynes, *J. Phys. Chem. B*, 2011, **115**, 10682–10691.
10. M. Benoit, A. Rodrigues, Q. Zhang, E. Fourre, K. D. O. Vigier, J.-M. Tatibouet and F. Jerome, *Angew. Chem. Int. Ed.*, 2011, **50**, 8964–8967.
11. R. Carrasquillo-Flores, M. Käldström, F. Schüth, J. A. Dumesic and R. Rinaldi, *ACS Catal.*, 2013, **3**, 993–997.
12. S. M. Hick, C. Griebel, D. T. Restrepo, J. H. Truitt, E. J. Buker, C. Bylda and R. G. Blair, *Green Chem.*, 2010, **12**, 468–474.
13. J. Hilgert, N. Meine, R. Rinaldi and F. Schüth, *Energy Environ. Sci.*, 2013, **6**, 92–96.
14. M. Käldström, N. Meine, C. Fares, R. Rinaldi and F. Schüth, *Green Chem.*, 2014, **16**, 2454–2462.
15. M. Käldström, N. Meine, C. Fares, F. Schüth and R. Rinaldi, *Green Chem.*, 2014, **16**, 3528–3538.
16. N. Meine, R. Rinaldi and F. Schüth, *ChemSusChem*, 2012, **5**, 1449–1454.
17. Q. Zhang and F. Jerome, *ChemSusChem*, 2013, **6**, 2042–2044.
18. J. Geboers, S. Van de Vyver, K. Carpentier, P. Jacobs and B. Sels, *Green Chem.*, 2011, **13**, 2167–2174.
19. H. Kobayashi, Y. Ito, T. Komanoya, Y. Hosaka, P. L. Dhepe, K. Kasai, K. Hara and A. Fukuoka, *Green Chem.*, 2011, **13**, 326–333.
20. H. Kobayashi, H. Matsuhashi, T. Komanoya, K. Hara and A. Fukuoka, *Chem. Commun.*, 2011, **47**, 2366–2368.
21. L.-N. Ding, A.-Q. Wang, M.-Y. Zheng and T. Zhang, *ChemSusChem*, 2010, **3**, 818–821.
22. X. Y. Wang and R. Rinaldi, *ChemSusChem*, 2012, **5**, 1455–1466.
23. J. Zakzeski, P. C. A. Bruijnincx, A. L. Jongerius and B. M. Weckhuysen, *Chem. Rev.*, 2010, **110**, 3552–3599.
24. E. Dorrestijn, L. J. J. Laarhoven, I. Arends and P. Mulder, *J. Anal. App. Pyrol.*, 2000, **54**, 153–192.
25. R. Parthasarathi, R. A. Romero, A. Redondo and S. Gnanakaran, *J. Phys. Chem. Lett.*, 2011, **2**, 2660–2666.
26. R. B. Santos, P. W. Hart, H. Jameel and H. M. Chang, *Bioresources*, 2013, **8**, 1456–1477.
27. S. Aziz and K. Sarkanen, *Tappi J.*, 1989, **72**, 169–175.
28. T. Vom Stein, P. M. Grande, H. Kayser, F. Sibilla, W. Leitner and P. Domínguez De María, *Green Chem.*, 2011, **13**, 1772–1777.
29. C. Schuerch, *J. Am. Chem. Soc.*, 1952, **74**, 5061–5067.
30. Y. Ni and Q. Hu, *J. Appl. Polym. Sci.*, 1995, **57**, 1441–1446.
31. Q. Wang, K. Chen, J. Li, G. Yang, S. Liu and J. Xu, *Bioresources*, 2011, **6**, 3034–3043.
32. W. Gordy and S. C. Stanford, *J. Chem. Phys.*, 1940, **8**, 170–177.
33. D. T. Balogh, A. A. S. Curvelo and R. Degroote, *Holzforschung*, 1992, **46**, 343–348.

34. A. L. Horvath, *J. Phys. Chem. Ref. Data*, 2006, **35**, 77–92.
35. E. I. Evstigneev, *Russ. J. Appl. Chem.*, 2011, **84**, 1040–1045.
36. J. Zakzeski and B. M. Weckhuysen, *ChemSusChem.*, 2011, **4**, 369–378.
37. T. Kleinert and K. v. Tayenthal, *Angew. Chem.*, 1931, **44**, 788–791.
38. P. Ferrini and R. Rinaldi, *Angew. Chem. Int. Ed.*, 2014, **53**, 8634–8639.
39. J. Wisniak, M. Hershkow, R. Leibowit and S. Stein, *J. Chem. Eng. Data*, 1974, **19**, 247–249.
40. T. E. Crozier and S. Yamamoto, *J. Chem. Eng. Data*, 1974, **19**, 242–244.
41. D. C. Engel, G. F. Versteeg and W. P. M. van Swaaij, *J. Chem. Eng. Data*, 1996, **41**, 546–550.
42. M. S. Wainwright, T. Ahn, D. L. Trimm and N. W. Cant, *J. Chem. Eng. Data*, 1987, **32**, 22–24.
43. E. Brunner, *J. Chem. Eng. Data*, 1985, **30**, 269–273.
44. E. Brunner, *Ber. Bunsen-Ges. Phys. Chem.*, 1979, **83**, 715–721.
45. V. R. Choudhary, M. G. Sane and H. G. Vadgaonkar, *J. Chem. Eng. Data*, 1986, **31**, 294–296.
46. J. V. H. d'Angelo and A. Z. Francesconi, *J. Chem. Eng. Data*, 2001, **46**, 671–674.
47. T. Katayama and T. Nitta, *J. Chem. Eng. Data*, 1976, **21**, 194–196.
48. K. Radhakrishnan, P. A. Ramachandran, P. H. Brahme and R. V. Chaudhari, *J. Chem. Eng. Data*, 1983, **28**, 1–4.
49. K. J. Kim, T. R. Way, K. T. Feldman and A. Razani, *J. Chem. Eng. Data*, 1997, **42**, 214–215.
50. J. F. Connolly and G. A. Kandalic, *J. Chem. Eng. Data*, 1986, **31**, 396–406.
51. M. Herskowitz, J. Wisniak and L. Skladman, *J. Chem. Eng. Data*, 1983, **28**, 164–166.
52. J. Park, R. L. Robinson and K. A. M. Gasem, *J. Chem. Eng. Data*, 1995, **40**, 241–244.
53. J. Park, R. L. Robinson and K. A. M. Gasem, *J. Chem. Eng. Data*, 1996, **41**, 70–73.
54. E. Brunner, *Ber. Bunsen-Ges. Phys. Chem.*, 1978, **82**, 798–805.
55. Z. K. Lopez-Castillo, S. N. V. K. Aki, M. A. Stadtherr and J. F. Brennecke, *Ind. Eng. Chem. Res.*, 2008, **47**, 570–576.
56. E. Brunner, *J. Chem. Thermodyn.*, 1980, **12**, 993–1002.
57. A. Jonasson, O. Persson and P. Rasmussen, *J. Chem. Eng. Data*, 1995, **40**, 1209–1210.
58. S. Raeissi, L. J. Florusse and C. J. Peters, *J. Chem. Eng. Data*, 2011, **56**, 1105–1107.
59. J. Kumelan, A. P. S. Kamps, D. Tuma and G. Maurer, *J. Chem. Eng. Data*, 2006, **51**, 11–14.
60. J. Kumelan, A. P.-S. Kamps, D. Tuma and G. Maurer, *J. Chem. Eng. Data*, 2006, **51**, 1364–1367.
61. Z. M. Zhou, Z. M. Cheng, D. Yang, X. Zhou and W. K. Yuan, *J. Chem. Eng. Data*, 2006, **51**, 972–976.
62. U. J. Jauregui-Haza, E. J. Pardillo-Fontdevila, A. M. Wilhelm and H. Delmas, *Latin Am. Appl. Res.*, 2004, **34**, 71–74.

63. H. Struebing, Z. Ganase, P. G. Karamertzanis, E. Siougkrou, P. Haycock, P. M. Piccione, A. Armstrong, A. Galindo and C. S. Adjiman, *Nature Chem.*, 2013, **5**, 952–957.
64. C. Reichardt, *Chem. Rev.*, 1994, **94**, 2319–2358.
65. E. Buncel and S. Rajagopal, *Acc. Chem. Res.*, 1990, **23**, 226–231.
66. J. P. Ceron-Carrasco, D. Jacquemin, C. Laurence, A. Planchat, C. Reichardt and K. Sraidi, *J. Phys. Org. Chem.*, 2014, **27**, 512–518.
67. L. Gilbert and C. Mercier, *Stud. Surf. Sci. Catal.*, 1993, **78**, 51–66.
68. U. K. Singh and M. A. Vannice, *Appl. Cat. A-Gen.*, 2001, **213**, 1–24.
69. Y. Marcus, *Chem. Soc. Rev.*, 1993, **22**, 409–416.
70. M. J. Kamlet, J. L. M. Abboud, M. H. Abraham and R. W. Taft, *J. Org. Chem.*, 1983, **48**, 2877–2887.
71. C. Reichardt and T. Welton, in *Solvents and Solvents Effects in Organic Chemistry*, Wiley-VCH, Weinheim, Germany, 4th edn, 2011.
72. U. Mayer, V. Gutmann and W. Gerger, *Monatsh. Chem.*, 1975, **106**, 1235–1257.
73. V. Gutmann, *Electrochim. Acta*, 1976, **21**, 661–670.
74. H. Takagi, T. Isoda, K. Kusakabe and S. Morooka, *Energy Fuels*, 1999, **13**, 1191–1196.
75. M. Bejblová, P. Zámostný, L. Červený and J. Čejka, *Appl. Catal. A: Gen.*, 2005, **296**, 169–175.
76. N. M. Bertero, A. F. Trasarti, C. R. Apesteguia and A. J. Marchi, *Appl. Catal. A: Gen.*, 2011, **394**, 228–238.
77. D. Al-Mawlawi and J. M. Saleh, *J. Chem. Soc. Faraday Trans. 1*, 1981, 77, 2965–2976.
78. D. Al-Mawlali and J. M. Saleh, *J. Chem. Soc. Faraday Trans. 1*, 1981, 77, 2977–2988.
79. U. K. Singh, S. W. Krska and Y. K. Sun, *Org. Proc. Res. Dev.*, 2006, **10**, 1153–1156.
80. L. J. Shorthouse, A. J. Roberts and R. Raval, *Surf. Sci.*, 2001, **480**, 37–46.
81. J. Geboers, X. Wang, A. B. De Carvalho and R. Rinaldi, *J. Mol. Catal. A: Chem.*, 2014, **388–389**, 106–115.
82. X. Wang and R. Rinaldi, *Angew. Chem. Int. Ed.*, 2013, **52**, 11499–11503.
83. X. Wang and R. Rinaldi, *Energy and Environ. Sci.*, 2012, **5**, 8244–8260.
84. J. Y. He, C. Zhao and J. A. Lercher, *J. Catal.*, 2014, **309**, 362–375.
85. G. C. A. Luijkx, N. P. M. Huck, F. van Rantwijk, L. Maat and H. van Bekkum, *Heterocycles*, 2009, 77, 1037–1044.
86. S. Nishimura, T. Itaya and M. Shiota, *Chem. Commun.*, 1967, 422–423.
87. A. F. M. Barton, *Handbook of Solubility Parameters*, CRC Press, pp. 153–157, 1983.
88. M. Mora-Pale, L. Meli, T. V. Doherty, R. J. Linhardt and J. S. Dordick, *Biotech. Bioeng.*, 2011, **108**, 1229–1245.
89. C. M. Hansen, "Hansen solubility parameters", PowerPoint presentation, http://hansen-solubility.com/, accessed on August 1, 2014.
90. Z. Lei, C. Dai and B. Chen, *Chem. Rev.*, 2014, **114**, 1289–1326.
91. J. P. Mikkola, T. Salmi and R. Sjoholm, *J. Chem. Technol. Biotechnol.*, 2011, **76**, 90–100.

CHAPTER 5

Hydrogenolysis of Cellulose and Sugars

PETER J. C. HAUSOUL, JENS U. OLTMANNS AND
REGINA PALKOVITS*

Lehrstuhl für Heterogene Katalyse und Technische Chemie, Institut für
Technische und Makromolekulare Chemie, RWTH Aachen University,
Worringerweg 1, 52074 Aachen, Germany
*Email: palkovits@itmc.rwth-aachen.de

5.1   Introduction

The first records on the hydrogenolysis of cellulose go back as far as 1925
when the early studies by Fierz-David and Hannig[1] and Bowen et al.[2] were
reported. Aiming to produce petroleum-like fuels from wood, they performed dry distillation of cellulosic materials either in the absence or
presence of a metal catalyst under high H_2 pressures and high temperatures.
In absence of a catalyst, the material subjected to dry distillation at 673 K
yielded mainly charcoal (32 wt%) and an aqueous distillate (39 wt%). These
results are similar to those achieved by Bergius concerning the carbonization
of wood to coal.[3] In contrast, the product distribution shifts dramatically in
the presence of nickel catalysts. Fierz-David and Hannig[1] showed only 1 wt%
of charcoal remaining after the reaction in addition to yields of distillate up
to of 82 wt%. The remaining weight was accountable as gaseous products.
The distillate contained a tar phase with a considerable fraction of ether-
soluble products (33 wt%), comprising organic acids, ketones, alcohols,
phenols and furans.[1]

RSC Energy and Environment Series No. 13
Catalytic Hydrogenation for Biomass Valorization
Edited by Roberto Rinaldi
© The Royal Society of Chemistry 2015
Published by the Royal Society of Chemistry, www.rsc.org

The seminal studies demonstrated the hydrogenation of biomass-based materials as a feasible approach for the production of liquid fuels and chemicals. In the following years, relevant research focused mainly on improving hydrocarbon yields through modification of the catalyst, improvement of the process conditions as well as utilization of different feedstocks.[4] Hence, Lindblad examined the activity of different catalysts (*e.g.* Co_2S, $Cu(OH)_2$, H_2MoO_4, and $ZnCl_2$) in the conversion of wood. Oil yields of up 40 wt% were obtained, which constituted a significant improvement, compared to the seminal works.[5] Routala reported similar results for the conversion of sawdust using catalysts, such as Cu, Fe, and $Co(OH)_2$, $Ni(HCO_2)_2$, $(NH_4)_2MoO_4$, and $(NH_4)_2CrO_4$.[6] Appell *et al.* studied the conversion of municipal waste, newspapers, and sewage sludge, using nickel catalysts at temperatures of 523 K.[7] Kaufmann *et al.* further developed this approach for the continuous conversion of newspaper slurries in mineral oil on a pilot-plant scale.[8] Working on a similar system, Gupta *et al.* proposed the use of temperatures and H_2 pressures above 673 K and 7.0 MPa, respectively, to enable the water-gas shift (WGS) reaction. Thereby, the need for hydrogen in the process was reduced.[9] Boocock *et al.* subjected poplar wood to Raney Ni suspended in water at 623 K under 10 MPa H_2. They found that poplar wood was fully converted into oily and gaseous products.[10] A patent granted to Johnson Matthey from 1983, describes the production of fuel oil from cellulosic materials (*e.g.* municipal and agricultural waste) using Ru or Cu catalysts under carbon monoxide and hydrogen pressures.[11]

Despite its promise, hydrogenation of cellulosic materials under pyrolytic conditions has not been utilized commercially mainly due to the high availability and therefore cheap nonrenewable resources (*i.e.* coal, oil and gas). Nevertheless, as already noted by Berl in 1944, processes for the production of synthetic fuels from biomass are vital assets in times of fuel shortage.[12] As a result of the limited petroleum reserves as well as the continuously increasing energy demand, catalytic processes for plant biomass conversion are currently gaining renewed importance as potential alternatives to their petrochemical counterparts.[13] Alongside the conversion towards liquid fuels, the hydrogenation of cellulose has also been extensively studied for the preparation of polyols (*e.g.* sorbitol, xylitol, glycerol, 1,2-propanediol and ethylene glycol). 1,2-Propanediol and ethylene glycol are traditional petrochemical-based compounds that are produced on a large scale for manufacturing polymers, antifreeze, cosmetics and food additives.

In 1928, Lautenschläger *et al.* described in a patent the use of nickel and copper catalysts for the hydrogenolysis of cellulosic materials towards polyols.[14] Aqueous slurries of cellulosic materials are subjected to Ni- or Cu-based catalysts at temperatures of 473–573 K under hydrogen pressures of 7–10 MPa. The main reaction products were glycerol and 1,2-propanediol. In 1949, Gürkan disclosed a process in which cellulosic materials are first dissolved in an ammoniacal CuO solution (Schweizer's reagent) and subsequently subjected to hydrogen pressure, leading to a cellulosic material impregnated with finely dispersed copper particles.[15] Further subjecting the resulting material in an aqueous solution of methanol to temperatures of 523–543 K under a hydrogen pressure of 6.9 MPa fully converts cellulose into a mixture of short-chain products (*e.g.* propanol, hydroxyacetone and 1,2-propanediol). These early examples clearly demonstrated the potential of cellulose as a replacement feedstock for the production of short-chain polyols.

In the same period, motivated by the high availability of cellulosic materials in agricultural and municipal waste, Russian scientists developed efficient methods for the production of hexitols, pentitols, and shorter chain polyols. The described methods are performed either as a two-stage process or a one-pot process.[16] In the two-stage process, purified cellulose is first hydrolyzed to glucose using concentrated mineral acids. After removal of the acids, the glucose solution was alkalinized and hydrogenated in the presence of Ni catalysts at 393 K under 6–15 MPa H_2, rendering high yields of hexitols. At temperatures of 493–513 K, glucose undergoes hydrogenolysis leading to glycerol, 1,2-propanediol or ethylene glycol, as major products. In the one-pot process, cellulose is directly converted into hexitols using Ru catalysts and small amounts of mineral acids (*e.g.* hydrochloric acid or phosphoric acid at 433 K). Although considerably more attractive from a technical point of view, the application of the latter method is limited to cases in which the catalytically active metal or the support does not dissolve in the acidic reaction medium (*i.e.* noble metals or carbon and polymeric supports, respectively). Also, the use of mineral acids and bases in the two-step processes is a serious drawback because their separation from the aqueous product stream poses difficulties in their recycling in the process.

Nowadays, research has been largely focused on the development and optimization of catalyst systems and conditions to facilitate efficient cellulose depolymerization as well as selective hydrogenation or hydrogenolysis. The more recent developments were comprehensively reviewed by Sels and coworkers,[17] and Palkovits and coworkers,[18] and are outlined in Section 5.4.

5.2 Mechanistic Understanding

Various fundamental studies on the hydrogenolysis of cellulose showed that the process can lead to a wide range of products.[19] In the presence of small amounts of mineral acids and at temperatures around 393–433 K, the reaction mainly produces sorbitol, mannitol, xylitol in addition to sorbitan and isosorbide. The formation of the latter products indicates that the dehydration of the sugar alcohols may also take place as a side reaction (Scheme 5.1). At temperatures around 523 K, the yields of hexitols/pentitols markedly decrease. Short-chain polyols (*e.g.* erythritol, glycerol and ethylene glycol) in addition to deoxygenated products (*e.g.* 1,2-butanediol and 1,2-propanediol) are formed. Furthermore, simple alcohols (*e.g.* methanol, 1- or 2-propanol) and gaseous products (*e.g.* methane and CO_2) are also produced. Overall, these observations clearly show that the cleavage of both C–C and C–O bonds is markedly enhanced at high temperatures. Conducive to

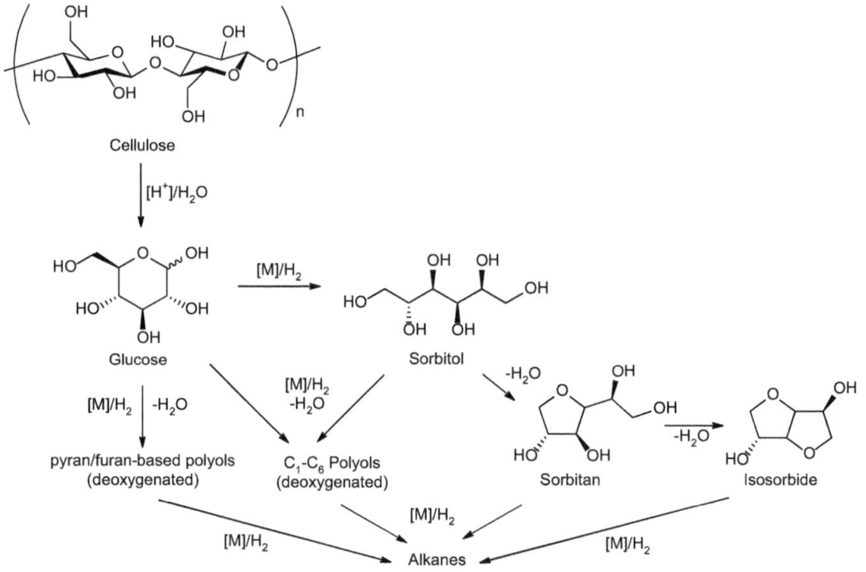

Scheme 5.1 Reaction network involved in the process for hydrogenolysis of cellulose.

this trend, further increasing the temperature or process duration leads to even more extensive fragmentation, and thus to the formation of hydrocarbons.

Scheme 5.1 shows that the hydrogenolysis of cellulose encompasses a large set of chemical reactions, all of which are strongly dependent on the employed catalyst and reaction conditions. Due to the simultaneous occurrence of these reactions as well as the relatively complex composition of product mixtures, the identification of the routes for the formation of specific products may then be difficult. To add to the complexity of the mechanism, some of the side reactions are also catalyzed by the added catalyst (*i.e.* metals, metal oxides, acids and bases), but they may also be caused by thermal degradation of the intermediates. Therefore, to understand the factors accounting for the product distribution obtained from the hydrogenolysis of cellulose, the next sections will cover the carbohydrate chemistry in aqueous medium (*i.e.* cellulose hydrolysis and glucose degradation), and subsequently the role of the hydrogenation/hydrogenolysis catalyst in the reaction network.

5.2.1 Hydrolysis of Cellulose

The first step in the efficient utilization of cellulose is the breakdown of its polymeric structure rendering glucose monomers. Unlike starch – an easy hydrolyzable, water-soluble biopolymer – cellulose is a thermally and chemically stable material, which is insoluble in water and many other common solvents. The main difference between starch and cellulose is the configuration of the glycosidic linkage. The glucose units are linked by 1,4-α-glycosidic bonds in starch, and by 1,4-β-glycosidic bonds in cellulose. By using a variety of spectroscopic and diffraction techniques, it has been established that the ordered structure of cellulose is the result of the strong hydrogen-bonding interactions between glucose units within the polymer chain as well as among different chains.[20] The chains are aligned in parallel forming cellulosic sheets, which are then stacked on top of one another. Due to the closely packed structure, the chemical conversion of cellulose is difficult. The reaction is, however, facilitated either by the addition of acids or by water autoprotolysis, which is enhanced at high reaction temperatures.[21] The mechanism of hydrolysis is proposed to proceed *via* the initial protonation of the glycosidic oxygen (Scheme 5.2). Subsequent dissociation of the acetal C–O bond leads to the formation of a cyclic carbonium ion and therefore to the cleavage of the polymer chain.[22] In the following step, the carbonium ion is hydrated to generate the protonated hemiacetal that in turn regenerates the proton species. An insightful DFT study on the mechanism of 1,4-β-glucans was recently reported by the research groups of Rinaldi and Thiel.[69]

In light of the application of cellulose hydrolysis for bioethanol production, various kinetic studies have been performed in order to develop models to predict the course of hydrolysis.[23] These studies showed that the glucose yield does not proceed to 100% but rather passes through a maximum (*i.e.* 60–65%) after which glucose undergoes slow decomposition into dehydration products. With the development of improved analytical techniques, insight into the processes occurring during the breakdown of cellulose has increased considerably. Recent studies on sub- and super-critical dissolution of cellulose showed that cellulose remains insoluble in water at the temperatures typically employed for the hydrogenolysis of cellulose (*i.e.* 373–523 K).[24] This finding implies that the hydrolysis primarily takes place on the surface of the cellulose particles. As a result, small oligomers and monomers are released into the solution (Figure 5.1).

As a consequence of the heterogeneous depolymerization mechanism, the process duration for the full conversion of cellulose slurries is much longer than that for cellulose fully dissolved in the reaction medium. Cellulose solubilizes in water under supercritical conditions.[24] However, water is an extremely corrosive medium under these conditions. Therefore, alternative reaction media that are able to dissolve cellulose under mild conditions, such as ionic liquids and molten salts, are currently being investigated as substitutes for water.[25]

5.2.2 Degradation of Glucose

Depending on the reaction conditions, sugars are prone to undergo several different reactions, such as epimerization, isomerization, dehydration and retro-aldol cleavage. These reactions are the primary degradation pathways observed for heated sugar solutions, and are typically accelerated by the

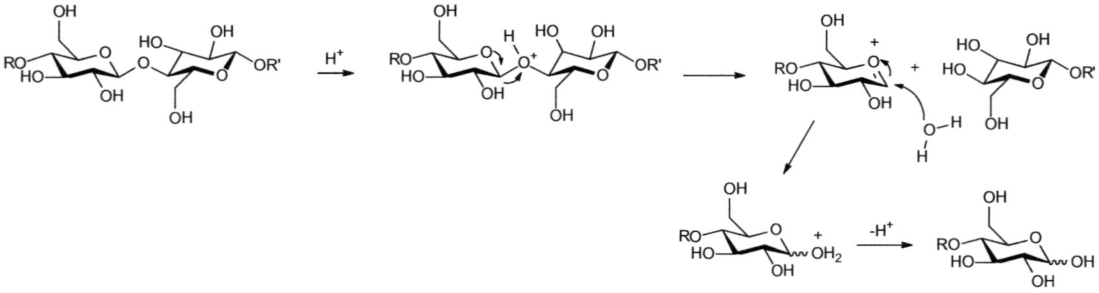

Scheme 5.2   Mechanism of the acid-catalyzed hydrolysis of the 1,4-β-glycosidic bonds.

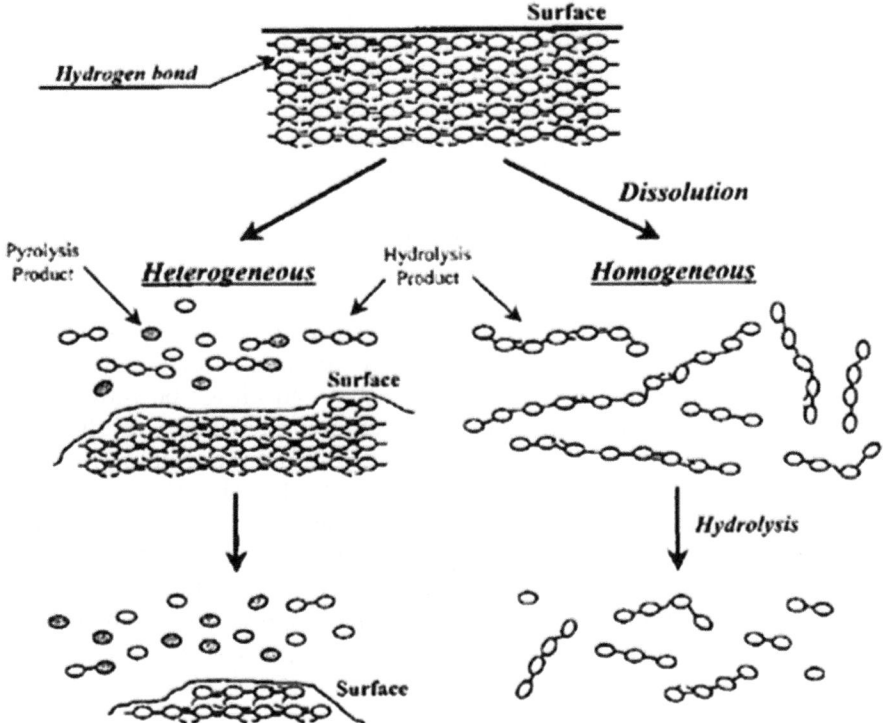

Figure 5.1 Reaction pathways starting with cellulose in heterogeneous and homo-geneous reactions.
(Reprinted (adapted) with permission from (M. Sasaki, Z. Fang, Y. Fukushima, T. Adschiri, K. Arai, *Ind. Eng. Chem. Res.*, 2000, **39**, 2883). Copyright (2013) American Chemical Society.

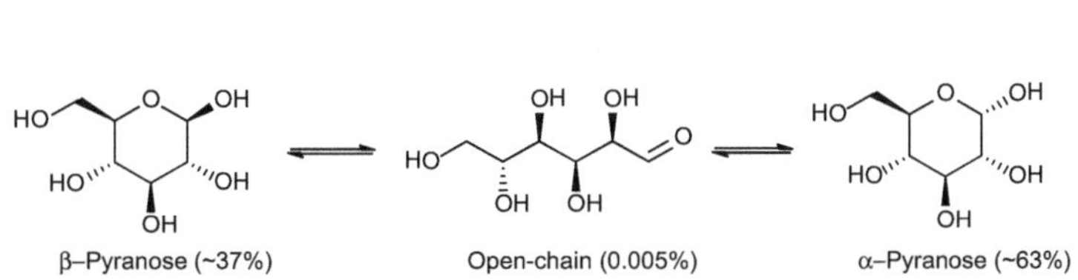

Scheme 5.3 Glucose epimerization and equilibrium composition in water at 303 K.[26]

presence of acids and bases. As discussed in the previous section, cellulose is a polymer exclusively made of glucose units that are in the β-pyranose form. Hydrolysis of the glycosidic bonds converts the acetal group to a hemiacetal, which can then be hydrolyzed to yield the open-chain form of the carbo-hydrate. Since this process is reversible, subsequent recyclization will render equilibrium compositions of the α/β-pyranose forms (Scheme 5.3).[26] It is noteworthy that the equilibrium compositions of the α/β-pyranose forms depend heavily upon the solvent.

The availability of the open-chain form of glucose enables the iso-merization of aldoses to ketoses *via* the aldose–ketose transformation (also known as Lobry de Bruyn–Alberda van Ekenstein transformation).[27] Al-though typically catalyzed by bases, this reaction can also be catalyzed by acids, albeit at a much slower rate.[28] Isotope exchange experiments have shown that besides the postulated enol/keto mechanism, which involves a 1,2-enediol species as an intermediate (Scheme 5.4), the reaction also pro-ceeds *via* a 1,2-hydride shift. The importance of this conversion is perceived in the hydrogenation of glucose that not only renders sorbitol but also mannose alongside fructose.

Under the acidic conditions typically employed for cellulose hydrolysis, fructose is easily dehydrated rendering 2-hydroxymethylfurfural (HMF).[29] HMF itself is also a reactive intermediate that can either be rehydrated, forming levulinic and formic acid, or undergo condensation resulting in water-insoluble, furan-based polymers called humins (Scheme 5.5). These reactions of HMF are irreversible and therefore lead to significant losses of glucose. Studies on the depolymerization of cellulose suggest that these processes should already occur while the glucose monomer is still attached as the reducing end of a cellulose oligomer/polymer.[23,24] Therefore, a more rapid depolymerization is beneficial for ultimate high yields of glucose.

Furthermore, studies on the decomposition of glucose and fructose in sub- and supercritical water showed the retro-aldol cleavage to also

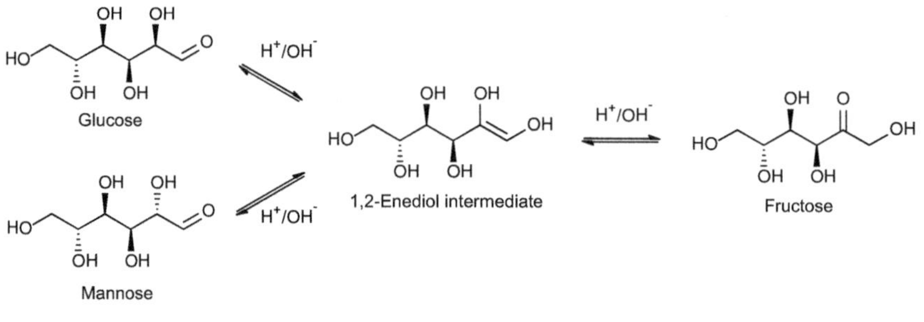

Scheme 5.4   Acid/base-catalyzed Lobry de Buyn–van Ekenstein isomerization of glucose to fructose and mannose.

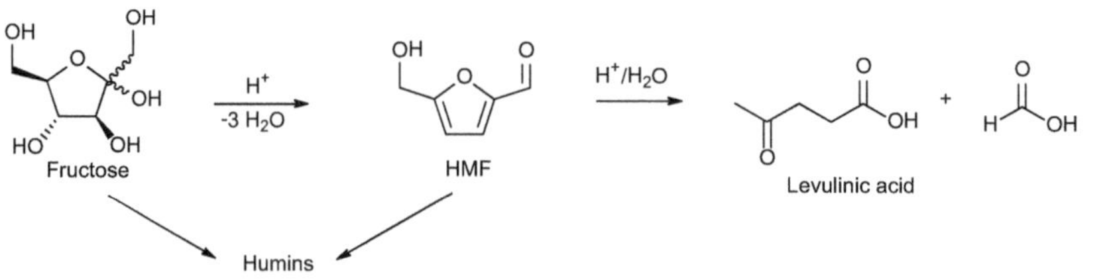

Scheme 5.5   Acid-catalyzed dehydration of fructose to HMF and sequential rehydra-tion to levulinic acid and formic acid.

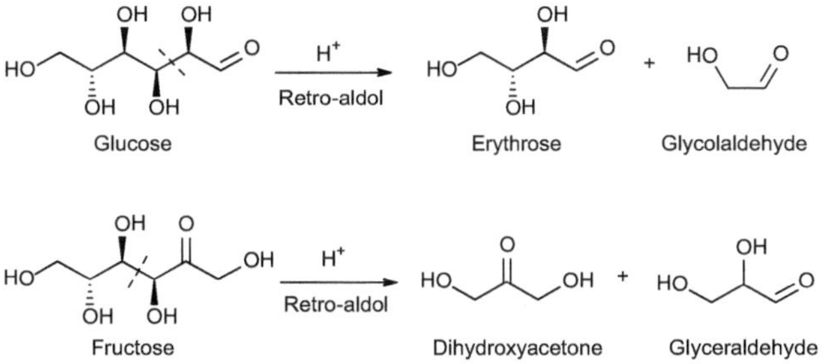

Scheme 5.6 Retro-aldol cleavage of glucose and fructose leading to short-chain sugars.

contribute to the sugar-degradation pathways at high temperatures (*i.e.* 473 K or higher).[30] As shown in Scheme 5.6, the retro-aldol cleavage of glucose renders erythrose and glycol aldehyde, whereas in the case with fructose, dihydroxyacetone and glyceraldehyde. Overall, retro-aldol reactions reduce the length of the carbohydrate chain *via* cleavage of C–C bonds.

5.2.3 Hydrogenolysis

The word "hydrogenolysis" is generally used to describe reductive catalytic processes in which carbon–carbon or carbon–heteroatom bonds are cleaved accompanied by the addition of hydrogen. It should be noted, however, that this description does not imply any mechanistic detail. Rather the term is used to describe the overall process, and this is especially the case for the hydrogenolysis of cellulose. As shown in the previous section, glucose is prone to undergo acid- and base-catalyzed reactions yielding a variety of products. In the presence of a suitable hydrogenation catalyst, these products can be hydrogenated to form their less reactive counterparts. As a consequence, the formation of the saturated products prevents the feed from degradation through side reactions, often improving the overall process yield. Nevertheless, depending on the specific nature of the catalyst, the feed may also undergo other side reactions catalyzed by the added metal catalyst.

5.2.3.1 Hydrogenation/Dehydrogenation

In the case of the hydrogenolysis of cellulose, a wide variety of transition-metal catalysts have been tested, although ruthenium-, nickel- and copper-based catalysts have been studied most extensively. This is mainly because they are relatively cheap, well known to efficiently catalyze the hydrogenation of glucose to sorbitol in addition to the hydrogenolysis of glycerol.[17,18] In contrast to glucose, sorbitol is a more stable compound. Should an effective hydrogenation catalyst be present in the hydrolysis of cellulose, the released glucose is directly converted into sorbitol effectively suppressing the formation of side products (*e.g.* humins). In fact, Sels and coworkers

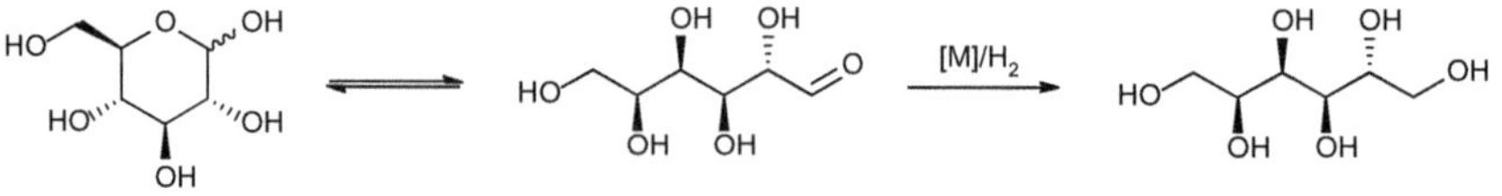

Scheme 5.7   Hydrogenation of glucose to sorbitol *via* open-chain form.

demonstrated the conversion of cellulose into hexitols at yields exceeding 90%.[31] The hydrogenation of glucose to sorbitol is generally believed to proceed *via* the open-chain form of the carbohydrate (Scheme 5.7). Based on kinetic studies, several different surface mechanisms have been proposed.[32] These mechanisms include hydrogenation of the unadsorbed carbonyl by interaction with the metal-hydride surface (Eley–Rideal) or *via* the adsorption of the carbonyl onto a vacant site of the metal-hydride surface (Langmuir–Hinschelwood).

At temperatures higher than 473 K, the product distribution of the hydrogenolysis reaction also includes short-chain polyols and gaseous products. This observation indicates the occurrence of C–C bond cleavage reactions. To account for the formation of these products, several reaction mechanisms, such as retro-aldol, retro-Michael, retro-Claisen and decarbonylation, have been proposed. Although some of these reactions can be envisaged to occur directly from the glucose released by cellulose hydrolysis, the low amounts of sugars together with the high amounts of polyols in reaction mixtures point towards an alternative explanation. Namely, that hydrogenation is sufficiently reversible to constantly resource the medium with intermediates containing carbonyl functions. Thus, the microreversibility of hydrogenation could also contribute to the controlled degradation of polyols.

5.2.3.2   C–C Bond Cleavage

Studying the hydrogenolysis of sorbitol to ethylene glycol, 1,2-propanediol and glycerol, Montassier and coworkers proposed the retro-aldol reaction to account for the C–C bond cleavage. As shown in Scheme 5.8, the reaction is initiated through a metal-catalyzed dehydrogenation, resulting in the formation of a β-hydroxyl carbonyl (Scheme 3).[33] The C–C bond cleavage is considered to proceed *via* a concerted mechanism, which converts the β-hydroxyl carbonyl to a 1,2-enediol and an aldehyde. Following isomerization and hydrogenation, the corresponding diols and alcohols are formed, respectively.

This mechanism was also suggested by Andrews and Klaeren to account for the selective formation of glycerol in the homogeneous $H_2Ru(PPh_3)_4$-catalyzed hydrogenolysis of fructose.[34] Later, Montassier *et al.* also postulated a retro-Michael pathway accounting for the formation of polyols with five or less carbon atoms that proceeds *via* a 1,5-dehydrogenated intermediate (Scheme 5.9).[35] Supposedly, this reaction occurs always in

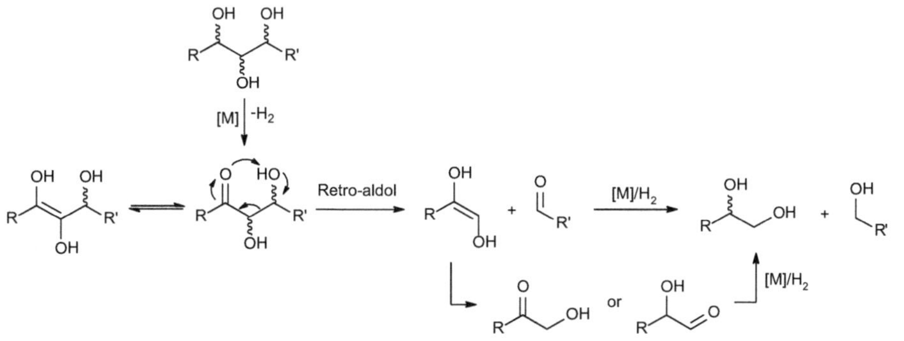

Scheme 5.8 Proposed retro-aldol cleavage of C–C bond in sorbitol.

competition with retro-aldol. Likewise, 1,3-dehydrogenated intermediates were suggested to account for the formation of CO_2 and ethylene glycol from glycerol *via* a retro-Claisen mechanism (Scheme 5.10). However, given the high hydrogen pressures used in these reactions, it appears unlikely that the proposed intermediates can be formed. In fact, using several model substrates, Wang *et al.* showed the retro-aldol cleavage to be the dominant pathway for the C–C bond cleavage under hydrogenolytic conditions.[36]

Shanks and coworkers studied the hydrogenolysis of hexitols using Ru/C catalysts and proposed decarbonylation reactions to account for the presence or absence of particular stereoisomers in the product distribution.[37] As shown in Scheme 5.11, the decarbonylation of aldehydes derived from dehydrogenation of primary alcohols yields $C_{(n-1)}$ polyols and CO. Unlike in the case of retro-aldol cleavage, which results in the loss of stereochemical information of three adjacent carbon atoms, decarbonylation only results in the loss of chirality at the carbon atom next to the aldehyde.

Recently, Li and Huber presented an indepth study on the Pt/SiO_2-Al_2O_3-catalyzed hydrodeoxygenation of sorbitol aiming at the production of hexane.[38] They concluded that the reaction pathway comprises four key steps: dehydration, hydrogenation, decarbonylation and retro-aldol reaction. In addition to C–C bond cleavage of sorbitol to glycerol, the dehydration of sorbitol and sorbitan leads to sorbitan and isosorbide, respectively, in the early process steps. Subsequent ring opening, dehydration and hydrogenation render 1,2,6-hexanetriol. Further conversion of the latter, in turn, forms hexanediol, hexanol, and eventually hexane. Moreover, Li and Huber also proposed the decarbonylation of these intermediates forming lighter alcohols or alkanes.[38]

5.2.3.3 C–O Bond Cleavage

The cleavage of C–O bonds is generally considered to proceed *via* a reaction sequence including dehydration, enol/keto tautomerization and

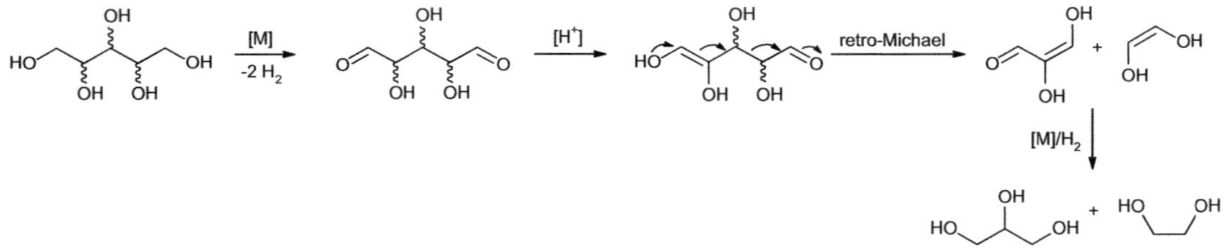

Scheme 5.9 Proposed retro-Michael cleavage proposed for the formation of glycerol and ethylene glycol from 1,5-dehydrogenated C$_5$-polyols.

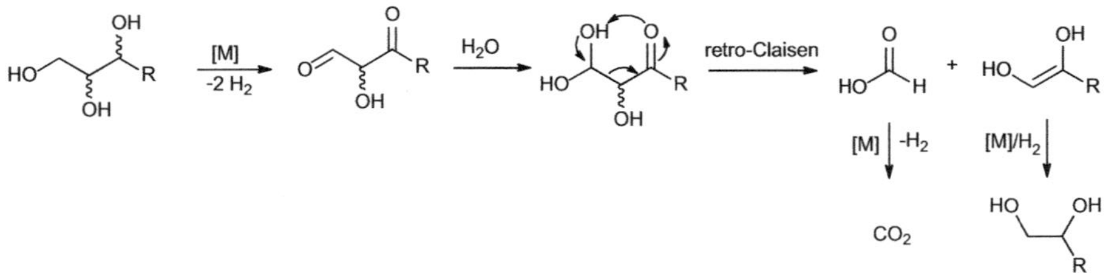

Scheme 5.10 Proposed retro–Claisen cleavage proposed for the formation CO$_2$ from 1,3-dehydrogenated C$_{3+}$-polyols.

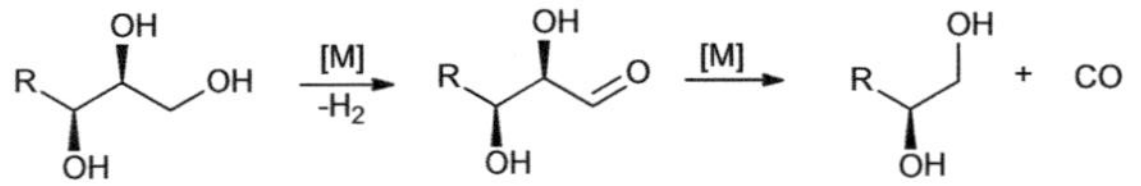

Scheme 5.11 Proposed decarbonylation pathway accounting for the reactions resulting in the cleavage of C–C bonds in dehydrogenated polyols through elimination of carbon monoxide.

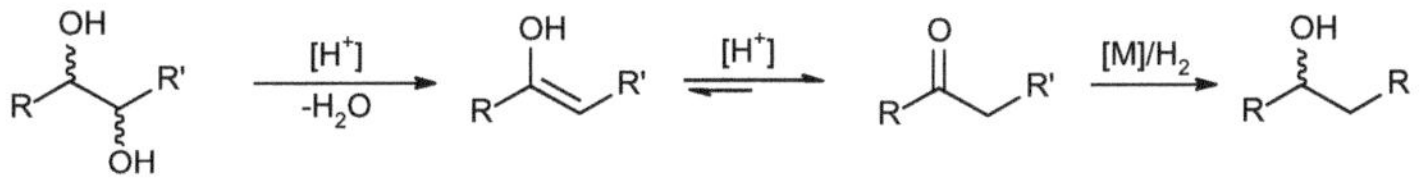

Scheme 5.12 Dehydration/hydrogenation pathway of polyols proposed for the overall deoxygenation.

hydrogenation (Scheme 5.12). As a result, the overall process of hydrodeoxygenation renders a large variety of dehydroxylation products. Depending on the reaction conditions, alkanes can also be obtained from the polyols.

5.3 Cellobiose as a Model Compound

To lend some mechanistic insights into the complex network of reactions involved in the hydrogenolysis of cellulose, cellobiose was used as a model compound in several studies. Cellobiose shares several structural features with cellulose (*e.g.* 1,4-β-glycosidic bond, and intramolecular H-bonding pattern around the glycosidic linkage). However, cellobiose is soluble in water. Therefore, the pronounced impact of the polymeric structure of cellulose upon the reaction performance cannot be accounted for in the experiments performed on cellobiose.

In 2006, a first mechanistic description of cellobiose hydrogenolysis was given by Kou and coworkers.[39] They subjected cellobiose to Ru, Rh, Pt and Pd nanoclusters stabilized by poly(*N*-vinyl-2-pyrrolidone) (PVP) for the production of C_6-alcohols from cellobiose in aqueous phase. Full conversion of cellobiose in the presence of Pd, Pt or Rh catalysts was achieved. Interestingly, glucose was the major product. Since the reaction was carried out at pH 2, the substrate was most likely converted through acid hydrolysis. Apparently, the metal nanoclusters exhibited almost no hydrogenation activity. In contrast, the use of Ru nanoclusters at neutral pH led to 88% conversion of cellobiose into 3-β-D-glucopyranosyl-D-glucitol (cellobitol) as the major product. Sorbitol, glucose, dideoxyhexitols and other polyols were formed as minor products. At pH 10, the conversion was lower (76%). However, at pH 2, full conversion of cellobiose into sorbitol was achieved. Kou and coworkers discuss two reaction mechanisms for the different pH environments based on these results. Under acidic conditions, cellobiose undergoes hydrolysis yielding glucose, which is sequentially hydrogenated to sorbitol (Scheme 5.13(a)). Under neutral or basic conditions, cellobiose undergoes either hydrogenolysis (forming dideoxyhexitol and glucose, which is then hydrogenated, leading to sorbitol) or hydrogenation (rendering cellobitol instead, Scheme 5.13(b)).

Under acidic conditions, the hydrolysis of cellobiose and subsequent hydrogenation of glucose rendering sorbitol has been confirmed in numerous studies.[40] However, Wang and coworkers identified cellobitol to be a possible intermediate involved in the production of sorbitol.[41] They

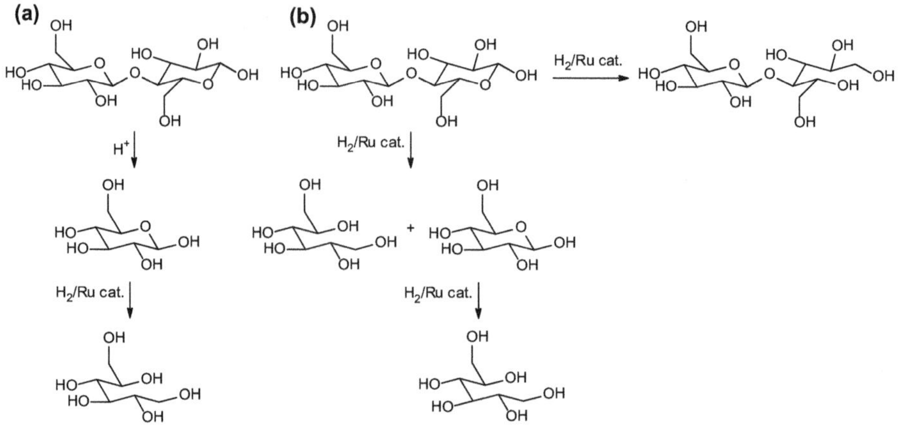

Scheme 5.13 Reaction mechanism of cellobiose hydrogenolysis under (a) acidic and (b) basic or neutral reaction conditions as proposed by Kuo and coworkers.[39]

reported the conversion of cellobiose into sorbitol, in the presence of Ru catalysts supported on carbon nanotubes, in neutral aqueous solutions. As such, they detected the formation of cellobitol and its subsequent hydrolysis forming sorbitol and glucose (Scheme 5.14). The authors suggested this pathway to be the predominant one under the reaction conditions studied.

More recently Makkee and coworkers investigated the conversion of cellobiose into sorbitol in aqueous $ZnCl_2$ with Ru/C as the catalyst and provided a detailed reaction mechanism.[42] They found two competing pathways for the formation of sorbitol. The reaction can proceed either *via* (1) the hydrogenation of cellobiose to cellobitol followed by its hydrolysis or through (2) hydrolysis of cellobiose and subsequent hydrogenation of glucose to sorbitol. The experimental data showed the pathway (1) to dominate under the reaction conditions.

Palkovits and coworkers presented a kinetic investigation of the catalytic conversion of cellobiose into sorbitol using silicotungstic acid and Ru/C as catalysts.[43] They confirmed a competition between the two presented pathways proposed for the formation of sorbitol from cellobiose. At 393 K, cellobitol was formed as the main product with 81% selectivity. At 443 K, the selectivity for cellobitol decreased to 1%, while the selectivity for sorbitol increased to 75%. Accordingly, the hydrolysis of the glycosidic linkage was identified to be a rate-determining step, independent of the reaction pathway. The authors concluded the cellobitol pathway to dominate at low reaction temperatures. Lower activation energies and higher rate constants could be determined for a hydrogenation and hydrolysis of cellobitol compared with the direct hydrolysis of cellobiose followed by hydrogenation of glucose. Thus, the possibility to selectively transform cellobiose to sorbitol *via* cellobitol as an intermediate was suggested.

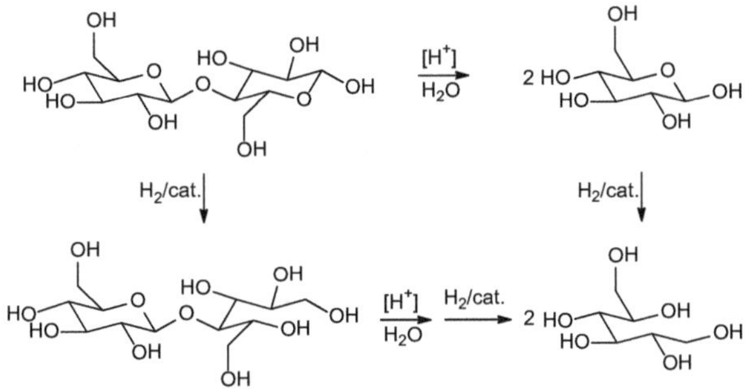

Scheme 5.14 Mechanism for sorbitol formation by hydrogenolysis of cellobiose including formation of cellobitol as an intermediate, as proposed by Wang and coworkers.[41]

5.4 Recent Advances in Hydrogenolysis of Cellulose

5.4.1 Application of Noble-Metal Catalysts in Aqueous Phase

The efficient hydrogenation of substrates is often possible with noble metals as they usually exhibit great activity and allow for high selectivity (for a detailed discussion refer to Chapter 3). Noble metals have therefore been extensively examined on the hydrogenolysis of cellulose. In 2006, Fukuoka and coworkers presented results on the catalytic conversion of cellulose into sugar alcohols in the presence of supported noble-metal catalysts.[40a,44] They screened Pt supported on γ-Al$_2$O$_3$, SiO$_2$-Al$_2$O$_3$ or HUSY in addition to Ru on HUSY as catalysts. All catalyst systems reached up to 30% yield of sorbitol after 24 h (463 K, initial H$_2$ pressure of 5 MPa at room temperature). The control tests with the metal-free supports led to as low as 3% yield of glucose. The authors suggested a spillover mechanism to explain cooperative effects between the metal catalyst and the support. In this mechanism, acid sites are formed *in situ* through the heterolytic dissociation of molecular hydrogen on the metal surface. The H$^+$ species spill over to the support surface.[45] Essayem and coworkers even suggested a dissociation of hydrogen on the Pt surface that leads to hydrogen atoms that are able to react either as H$^+$ or as H$^-$ species, depending on the reacting substrate.[46]

In 2007, Liu and coworkers reported the conversion of cellulose into polyols catalyzed by reversibly formed acids and Ru/C in hot water.[19a] The hydrogenolysis of cellulose was carried out at 518 K and 6 MPa H$_2$ within a duration of 5 or 30 min. The high temperature allowed for a fast hydrolysis of cellulose generating glucose that then undergoes hydrogenation to sorbitol and mannitol. A 40% yield of hexitols, at full cellulose conversion, was achieved. However, side reactions also occurred, leading to a vast variety of products (*i.e.* sorbitan, xylitol, erythritol, glycerol, propylenediol, ethylene glycol, methanol and methane). In this manner, the process suffers from rather low carbon efficiency because of the formation of humins or overreaction leading to other degradation products. Moreover, water was found to be essential for the reaction since no conversion was achieved in the experiments performed in dioxane or ethanol. However, in a solution of water and ethanol, the hydrogenolysis takes place, but to a limited conversion of 10%. The authors claimed that *in situ* formation of acid sites is responsible for the fast hydrolysis due to a shift of the autoprotolysis equilibrium of water at high temperatures. This statement leads to the assumption that a high dielectric constant is essential for the reaction. In Chapter 11, a detailed discussion on the effect of dielectric constant of pressurized hot water upon the reaction mechanisms found for hydrothermal processing of biomass is provided.

Wang and coworkers investigated several metals (*e.g.* Fe, Co, Ni, Pd, Pt, Rh, Ru, Ir, Ag and Au) and support materials (SiO_2, CeO_2, MgO, Al_2O_3 and CNT). They identified Ru/CNT to reach superior yields in the conversion of cellulose into sorbitol.[47] With this catalyst system at 458 K and an initial H_2 pressure of 5 MPa (measured at room temperature), a 36% yield of sorbitol from untreated cellulose (85% crystallinity) was obtained after 24 h. An improvement was achieved through pretreatment of cellulose in 85% H_3PO_4 at 323 K for 40 min in order to reduce crystallinity. A 69% yield of sorbitol could be reached starting with pretreated cellulose (33% crystallinity) at 458 K and under 5 MPa H_2 (measured at room temperature).

5.4.2 Application of Noble-Metal Catalysts and Acids in Aqueous Phase

The addition of acids has been investigated to facilitate the hydrolysis of cellulose under mild reaction conditions. In 2010, Palkovits *et al.* used a combination of mineral acids and different hydrogenation catalysts for the hydrogenolysis of cellulose.[19d] They compared Pt, Pd and Ru supported on activated carbon in the presence of phosphoric or sulfuric acid. Due to the use of mineral acids, a reaction temperature of 433 K under an initial H_2 pressure of 5 MPa (measured at room temperature) suffices to achieve high conversion. Ru/C combined with sulfuric acid led to high to full conversion of cellulose. Yields of C_4 to C_6 sugar alcohols up to 60% were achieved. In subsequent publications, Palkovits *et al.* identified heteropoly acids as efficient acid catalysts for hydrogenolysis of cellulose.[48] In combination with Ru/C, $H_3PW_{12}O_{40}$ and $H_4SiW_{12}O_{40}$ allowed for sugar alcohols yields above 80% within 7 h. Interestingly, this catalyst system could be directly applied to spruce as a real biomass feedstock. Full conversion of cellulose and hemicellulose of the wooden biomass were reached; yields of up to 67% of polyols were obtained.

Sels and coworkers reported an efficient catalytic conversion of concentrated cellulose feeds with heteropoly acids and Ru/C.[49] Under optimized reaction conditions ($[H^+] = 1.22 \times 10^{-2}$ mol L^{-1}, Ru/C $= 0.25$ g, 463 K, 9.5 MPa H_2), a 100% selectivity for hexitols including 85% sorbitol and 15%

sorbitan was obtained. Full conversion was reached beginning the experiment with ball milled cellulose. Additionally, they investigated insoluble cesium salts of heteropoly acids in combination with Ru/C. Compared to the heteropoly acids, higher activities were achieved. Moreover, because of the fact that cesium salts of the heteropoly acids can be recrystallized at room temperature, the recycling of the acid catalyst is facilitated.[50] In a subsequent work, the use of Ru-loaded zeolites and trace amounts of mineral acids was reported.[31] Hexitol yields of up to 90% were obtained with concentrations of mineral acids as low as 35–177 ppm in combination with Ru-loaded zeolites (USY and MOR) and ball-milled cellulose as the substrate.

While tungsten-based heteropoly acids catalyze the reaction quite well, Zhang and coworkers demonstrated a different approach for ethylene glycol production from cellulose through the discovery of a temperature-controlled phase-transfer catalyst based on tungsten.[51] Tungstic acid combined with Ru/C was used as a catalytic system at 518 K and under 6 MPa H_2 for 0.5 h. Through an optimization of the metal-to-acid ratio, yields of ethylene glycol above 50% were achieved. Since full cellulose conversion was reached, only moderate carbon efficiency of the transformation could be achieved. As byproducts, propanediol, glycerol, 1,2-butanediol, erythritol, mannitol, and sorbitol were detected. While H_2WO_4 is insoluble in water at room temperature, it becomes soluble at elevated temperatures. Therefore, the system is homogenously catalyzed by tungsten acid under the reaction conditions. As the catalyst precipitates after cooling of the reaction mixture, it can be filtered off and reused. The authors demonstrated exceptional reusability of more than 20 cycles with a quantity of tungsten of 30 ppm leached per reaction run.

In contrast to this homogenously catalyzed system, Xu and coworkers investigated Brønsted acid-promoted ruthenium catalysts.[52] Sulfonic acid functionalized mesoporous silica (MCM-41) was combined with Ru/C in a one-pot process at 503 K and under 6 MPa H_2 for 40 min, yielding mainly ethylene glycol and propanediol. However, the catalyst suffered from rapid deactivation. After the first use cycle, irreversible changes of the mesoporous structure of the catalyst in addition to loss of acidic groups resulted in catalyst deactivation. The same catalyst system was used in a sequential process under similar conditions in which MCM-41 was filtered off and exchanged for Ru/C. Interestingly, product selectivity shifted to γ-valerolactone and hexitols. However, catalyst lifetime was increased by six times compared with the original durability. With this sequential process, up to 40% selectivity to GVL was achieved at almost full conversion of cellulose.

In 2013, Rinaldi and coworkers reported a unique approach combining a mechanocatalytic depolymerization of cellulose[53a] with a subsequent hydrogenolysis.[53b] To cope with the insolubility of cellulose and the difficulties during hydrolysis of the polymer, they introduced the solid-state mechanocatalytic depolymerization by ball milling cellulose impregnated with low quantities of H_2SO_4 or HCl.[53a] Directly after the solid-state reaction, hydrogenation of the water-soluble product was performed in the presence of H_2O and Ru/C at 423 K under an initial H_2 pressure of 5 MPa (measured at room temperature). The overall process duration was less than 4 h. In this approach, yields of up to 94% hexitols at full cellulose conversion were obtained.[53b] The authors demonstrated six successful recycling runs of the catalyst system and estimated this value to be much higher as the catalyst was still highly active. Additionally, no corrosion of the ball mill was detected even after 300 experiments. This is explained by a strong affinity of the sulfuric acid towards cellulose reducing the acid activity on the stainless steel surface. However, downstream acid handling is still a problem, and thus new solutions for an efficient acid recovery are still needed, as also found for most of the aforementioned systems.

5.4.3 Catalysis in Nonaqueous Media

The insolubility of cellulose in water and most of the common solvents poses a challenge for the efficient conversion of cellulose through the hydrogenolysis approach. Solid catalysts exhibit much less activity towards solid substrates due to the inefficient solid–solid interactions. To overcome this problem, several research groups chose to examine the use of solvents other than water for the hydrogenolysis. However, there are just a few suitable solvents for cellulose that are also compatible with hydrogenation catalysts. One of the few possibilities is the use of ionic liquids. Zhu *et al.* converted cellulose with ionic liquid-stabilized ruthenium nanoparticles and a reversible binding agent in 2010.[25a] The ionic liquid was functionalized with boronic acid as an acid catalyst, while Ru nanoparticles were used as the hydrogenation catalyst. The reaction was carried out at only 353 K under an initial H_2 pressure of 1 MPa (measured at room temperature). As cellulose was completely dissolved in the ionic liquid, near to full conversion of cellulose into glucose was achieved after 5 h in the experiment performed in absence of the Ru catalyst. Experiments combining Ru and the functionalized ionic liquid produced high amounts of sorbitol under the same reaction conditions. Different hydrogen donors, namely pure hydrogen, sodium formate and formic acid, were successfully applied. Yields of sorbitol of up to 94% were reported.

De Vos and coworkers investigated the activity of $HRuCl(CO)(PPh_3)_3$ as a molecular catalyst combined with bases. 1-Butyl-3-methylimidazolium chloride was used as the solvent for cellulose.[25b] Full conversion of cellulose could only be obtained after 48 h at 423 K under an initial H_2 pressure of 3.5 MPa (measured at room temperature) when the catalyst was combined with Pt/C or Ru/C. Interestingly, the main product was glucose rather than sorbitol with yields of up to 76%. As byproducts, glucose dimers, C_6 sugars, sorbitol, mannitol, levoglucosan, shorter sugars and alcohols in addition to HMF were detected. The product selectivity could strongly be influenced by the ratio of molecular-to-solid catalyst. With Ru/C as the solid catalyst, sorbitol was the main product obtained at yields of up to 74%.

Apart from high costs, the main challenge while applying ionic liquids to the conversion of cellulose concerns an efficient product separation from the solvent, as most of the polar products are highly soluble in ionic liquids. Moulijn and coworkers addressed this challenge by dissolving cellulose in a $ZnCl_2$ molten hydrate.[25c] With addition of minor quantities of HCl, the hydrolysis of cellulose to glucose was facilitated. The system could even be used under acid-free conditions at 373 K without a H_2 pressure. HCl could be removed as a gas after the reaction. This approach led to full conversion of cellulose into glucose with no significant formation of byproducts. Next, the hydrogenation of glucose to sorbitol was carried out for 60 min at 373 K under an initial H_2 pressure of 5 MPa (measured at room temperature) with Ru/C as a catalyst to nearly quantitative yield of sorbitol. Thus, $CuCl_2$ or $NiCl_2$ was added to the system at 453 K and 4 MPa H_2 pressure in order to catalyze the dehydration of sorbitol to isosorbide. First, sorbitol is dehydrated to either 1,4-, 2,5- or 3,6-anhydrosorbitol. However, only the 1,4- and 3,6-anhydrosorbitol isomers can be further dehydrated to isosorbide. Therefore, selectivity to isosorbide should be limited by the formation of 2,5-anhydrosorbitol. Surprisingly, almost no 2,5-anhydrosorbitol is formed under the described reaction conditions, thus leading to a selective dehydration of sorbitol to isosorbide. This is explained by an interaction of sugar hydroxyls with $ZnCl_2$ salt hydride species, which has been reported previously for sorbitol dehydration in pyridinium chloride.[54] Isosorbide can be extracted with xylene at elevated temperatures. Subsequent precipitation is also possible at lower temperatures resulting in a separation of isosorbide from the $ZnCl_2$ molten hydrate.[55]

5.4.4 Catalysis by Base-Metal Catalysts

While noble-metal catalysts usually exhibit quite high activities, major drawbacks are their rareness and high cost. These drawbacks pose major challenges when facing a worldwide application of a working catalyst system to replace integral platforms of today's oil industry. Therefore, base-metal catalysts seem to be much more feasible, as they are less rare and therefore less expensive. Even though they exhibit less activity in hydrogenation, this problem can be solved by prolonging the process duration, increasing the reaction temperature or using higher amounts of catalyst.

A base-metal catalytic system was introduced by Zhang and coworkers.[56] Promoted tungsten carbide supported on activated carbon was examined in the hydrogenolysis of cellulose to ethylene glycol. Yields of ethylene glycol up to 61%, at 518 K, 6 MPa H_2 and a reaction duration of 0.5 h, were reported. The addition of Ni had a stabilizing effect on W_2C, and the catalyst proved to be a low cost alternative to Pt and Ru catalysts. The catalytic hydrogenation of corn stalk to ethylene glycol and 1,2-propylene glycol was also showed.[57]

Furthermore, Zhang and coworkers applied the carbon supported Ni-W_2C catalyst system to the hydrocracking of raw wooden biomass.[58] While they also tested noble-metal-based tungsten carbide catalysts, Ni exhibited competitive yields and activity. The conversion of cellulose, hemicellulose and lignin could also be performed as a one-pot reaction. Cellulose and hemicellulose resulted in ethylene glycol and other diols with a 75.6% yield at 508 K (6 MPa H_2), while lignin was converted into monophenols with yields of 46.5%. However, the chemical composition and structure of the biomass exerts a notable effect upon the catalytic activity and product distribution.

The same research group investigated the effects of the preparation methods of nickel-promoted tungsten carbide on its catalytic activity.[59] Two preparation methods were compared: (1) the postimpregnation method depositing Ni as nickel nitrate on previously prepared W_2C/C catalysts; and (2) the coimpregnation of Ni and W on activated carbon at the same time prior to the carburization procedure. The postimpregnation method results in a better dispersion of active species on the catalyst. They propose the W_2C particles to be unstable in aqueous solution during impregnation, resulting in partial dissolution of the particles and redispersion of W species. The improved dispersion achieved through the postimpregnation method led to an efficient catalyst for the cellulose conversion. A 73% yield of ethylene glycol at full conversion of cellulose was obtained at 518 K and an initial H_2 pressure of 6 MPa (measured at room temperature) after 30 min. This yield is 12% higher than that achieved in the presence of catalysts obtained from coprecipitation.

Apart from tungsten carbide based catalysts, Zhang and coworkers identified bimetallic tungsten catalysts to be active for hydrogenolysis.[60] Bimetallic catalysts comprising tungsten and a metal from groups 8, 9 or 10 of the periodic table, both supported on activated carbon, delivered promising results concerning the yields and selectivity to ethylene glycol. Notable combinations were Ni–W, Pd–W, Pt–W, Ru–W and Ir–W. The Ni–W catalyst exhibited superior selectivity to ethylene glycol with a 75.4% yield of ethylene glycol at full conversion at 518 K and an initial H_2 pressure of 6 MPa (measured at room temperature) after 30 min.

Additionally, the same group explored Raney Ni and tungstic acid as a low-cost binary catalyst for cellulose conversion into ethylene glycol.[61] This work is based on the previously discussed publication using Ru/C and tungstic acid as a phase-transfer catalyst.[51] Thus, they examined tungsten compounds in combination with Raney Ni and compared the ethylene glycol yields obtained in the presence of these catalysts. They found the following yield order: $H_4SiW_{12}O_{40} < H_3PW_{12}O_{40} < WO_3 < H_2WO_4$. Apparently, WO_3 and H_2WO_4 dissolved partially in the reaction medium forming the active species H_xWO_3 under reaction conditions. Therefore, it was stated that the C–C bond cleavage of cellulose with a tungsten catalyst is homogenously catalyzed. By this approach, yields of up to 65% of ethylene glycol were obtained after 30 min at 518 K and 6 MPa of H_2 pressure (measured at room temperature), since Raney Ni is highly active for hydrogenation but rather inert for degradation of ethylene glycol. Furthermore, the catalyst could be reused at least 17 times.

In other publications, Zhang and coworkers used bifunctional nickel phosphide catalysts, leading to a 48.4% sorbitol yield.[62] Acidic and metallic acid sites on the support material are proposed to account for the catalytic activity. They also investigated Ni-based bimetallic catalysts on mesoporous carbon (MC).[63] Ni was combined with Pt, Pd, Ru, Rh and Ir. A hexitol yield of 59.8% was achieved in the presence of a catalyst comprising 1% Rh and 5% Ni on MC. However, the studied bimetallic catalysts led to comparable yields of hexitols. It is noteworthy that mesoporous carbon showed superior performance as a support material, compared to activated carbon. In fact, the catalyst supported on MC exhibited two to four times higher yields under the same reaction conditions (518 K, 6 MPa H_2 pressure, 0.5 h). A plausible explanation is that the mesoporous carbon material shows, on its own, an elevated catalytic activity for the hydrolysis of cellulose, whereas activated carbon seems to be an inert material for this reaction. Additionally, Zhang and coworkers reported that mesoporous carbon shows an uptake of both cellulose and glucose up to three times higher than that found for activated carbon. As a result, sugar intermediates should be more readily adsorbed throughout the course of cellulose conversion.

Further research regarding Ni-based catalysts was carried out by Tanksale and coworkers.[64] They investigated promoted nickel catalysts supported on mesoporous alumina. With an atom ratio Ni-to-Pt of 22 : 1, the catalyst enabled a 32.4% hexitol yield, compared to 5% yield with Ni catalysts. Ni–Pt on mesoporous β-zeolite provided slightly better yields (36.6%, at 473 K and 5 MPa H_2 pressure for 6 h). The protonation of water by the heterolytic cleavage of hydrogen molecules on the Pt surface was postulated for mechanistic implication since a spillover of the H^+-species to Ni generates *in situ* acid sites.

Mu and coworkers explored several supports (*e.g.* alumina, Kieselguhr, TiO_2, SiO_2, activated carbon, ZnO, ZrO_2 and MgO) in combination with nickel for the conversion of microcrystalline cellulose.[65] They identified the support to play a crucial role in the product distribution. Ni/ZnO exhibited superior properties. A 70.4% alkanediol yield at full conversion of microcrystalline cellulose was achieved (518 K, 6 MPa of H_2 pressure (measured at room temperature) for 2 h). The same group also investigated the hydrogenolysis of cellulose in the presence of Ni-Cu/ZnO catalysts.[66] Several Cu-to-Ni ratios were compared. They found that Ni and Cu exert synergetic effects upon the dehydroxylation. A Ni-to-Cu ratio of 2 : 3 was identified to provide a catalyst with high selectivity to 1,2-alkanediols.

Sels and coworkers reported that reshaped Ni nanoparticles at the tip of carbon nanofibers are active for the conversion of cellulose.[19c] Ni nanoparticles were deposited at the tip of carbon nanofibers rather than on the bulky material, which led to an easier contact between metal and solid substrate. A 92% conversion of cellulose with a 50% yield of sorbitol (56.5% hexitols) was demonstrated at 503 K (6 MPa H_2 pressure) after 4 h. The high yield of hexitols was explained by a reshaping of the crystal structure of Ni particles induced *via* synthesis, which suppressed undesired C–C and C–O bond cleavage.

There have been few other publications regarding the use of base metals other than nickel. Palkovits and coworkers investigated copper-based catalysts for an efficient valorization of cellulose.[67] Based on promising results with a commercially available catalyst for methanol synthesis, copper-based ZnO/Al_2O_3 catalysts were synthesized. At temperatures of 518 K for 3 h (initial H_2 pressure of 5 MPa, measured at room temperature), a 95% yield of liquid-phase products was achieved. The product mixture comprised C_1–C_3 compounds (67.4%) with 1,2-propanediol (15.4%) and methanol (27.1%) as major products. XPS measurements showed Cu^0 to be the catalytically active species.

Liang and coworkers reported highly efficient CuCr catalysts.[68] At full conversion of a feed of up to 15 wt% cellulose in water and $Ca(OH)_2$ as an additive, 1,2-propanediol and ethylene glycol yields of 42.6% and 31.6%, respectively, were obtained. However, no explanation of the promoting effect of $Ca(OH)_2$ was provided. Apparently, a change in pH has a strong effect on the product selectivity as the different intermediates are formed at different rates depending on the pH of the reaction medium.

5.5   Concluding Remarks

While a broad selection of catalyst systems has been investigated until now, major breakthroughs are still awaited. Some publications, however, show results with great potential. Hydrolysis of cellulose, for example, is able to yield up to 95% of glucose, opening up new horizons for selective two-step processes. Regarding direct hydrogenolysis, promising results include the conversion of cellulose into hexitols with up to 100% selectivity. However, the catalyst systems still suffer from poor catalyst recovery, as additional soluble acids were used. Additionally, full conversion could only be reached with ball-milled cellulose. On a large scale, these pretreatments are usually very time- and energy-demanding processes. However, preliminary analysis shows the mechanocatalytic depolymerization of lignocellulose to be more energetically efficient than conventional milling.[53a,70] Moreover, various systems without the addition of acid or base have been discussed, although the catalyst is then usually less selective, unstable or otherwise not yet suitable for a long-term continuous process. To influence this behavior, knowledge-driven catalyst development needs to be pursued. Only a basic understanding of the mechanism of the various reactions has been established yet. However, a fundamental comprehension of all factors influencing the cellulose conversion and product selectivity awaits discovery. Nevertheless, current attempts to gain deeper insight into the complexity of involved reaction pathways and the extensive dominating reaction mechanisms will provide the possibility to enable knowledge-based and driven catalyst development.

References

1. H. E. Fierz-David and M. Hannig, *Helv. Chim. Acta*, 1925, **8**, 900.
2. A. R. Bowen, H. G. Shatwell and A. W. Nash, *J. Soc. Chem. Ind.*, 1925, **44**, 507.
3. F. Bergius, *J. Soc. Chem. Ind.*, 1913, **32**, 462.
4. A. H. Weiss, *Text. Res. J.*, 1972, **42**, 526.
5. A. Lindblad, *Ing. Vetenskaps Akad. Handl.*, 1931, **107**, 7.
6. O. Routala, *Acta Chem. Fennica*, 1939, **3**, 115.
7. H. R. Appell, I. Wender and R. D. Miller, *Prepr. Am. Chem. Soc. Div. Fuel Chem.*, 1969, **4**, 39.
8. J. A. Kaufman, D. V. Gupta, T. S. Szatkowski and A. H. Weiss, *Chem. Ing. Tech.*, 1974, **14**, 609.
9. D. V. Gupta, W. L. Kranich and A. H. Weiss, *Ind. Eng. Chem. Process Des. Dev.*, 1976, **15**, 256.
10. (a) D. G. B. Boocock, D. Mackay, M. Mcpherson, S. Nadeau and R. Thurier, *Can. J. Chem. Eng.*, 1979, **57**, 98; (b) D. Beckman and D. G. Boocock, *Can. J. Chem. Eng.*, 1983, **61**, 80.
11. G. C. Bond, A. J. Bird, US Patent 4 396 786, 1983.
12. E. Berl, *Science*, 1944, **99**, 309.
13. G. W. Huber, S. Iborra and A. Corma, *Chem. Rev.*, 2006, **106**, 4044.
14. K. L. Lautenschläger, M. Bockmühl and G. Ehrhart, US Patent 1 915 431, 1933.
15. H. H. Gürkan, US Patent 2 488 722, 1949.
16. (a) V. I. Sharkov, *Angew. Chem. Int. Ed.*, 1963, **2**, 405; (b) V. I. Sharkov, *Chem. Ing. Tech.*, 1963, **35**, 494.
17. S. Van de Vyver, J. Geboers, P. A. Jacobs and B. F. Sels, *ChemCatChem*, 2011, **3**, 82.
18. A. M. Ruppert, K. Weinberg and R. Palkovits, *Angew. Chem. Int. Ed.*, 2012, **51**, 2564.
19. (a) C. Luo, S. Wang and H. Liu, *Angew. Chem. Int. Ed.*, 2007, **46**, 7636; *Angew. Chem.*, 2007, **119**, 7780; (b) L.-N. Ding, A.-Q. Wang, M.-Y. Zheng and T. Zhang, *ChemSusChem*, 2010, **3**, 818; (c) S. Van de Vyver, J. Geboers, M. Dusselier, H. Schepers, T. Vosch, L. Zhang, G. Van Tendeloo, P. A. Jacobs and B. F. Sels, *ChemSusChem*, 2010, **3**, 698; (d) R. Palkovits, K. Tajvidi, J. Procelewska, R. Rinaldi and A. Ruppert, *Green. Chem.*, 2010, **12**, 972.
20. (a) A. Sarko and R. Muggli, *Macromolecules*, 1974, 7, 486; (b) A. C. O'Sullivan, *Cellulose*, 1997, **4**, 173; (c) Y. Nishiyama, P. Langan and H. Chanzy, *J. Am. Chem. Soc.*, 2002, **124**, 9074.
21. (a) L. Meiler and H. Scholler, US Patent 1,641,771, 1927; (b) E. E. Harris and E. Beglinger, *Ind. Eng. Chem.*, 1946, **38**, 890.
22. (a) J. N. Bemiller, *Adv. Carbohyd. Res.*, 1967, **22**, 25; (b) Y. V. Moiseev, N. A. Khaltruniskii and G. E. Zaikov, *Carbohyd. Res.*, 1976, **51**, 23; (c) R. Rinaldi and F. Schüth, *ChemSusChem*, 2009, 2, 1096.
23. (a) H. Lüers, *Z. Angew. Chem.*, 1930, **43**, 455; (b) J. F. Saeman, *Ind. Eng. Chem.*, 1945, **37**, 43; (c) A. H. Conner, B. F. Wood, C. G. Hill Jr. and

J. F. Harris, *J. Wood Chem. Technol.*, 1985, **5**, 461; (d) J. Bouchard, N. Abatzoglou, E. Chornet and R. P. Overend, *Wood Sci. Technol.*, 1989, **23**, 343; (e) J. Bouchard, R. P. Overend and E. Chornet, *J. Wood Chem. Technol.*, 1992, **12**, 335.

24. (a) M. Sasaki, B. Kabyemela, R. Malaluan, S. Hirose, N. Takeda, T. Adschiri and K. Arai, *J. Supercrit. Fluids*, **13**, 261; (b) M. Sasaki, Z. Fang, Y. Fukushima, T. Adschiri and K. Arai, *Ind. Eng. Chem. Res.*, 2000, **39**, 2883.

25. (a) Y. Zhu, Z. N. Kong, L. P. Stubbs, H. Lin, S. Chen, E. V. Anslyn and J. A. Magire, *ChemSusChem*, 2010, **3**, 67; (b) I. A. Ignatyev, C. Van Doorslaer, P. G. N. Mertens, K. Binnemans and D. E. De Vos, *ChemSusChem*, 2010, **3**, 91; (c) R. M. de Almeida, J. Li, C. Nederlof, P. O'Conner, M. Makkee and J. A. Moulijn, *ChemSusChem*, 2010, **3**, 325; (d) R. M. de Almeida, S. Daamen, J. A. Moulijn, M. Makkee, P. O'Conner, EP 21,000,972A1, 2009.

26. (a) S. J. Angyal, *Angew. Chem. Int. Ed.*, 1969, **8**, 157; (b) F. Franks, *Pure Appl. Chem.*, 1987, **59**, 1189; (c) H. E. van Dam, A. P. G. Kieboom and H. van Bekkum, *Starch*, 1986, **38**, 95.

27. (a) C. A. Lobry de Bruyn and W. Alberda van Ekenstein, *Recl. Trav. Chim. Pays-Bas*, 1895, **14**, 203; (b) C. A. Lobry de Bruyn and W. Alberda van Ekenstein, *1896 Recl. Trav. Chim. Pays-Bas*, 1896, **15**, 92.

28. (a) J. C. Sowden and R. Schaffer, *J. Am. Chem. Soc.*, 1952, **74**, 505; (b) D. W. Harris and M. S. Feather, *Carbohyd. Res.*, 1973, **30**, 359.

29. (a) J. Lewkowski, *Arkivoc*, **2001**, 17; (b) M. L. Mednick, *J. Org. Chem.*, 1962, **27**, 398; (c) M. J. Antal, Jr., W. S. L. Mok and G. N. Richards, *Carbohyd. Res.*, 1990, **199**, 91.

30. (a) B. M. Kabyemela, T. Adschiri, R. M. Malaluan and K. Arai, *Ind. Eng. Chem. Res.*, 1999, **38**, 2888; (b) B. M. Kabyemela, T. Adschiri, R. M. Malaluan and K. Arai, *Ind. Eng. Chem. Res.*, 1997, **36**, 1552; (c) B. M. Kabyemela, T. Adschiri, R. M. Malaluan and K. Arai, *Ind. Eng. Chem. Res.*, 1997, **36**, 2025.

31. J. Geboers, S. Van de Vyver, K. Carpentier, P. Jacobs and B. Sels, *Chem. Commun.*, 2011, **47**, 5590.

32. (a) J. Wisnlak and R. Simon, *Ind. Eng. Chem. Prod. Res. Dev.*, 1979, **18**, 50; (b) E. Crezee, B. W. Hoffer, R. J. Berger, M. Makkee, F. Kapteijn and J. A. Moulijn, *Appl. Catal. A*, 2003, **251**, 1; (c) N. Dchamp, A. Gamez, A. Perrard and P. Gallezot, *Catal. Today*, 1995, **24**, 29; (d) B. W. Hoffer, E. Crezee, P. R. M. Mooijman, A. D. van Langeveld, F. Kapteijn and J. A. Moulijn, *Catal. Today*, 2003, **79**, 35; (e) P. H. Brahme and L. K. Doraiswamy, *Ind. Eng. Chem. Process Des. Dev.*, 1976, **15**, 130.

33. (a) D. K. Sohounloue, C. Montassier and J. Barbier, *React. Kinet. Catal. Lett.*, 1983, **22**, 391; (b) C. Montassier, J. C. Menezo, L. C. Hoang, C. Renaud and J. Barbier, *J. Mol. Catal. A Gen.*, 1991, **70**, 99.

34. M. A. Andrews and S. A. Klaeren, *J. Am. Chem. Soc.*, 1989, **111**, 4131.

35. (a) C. Montassier, D. Giraud, J. Barbier and J. P. Boitiaux, *Bull. Soc. Chim. Fr.*, 1989, **2**, 148; (b) J. M. Montassier, P. Dumas, Granger and J. Barbier,

Appl. Catal. A, 1995, **121**, 231; (c) C. Montassier, D. Giraud and J. Barbier, *Stud. Surf. Sci. Catal.*, 1988, **41**, 165; (d) C. Montassier, J. C. Menezo, J. Moukolo, J. Naja and J. Barbier, *Stud. Surf. Sci. Catal.*, 1991, **59**, 223.

36. K. Wang, M. C. Hawley and T. D. Furney, *Ind. Eng. Chem. Res.*, 1995, **34**, 3766.
37. K. L. Deutsch, D. G. Lahr and B. R. Shanks, *Green Chem.*, 2012, **14**, 1635.
38. N. Li and G. W. Huber, *J. Catal.*, 2010, **270**, 48.
39. N. Yan, C. Zhao, C. Luo, P. J. Dyson, H. Liu and Y. Kou, *J. Am. Chem. Soc.*, 2006, **128**, 8714.
40. (a) A. Fukuoka and P. L. Dhepe, *Angew. Chem. Int. Ed.*, 2006, **45**, 5161; (b) G. F. Liang, C. Y. Wu, L. M. He, J. Ming, H. Y. Cheng, L. H. Zhuo and F. Y. Zhao, *Green Chem.*, 2011, **13**, 839; (c) P. L. Dhepe and A. Fukuoka, *ChemSusChem*, 2008, **1**, 969.
41. W. Deng, M. Liu, X. Tan, Q. Zhang and Y. Wang, *J. Catal.*, 2010, **271**, 22.
42. J. Li, H. S. M. P. Soares, J. A. Moulijn and M. Makkee, *Catal. Sci. Technol.*, 2013, **3**, 1565.
43. L. Negahdar, J. U. Oltmanns, S. Palkovits and R. Palkovits, *Appl. Catal. B*, 2014, **147**, 677.
44. P. L. Dhepe and A. Fukuoka, *Catal. Surb. Asia*, 2007, **11**, 186.
45. H. Hattori and T. Shishido, *Catal. Surv. Jpn.*, 1997, **1**, 205.
46. V. Jollet, F. Chambon, F. Rataboul, A. Cabiac, C. Pinel, E. Guillon and N. Essayem, *Green Chem.*, 2009, **11**, 2052.
47. W. Deng, X. Tan, W. Fang, Q. Zhang and Y. Wang, *Catal. Lett.*, 2009, **133**, 167.
48. R. Palkovits, K. Tajvidi, A. M. Ruppert and J. Procelewska, *Chem. Commun.*, 2011, **47**, 576.
49. J. Geboers, S. Van de Vyver, K. Carpentier, K. de Blochouse, P. Jacobs and B. Sels, *Chem. Commun.*, 2010, **46**, 3577.
50. J. Geboers, S. Van de Vyver, K. Carpentier, P. Jacobs and B. Sels, *Green Chem.*, 2011, **13**, 2167.
51. Z. Tai, J. Zhang, A. Wang, M. Zheng and T. Zhang, *Chem. Commun.*, 2012, **48**, 7052.
52. Z. Wu, S. Ge, C. Ren, M. Zhang, A. Yip and C. Xu, *Green Chem.*, 2012, **14**, 3336.
53. (a) N. Meine, R. Rinaldi and F. Schüth, *ChemSusChem*, 2012, **5**, 1449–1454; (b) J. Hilgert, N. Meine, R. Rinaldi and F. Schüth, *Energy Environ. Sci.*, 2013, **6**, 92.
54. A. Duclos, C. Fayet and J. Gelas, *Synthesis*, 1994, 1087.
55. W. C. Brinegar, M. Wohlers, M. A. Hubbard, E. G. Zey, G. Kvakovszky, T. H. Shockley, R. Roesky, U. Dingerdissen and W. Kind (E. I. du Pont de Nemours and Company), WO 00/14081, 2000.
56. N. Ji, T. Zhang, M. Zheng, A. Wang, H. Wang, X. Wang and J. G. Chen, *Angew. Chem. Int. Ed.*, 2008, **47**, 8510.
57. J. Pang, M. Zheng, A. Wang and T. Zhang, *Ind. Eng. Chem. Res.*, 2011, **50**, 6601.

58. C. Li, M. Zheng, A. Wang and T. Zhang, *Energy Environ. Sci.*, 2012, **5**, 6383.

59. N. Ji, M. Zheng, A. Wang, T. Zhang and J. G. Chen, *ChemSusChem*, 2012, **5**, 939.

60. M. Zheng, A. Wang, N. Ji, J. Pang, X. Wang and T. Zhang, *ChemSusChem*, 2010, **3**, 63.

61. Z. Tai, J. Zhang, A. Wang, J. Pang, M. Zheng and T. Zhang, *ChemSusChem*, 2013, **6**, 652.

62. L. Ding, A. Wang, M. Zheng and T. Zhang, *ChemSusChem*, 2010, **3**, 818.

63. J. Pang, A. Wang, M. Zheng, Y. Zhang, Y. Huang, X. Chen and T. Zhang, *Green Chem.*, 2012, **14**, 614.

64. A. Shrotri, A. Tanksale, J. N. Beltramini, H. Gurav and S. V. Chilukuri, *Catal. Sci. Technol.*, 2012, **2**, 1852.

65. X. Wang, L. Meng, F. Wu, Y. Jiang, L. Wang and X. Mu, *Green Chem.*, 2012, **14**, 758.

66. X. Wang, F. Wu, S. Yao, Y. Jiang, J. Guan and X. Mu, *Chem. Lett.*, 2012, **41**, 476.

67. K. Tajvidi, K. Pupovac, M. Kükrek and R. Palkovits, *ChemSusChem*, 2012, **5**, 2139.

68. Z. Xiao, S. Jin, M. Pang and C. Liang, *Green Chem.*, 2013, **15**, 891.

69. C. Loerbroks, R. Rinaldi and W. Thiel, *Chem. Eur. J.*, 2013, **19**, 16282.

70. (a) M. Käldström, N. Meine, C. Fàres, R. Rinaldi and F. Schüth, *Green Chem.*, 2014, **16**, 2454; (b) M. Käldström, N. Meine, C. Fàres, F. Schüth and R. Rinaldi, *Green Chem.*, 2014, **16**, 3528–3538; (c) F. Schüth, R. Rinaldi, N. Meine, M. Käldström, J. Hilgert and M. D. Kaufman Rechulski, *Catal. Today*, 2014, **234**, 24.

CHAPTER 6

Hydrodeoxygenation of Lignocellulose-Derived Platform Molecules

KONSTANTIN HENGST, MARTIN SCHUBERT, WOLFGANG KLEIST AND JAN-DIERK GRUNWALDT*

Institute for Chemical Technology and Polymer Chemistry (ITCP) and Institute of Catalysis Research and Technology (IKFT), Karlsruhe Institute of Technology (KIT), D-76131, Karlsruhe, Germany
*Email: grunwaldt@kit.edu

6.1 Introduction

Two main routes have been established for the utilization of plant biomass in chemical processes.[1-3] In the first route, biomass is gasified at high temperatures and its complex carbon backbone is decomposed into synthesis gas (CO and H_2, Figure 6.1).[1,4] Gasification of biomass has been examined on the pilot to the industrial scale, e.g. the Bioliq process,[5a] Chemrec's gasification,[5b] or the Carbona/Haldor Topsøe gasification plant.[5c] An advantage is that the whole biomass is used and clean fuels or fuel additives can be obtained. A fundamental disadvantage of the gasification process is that synthetic fuels or chemicals have to be built up from the bottom (e.g. by Fischer–Tropsch, methanol or dimethyl ether syntheses). Consequently, insertion of functional groups is quite elaborate within this approach, since they need to be reintroduced into the hydrocarbon or alcohol backbone. In the second route, a structural use of plant biomass is

RSC Energy and Environment Series No. 13
Catalytic Hydrogenation for Biomass Valorization
Edited by Roberto Rinaldi
© The Royal Society of Chemistry 2015
Published by the Royal Society of Chemistry, www.rsc.org

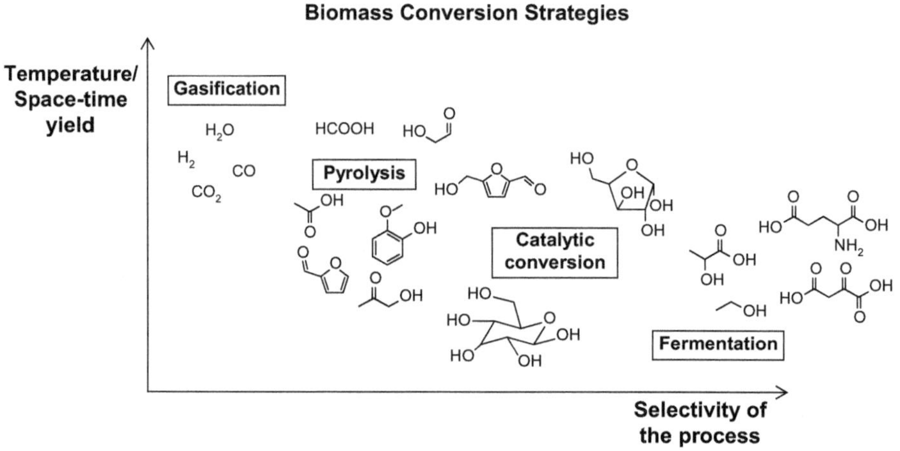

Figure 6.1 Approaches for the utilization of plant biomass and platform (bio)chemicals.

achieved by thermochemical treatment (*e.g.* flash pyrolysis), selective chemical conversion or fermentation performed in the liquid phase at lower temperatures. The processes are carried out at lower temperatures than gasification processes (Figure 6.1).[2,6,7] Whereas flash pyrolysis results in a bio-oil that needs to be upgraded by hydrodeoxygenation (HDO) and fractionated[8,9] (as will be discussed in Chapter 7), the selective defunctionalization of biomass is an important pathway to pave the way for tailor-made biofuels and chemicals. This route has been exploited by many research consortia, *e.g.* "Tailor-Made Fuels from Biomass" (Aachen/Germany),[5d] "CatchBio" in the Netherlands,[5e] or "Catalysis for Sustainable Energies (CASE)" in Denmark,[5f] and several other initiatives worldwide.

The selective structural use of biomass is currently present in several industrial processes for the large-scale production of many platform chemicals (*e.g.* bioethanol, citric acid, lactic acid, *etc.*). It is noteworthy that the space–time yield of the enzymatic and fermentative processes is rather low compared to chemical processes performed either in gas or liquid phase. Accordingly, chemical and thermochemical conversions of biomass may still offer more attractive alternatives for large-scale operations.[8,9] Prior to the chemical conversion of lignocellulose, the fractionation of its major components – cellulose, hemicellulose and lignin – is required for the efficient catalytic conversion into desired products (Figure 6.2).[1,3,10] Alternatively, the biomass is pyrolyzed and converted by HDO, forming a less-complex reaction mixture that can be fractionized similarly to crude oil.[8,9]

Hydrolysis of lignocellulose leads first to sugars, which can subsequently be transferred into a number of platform molecules. Some of these chemicals have been outlined in a report by the *National Renewable Energy Laboratory* (NREL)[10] and are depicted in Figure 6.3. These platform chemicals show high oxygen contents and structural similarity to compounds that are

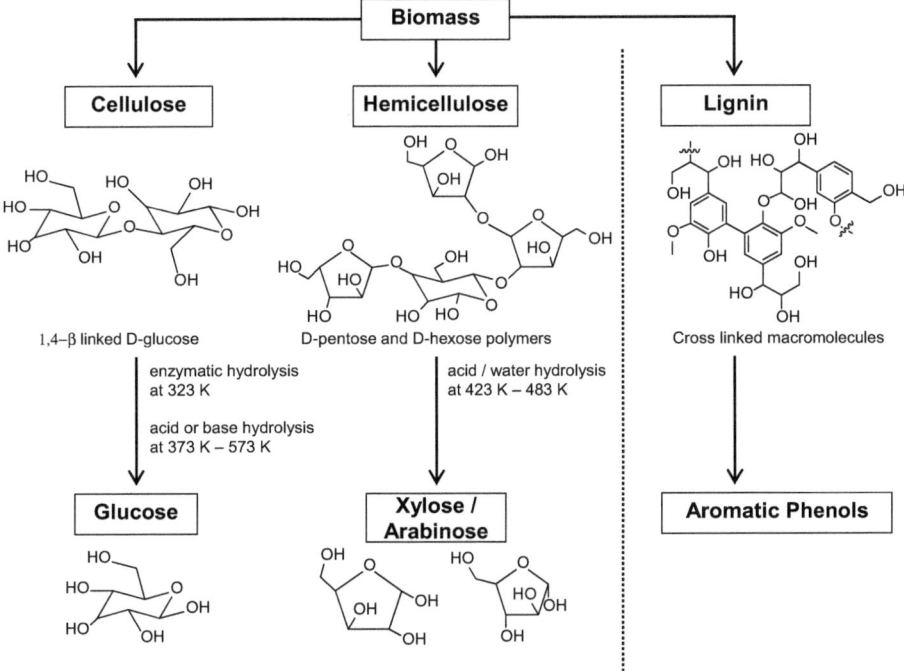

Figure 6.2 Components of lignocellulose and depolymerization products.

found in the pyrolysis oil after thermal treatment. Therefore, selective removal of the oxygen functionalities by decarboxylation, dehydration, ketonization or HDO reactions is crucial for the synthesis of various commodity chemicals or, in case of pyrolysis oil, to receive a less complex mixture.

This chapter focuses on the general aspects related to HDO of biomass-derived platform molecules to speciality chemicals, synthetic fuels and fuel additives taking 5-hydroxymethylfurfural (HMF) and levulinic acid (LA) as representative examples. HMF and LA can be obtained at relatively high purity degree from lignocellulosic feedstocks. Most importantly, unlike other top-platform molecules, only de- and rehydration reactions are needed to produce HMF and LA. Accordingly, no molecular hydrogen is needed for the deoxygenation of cellulose and hemicellulose rendering HMF and LA. This fact is important at the current industrial development, since molecular hydrogen is still mostly obtained from nonrenewable resources (*e.g.* natural gas, as indepth discussed in Chapter 1). In addition, these two sample compounds may be considered also as representatives of organic ketones, alcohols or acids that are typical constituents of pyrolysis oils produced *via* pyrolysis processes.

6.2 Effective H/C Ratio and its Importance

Typically, platform chemicals and fuels derived from biomass are much more functionalized than hydrocarbons. This fact accounts for the lower

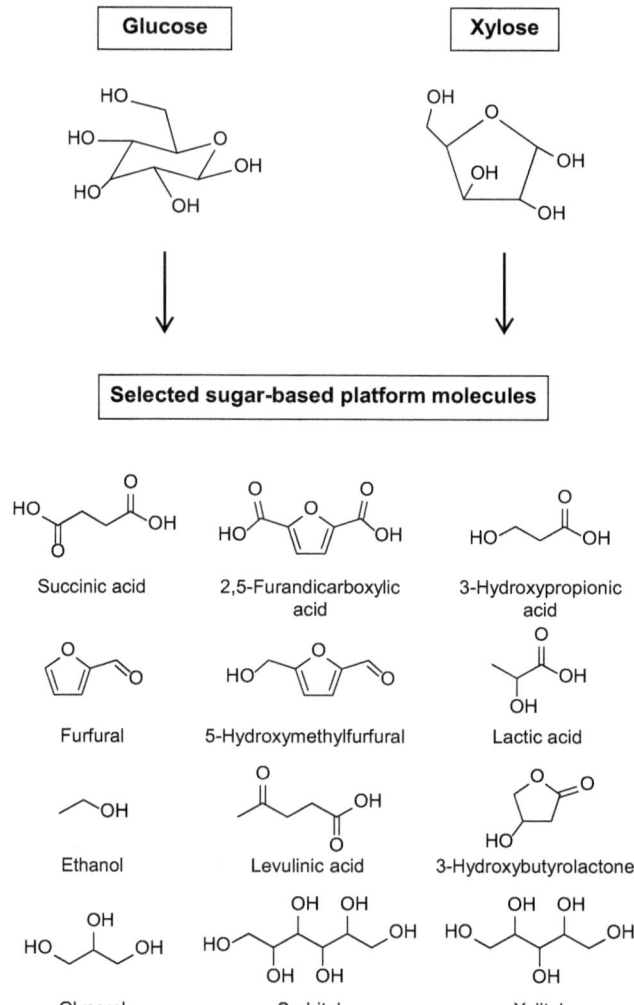

Figure 6.3 Biomass platform molecules from sugar as examples discussed in ref. 10.

energy density of first-generation biofuels compared to petroleum-based fuels. The degree of functionalization of a molecule can be compared by its "mean oxidation number of organic carbon" (MOC) or by the "effective H/C ratio" (*i.e.* effective H/C ratio $=-$ MOC).

According to Vogel *et al.*,[11] a high oxygen content corresponds to a high "mean oxidation number of organic carbon" (MOC), which is defined as:

$$MOC = \frac{\sum\limits_{i=1}^{n} OC_i}{n} \qquad (6.1)$$

where OC_i stands for the formal oxidation state of the *i*th carbon atom in the organic molecule, and *n* corresponds to the number of carbon atoms in the molecule. Since the MOC concept does not differentiate between heteroatoms, different compounds may be classified by the same MOC value.[15] A more convenient manner to classify different feedstocks and biofuels is by using their effective H/C ratios, as proposed by Vennestrøm *et al.*[12]

Considering the transformation of plant biomass into biofuels, a high effective H/C ratio is related to a high energy content per carbon (*i.e.* heating value or combustion enthalpy). For example, the combustion enthalpies per carbon atom for methane and octane are –890.4 and –683.8 kJ/mol, respectively, compared to –485.4 kJ/mol for LA and –463.5 kJ/mol for HMF. In comparison, the effective H/C ratio decreases in the same order found for the combustion enthalpies (*i.e.* the effective H/C ratio is 4 for methane, 2.25 for octane, 0.4 for LA, and zero for HMF). Transportation fuels exhibit effective H/C ratios in the range from 1 to 2.3.[12] While the use of effective H/C ratios is a good descriptor to group several compounds regarding their overall functionalization, any generalization always shows some limitations. Although carbohydrates show a much lower energy content than high-ranking coals, the classification by effective H/C ratio clusters carbohydrates close to high-ranking coals.

The concept of the effective H/C ratio underlines that subsequent de-oxygenation of the biomass streams is mandatory for the production of biofuels. In practice, biomass streams can be upgraded by catalytic HDO in order to increase their effective H/C ratio, and consequently their energy content. Moreover, deoxygenation of fast pyrolysis oil not only increases its energy content but also its chemical stability, as will be addressed in Chapter 7.

Figure 6.4 classifies resources, platform and intermediates, and target chemicals according to their effective H/C ratios, and reveals another important aspect of the analysis based on the effective H/C ratio. The horizontal axis represents the degree of processing. The second vertical axis semiquantitatively correlates the effective H/C ratio of the respective substances with the energy content per carbon atom. This correlation is helpful when discussing upgrading of biomass platform molecules *via* HDO for the production of biofuels.

A large difference between the effective H/C ratio of a resource and a target chemical is, in most cases, accompanied by a need for complex processing.[12] Moreover, this implies that substantial amounts of energy (and most likely losses of energy and/or product) would be involved throughout the process chain. In this context, carbohydrates have effective H/C ratios comparable to many highly functionalized platform or target chemicals. In some scenarios, plant biomass should be a more suitable feedstock than conventional resources (*e.g.* natural gas, oil or coal), because fewer, or even no, HDO steps may be necessary. Conversely, a process initiated with conventional resources may need many steps in order to introduce the desired functionalities.[12]

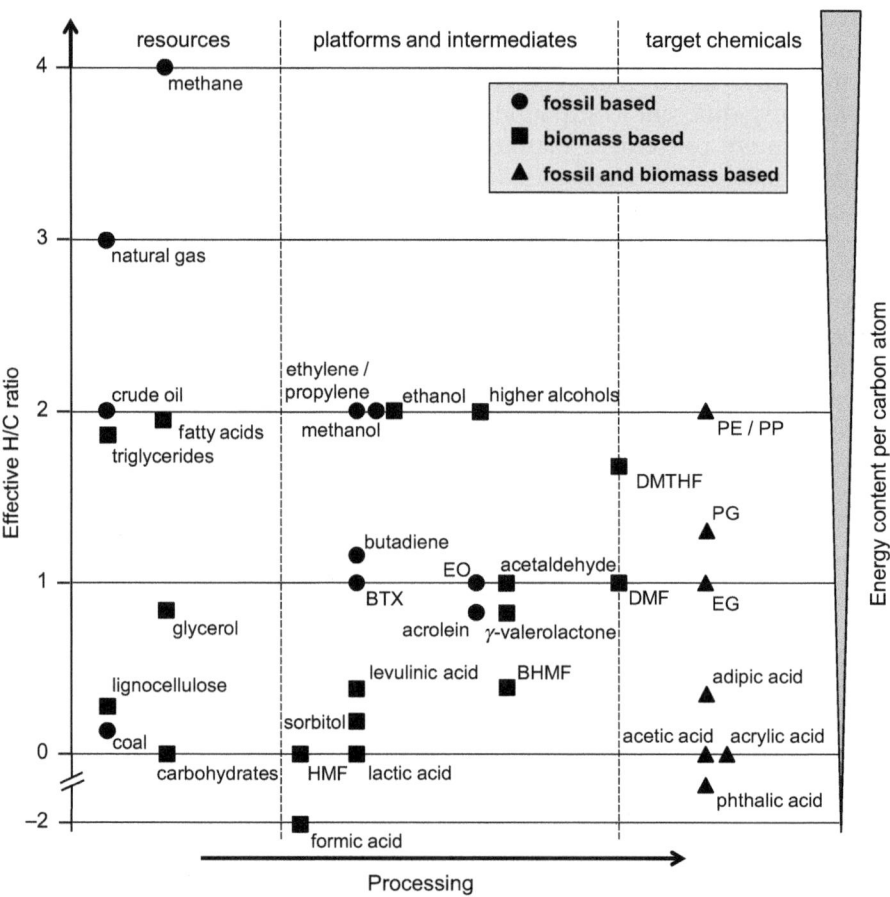

Figure 6.4   Effective H/C ratio versus degree of processing. Abbreviations: HMF –
5-hydroxymethylfurfural, BTX – benzene, toluene, xylene, EO – ethylene
oxide, BHMF – 2,5-bis(hydroxymethyl)furfural, DMF – 2,5-dimethylfuran,
DMTHF – 2,5-dimethyltetrahydrofuran, EG – ethylene glycol, PG – propylene
glycol, PE – polyethylene, PP – polypropylene.

6.3   (Hydro)deoxygenation of 5-Hydroxymethylfurfural

HMF represents a biomass platform molecule available from hexoses but
also an important constituent of pyrolysis oil from flash pyrolysis, which
exhibits an aldehyde and a hydroxymethyl group on the furanic ring at the
positions 2 and 5, respectively. These functionalities are very reactive.
Accordingly, HMF can serve as a platform for the production of several
chemical intermediates and fuel additives (Figure 6.5). For biofuels, 2,5-
dimethylfuran (DMF) and 2,5-dimethyltetrahydrofuran (DMTHF) are
potential candidates due to their high energy density, low volatility, and high

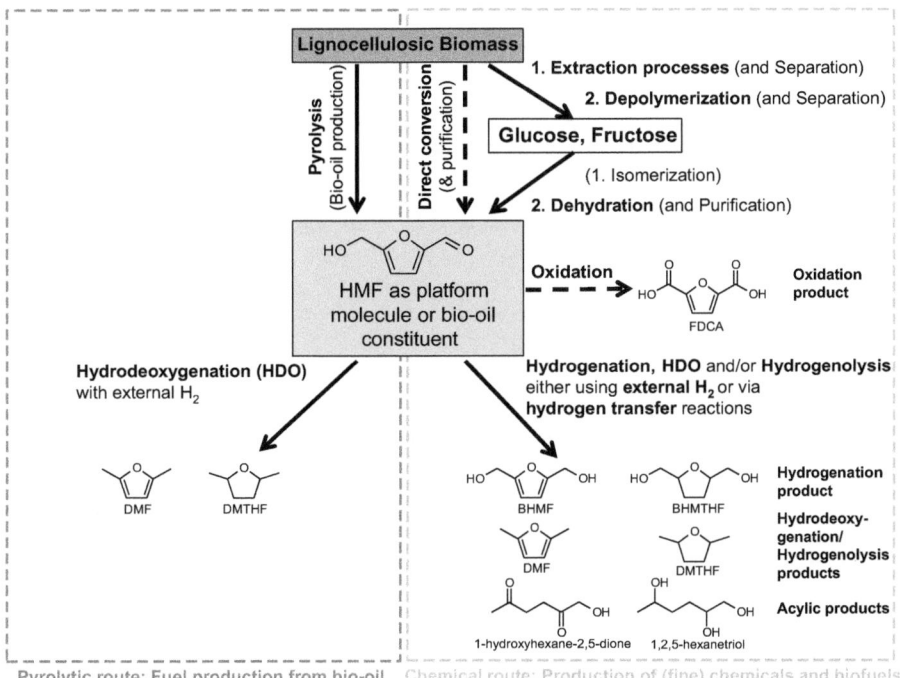

Figure 6.5 Transformation pathways of lignocellulosic biomass to HMF and its conversion into selected valuable chemicals and fuel components.

research octane number (RON) of 119 (DMF) and 82 (DMTHF). Furthermore, DMTHF has been explored as a biobased solvent and substitute for petroleum-derived THF.[13,14] HMF is also a platform molecule for the production of value-added chemicals through oxidation (*e.g.* 2,5-furandicarboxylic acid, FDCA; and 2,5-bis(hydroxylmethyl)furan, BHMF), hydrogenation (*e.g.* 2,5-bis(hydroxymethyl)tetrahydrofuran, BHMTHF) and hydrogenolysis (*e.g.* 1-hydroxyhexane-2,5-dione and hexane-1,2,5-triol). FDCA is proposed as a renewable alternative to terephthalic acid as a monomer for polyesters.[10,12,15] The hydrogenation products BHMF and BHMTHF are valuable intermediates for polymer production.[10,15,16] The hydrogenolysis products 1-hydroxyhexane-2,5-dione and 1,2,5-hexanetriol are specialty chemicals used in the flavor and polymer industries.[17]

6.3.1 Hydrogenation and Hydrodeoxygenation Using Molecular Hydrogen

Hydrogenation and hydrodeoxygenation of HMF using externally supplied molecular hydrogen have been extensively explored under different conditions (see for example Tables 6.1 and 6.2). The product distribution depends upon temperature, H_2 pressure, catalysts and solvents. In the recent literature, there is a consensus that HMF especially over noble-metal based

Table 6.1 Selected results for the production of 2,5-dimethylfuran *via* hydrogenolysis / hydrogeoxygenation of HMF with molecular hydrogen.

Entry	Feedstock	Reactor[f]	Catalyst	Solvent	T [K]	p_{H_2} [MPa]	t [h] WHSV [h^{-1}]	Conv. [%]	Yield [%]			Ref.
									BHMF	DMF	DMTHF	
1	HMF	B	Cu–Ru/C (3:1)[h]	1-BuOH	493	0.68	10	100	n.a.	71	n.a.	13
2	HMF[a]	B	Cu–Ru/C (3:1)[h]	1-BuOH	493	0.68	10	100	n.a.	61	n.a.	13
3	HMF	B	CuCrO$_4$	1-BuOH	493	0.68	10	100	n.a.	61	n.a.	13
4	HMF[a]	B	CuCrO$_4$	1-BuOH	493	0.68	10	94	13	6	n.a.	13
5	HMF	F[g]	Cu–Ru/C (3:2)[h]	1-HexOH	493	1.72	0.15	100	n.a.	77	n.a.	13
6	HMF	F[g]	Cu–Ru/C (3:2)[h]	1-HexOH	493	0.68	0.98	100	n.a.	78	n.a.	13
7	HMF	F[g]	Cu–Ru/C (3:1)[h]	1-BuOH	493	1.72	0.15	100	n.a.	72	n.a.	13
8	HMF[b]	F[g]	Cu–Ru/C (3:1)[h]	1-BuOH	493	1.72	0.17	100	n.a.	72	n.a.	13
9	Glucose	B	Sn–Ru/C	GVL or γ-hexalactone	n.a.	n.a.	n.a.	n.a.	n.a.	25–27	n.a.	26
10	HMF	B	10% Pd/C	1-PrOH	333	0.1	2.5	100	0	35.7	0.5	19
11	HMF	B	5% Pd/Al$_2$O$_3$	1-PrOH	333	0.1	8.25	19.2	14.1	0	0	19
12	HMF	B	5% Pd/Al$_2$O$_3$	1-PrOH[i]	333	0.1	7	100	0	28.3	0.5	19
13	HMF	B	10% Pd/C	1,4-dioxane	333	0.1	12.33	90	79.9	4.8	0	19
14	HMF[c]	B	Cu–Ru/C	1-BuOH	493	0.68	10	n.a.	n.a.	49	n.a.	33
15	HMF	B	10% Pd/C	EMIMCl	393	6.2	1	19	n.a.	2.5	n.a.	20
16	HMF	B	10% Pd/C	EMIMCl-acetonitrile	393	6.2	1	47	0.5	15	n.a.	20
17	HMF[d]	B	10% Pd/C	EMIMCl-acetonitrile	393	6.2	1	44	0.8	12	n.a.	20
18	HMF	B	Ru/C	1-BuOH	533	3.9	1.5	99.8	n.a.	60.3	n.a.	25
19	HMF[e]	B	Ru/C	1-BuOH	533	3.9	1.5	93.4	n.a.	32.7	n.a.	25

Abbreviations: HMF – hydroxymethylfurfural, DMF – 2,5-dimethylfuran, BHMF – 2,5-bis(hydroxymethyl)furan, DMTHF – 2,5-dimethyltetrahydrofuran, EMIMCl – 1-ethyl-3-methylimidazolium chloride, 1-BuOH – 1-butanol, 1-PrOH – 1-propanol, 1-HexOH – 1-hexanol, n.a. – not available/not mentioned in the respective article.

[a] HMF precontacted with a saturated aqueous solution of NaCl (26 mmol L^{-1} NaCl) and purified by evaporation of 25% of the mass, leaving 1.6 mmol/L in the HMF phase.
[b] obtained from fructose.
[c] HMF produced from untreated corn stover in *N,N*-dimethylacetamide-LiCl and purified with ion-exclusion chromatography.
[d] HMF obtained upon glucose dehydration in EMIMCl-acetonitrile using 12-molybdophosphoric acid.
[e] a 1:1 HMF and furfural mixture was obtained from glucose and xylose dehydration in a biphasic water – 1-butanol reaction mixture using SO_4^{2-}/ZrO_2-TiO_2.
[f] B = batch, F = flow.
[g] vapor phase hydrogenolysis.
[h] molar ratio Cu:Ru.
[i] conc. HCl added.

Table 6.2 Selected results for the production of 2,5-dimethylfuran *via* hydrogenolysis/hydrogeoxygenation of HMF or more complex starting materials, such as fructose and glucose, cellulose or agar *via* transfer hydrogenation.

Entry	Feedstock	Reactor[a]	Catalyst	Solvent	H_2 (Leerzeichen) donor	T [K]	p [MPa]	t [h]	Conv. [%]	Yield [%] BHMF	DMF	DMTHF	Ref.
1	HMF	B	Cu/PMO	scMeOH	scMeOH	533	10.7[h]	3	100	n.a.	48	10	14
2	HMF[a]	B	Cu/PMO	scMeOH	scMeOH	573	10.7[h]	3	100	n.a.	12	10	14
3	HMF	B	5% Pd/C[f]	THF	FA	343[g]	0.1	15	100	-	>95	-	31
4	Fructose[b]	B	5% Pd/C[f]	THF	FA	343[g]	0.1	15	100	-	51	-	31
5	Fructose[c]	B	5% Ru/C[f]	THF	FA	348[g]	0.1	15	n.a.	n.a.	30	n.a.	32
6	Fructose[c]	B[e]	5% Ru/C[f]	THF	FA	348	i	0.75	n.a.	n.a.	32	n.a.	32
7	α-Cellulose[c]	B[e]	5% Ru/C[f]	THF	FA	348	i	0.75	n.a.	n.a.	16	n.a.	32
8	Agar[c]	B[e]	5% Ru/C[f]	THF	FA	348	i	0.75	n.a.	n.a.	27	n.a.	32

Abbreviations: HMF – 5-hydroxymethylfurfural, DMF – 2,5-dimethylfuran, BHMF – 2,5-bis(hydroxymethyl)furan, DMTHF – 2,5-dimethyltetrahyxrpofuran, scMeOH – supercritical methanol, THF – tetrahydrofuran, n.a. – not available/not mentioned in the article.

[a]HMF derived from dehydration of glucose and purified by extraction with methyl isobutyl ketone (which was removed in vacuum after extraction leaving a brown sticky material, *i.e.* crude HMF).

[b]one-pot synthesis of DMF from fructose performed in two-steps: 1st, dehydration of fructose in formic acid, and 2nd, hydrogenolysis of the obtained material according to the parameters given in the table.

[c]multistep one-pot synthesis, only the hydrogenolysis step is given in the table, previous steps include formation of HMF from fructose in formic acid, or even depolymerization of cellulose or agar in an ionic liquid catalyst system [DMA][+][CH₃SO₃][−]/DMA-LiCl (DMA = *N,N*-dimethylacetamide).

[d]B = batch.

[e]Microwave heating.

[f]H₂SO₄ as cocatalyst.

[g]reflux.

[h]estimated.

[i]autogenic vapor pressure.

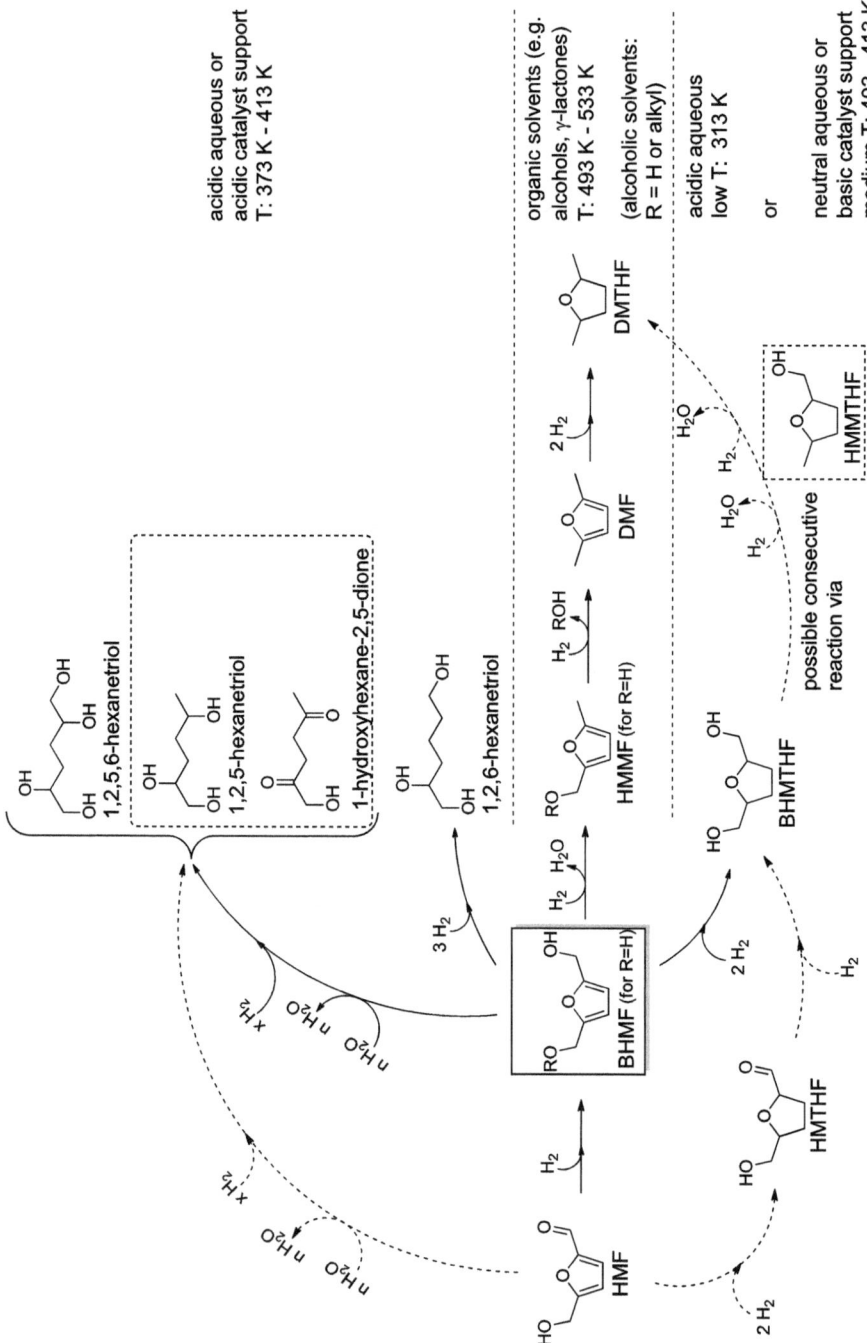

catalysts first undergoes hydrogenation forming a diol intermediate (BHMF). This reduction initiates the reaction chain occurring in the presence of noble-metal catalysts (see Scheme 6.1).[13,15,16,18,19] As discussed in Chapter 4, solvent effects play major roles in the control of the product distribution obtained by heterogeneously catalyzed reductions. For instance, HMF conversion takes place *via* a hemiacetal or acetal intermediate, when the reaction is performed in an alcoholic solvent. As a result, not BHMF but a 2-alkoxy-5-hydroxymethylfuran product is obtained as the first intermediate (Scheme 6.1).[19] In an electrolyte (*e.g.* 1-ethyl-3-imidazolium chloride dissolved in acetonitrile), the conversion of HMF proceeds *via* 5-methylfurfural as first intermediate using a Pd/C catalyst (Table 6.1, entries 16 and 17).[20] In water[16] or diethyl carbonate[21] full saturation of HMF (rendering BHMTHF) is favored on mono- or bimetallic Pd-based catalysts. These conversions proceeded to a remarkable extent via the ring-hydrogenated, 5-hydroxymethyl-2,3,4,5-tetrahydro-2-furaldehyde (HMTHF). The main reaction pathway, however, still includes BHMF as the first intermediate product (Scheme 6.1).

BHMF holds promise as a specialty chemical for several applications,[10,15,18] especially in the production of polymers. High to quantitative yield of BHMF (95–100%), was reported for the conversion of HMF in the presence of copper chromite, Pt/C, Ru/C or PtO$_2$ performed in water under H$_2$ pressure of 7 MPa at temperatures of 373–413 K at reaction durations of 30–60 min.[17] Furthermore, Ohyama *et al.*[18] found Au supported on γ-Al$_2$O$_3$ (Au cluster size of 0.88 ± 0.30 nm corresponding to an Au$_{13}$ cluster) to be a suitable catalyst for the selective hydrogenation of HMF to BHMF in aqueous solution at 393 K and under a H$_2$ pressure of 6.5 MPa. The obtained BHMF yield was 96% under these conditions. More recently, a selectivity of 99% was achieved over Ir–ReO$_x$/SiO$_2$ catalysts.[22] Dumesic's group[23] examined the conversion of HMF into BHMF in a biphasic 1-butanol/water system. They reported BHMF yields of 81–94% at full HMF conversion in the presence of Ru supported on metal oxides (*e.g.* CeO$_x$, γ-Al$_2$O$_3$, or magnesia–zirconia) under a H$_2$ pressure of 2.75 MPa at 403 K for 2 h. Prolonged reaction durations decreased the selectivity to BHMF due to further hydrogenation yielding BHMTHF (Scheme 6.1).

BHMTHF is a key intermediate for the production of 1,6-hexanediol from renewable resources, which is utilized in the manufacturing of several polymers.[16,24] Note that the hydrogenation of HMF (or BHMF) to BHMTHF increases the effective H/C ratio, but does not decrease the O/C ratio (*i.e.* it does not deoxygenate HMF). Full saturation of BHMF to BHMTHF is

---

Scheme 6.1   Hydrogenation pathways of 5-HMF under varying conditions. Solid arrows represent the main reaction pathways, dashed arrows correspond to minor reaction pathways. Hydrodeoxygenation of BHMTHF to DMTHF *via* HMMTHF may occur, but was not explicitly mentioned in the literature. Note that the reaction path *via* 5-methylfurfural found in the presence of ionic liquids[20] is not displayed in this figure. Abbreviations are given in the text.

possible in the presence of bimetallic Ni–Pd supported on silica in an acidic aqueous medium at 304 K, under a H_2 pressure of 8 MPa, for 2 h.[16] The catalyst with a Ni:Pd ratio of 7:1 showed the best catalytic performance, achieving a 96% BHMTHF yield (Scheme 6.1). The hydrogenation of BHMF to BHMTHF can also be carried out with Pd/C, Raney Ni, or Ru supported on carbon, CeO_x, γ-Al_2O_3 or magnesia-zirconia (hence, neutral or slightly basic catalyst supports) under more severe conditions (at 403–413 K under H_2 pressures of 2.5–7 MPa).[17,18,23]

Using Pt/C or Ru/C, acyclic products (*e.g.* 1-hydroxyhexane-2,5-dione and "hexane-polyols") are obtained with selectivities of up to 41% (sum of all acyclic products) at full HMF conversion[23] in the presence of an acid (*e.g.* acidic impurities in the feedstock, acidic reaction medium or acidic catalyst support), at moderate temperatures (373–403 K) and prolonged reaction times.[17,23] Note that these acyclic products are formed from BHMF,[23] the first hydrogenation product of HMF (Scheme 6.1). A mechanism for the acid-catalyzed ring-opening of BHMF was proposed by Dumesic's group.[23] According to this mechanism, the acyclic products 1-hydroxyhexane-2,5-dione, 1,2,5-hexanetriol and 1,2,5,6-hexanetetraol are formed by hydration-dehydration and subsequent hydrogenation steps of BHMF. In contrast, no water is needed for the ring opening of BHMF and subsequent hydrogenation to form 1,2,6-hexanetriol. Alternatively, 1-hydroxyhexane-2,5-dione and 1,2,5-hexanetriol may also form from HMF directly by hydration (and subsequent hydrogenation of the intermediate).[17,19] This pathway may also lead to levulinic and formic acid as side products. This reaction path leads to an increased effective H/C ratio, but not to deoxygenation.

For the hydrodeoxygenation of BHMF to DMF (effective H/C ratio of 1) mostly organic solvents, such as alcohols[13,19,25] or γ-lactones,[26] were used. The reaction pathway probably occurs *via* 2-hydroxymethyl-5-methylfuran (HMMF) that is also observed as a byproduct.[13] The best yields of DMF are achieved using Ru/C,[25] bimetallic Cu–Ru/C[13] or Sn–Ru/C[26] at temperatures of 493–533 K under H_2 pressures of 0.7–3.9 MPa (25–78%, Table 6.1, entries 1, 5, 6, 9, 18, and 19). Copper chromite was also used to produce DMF.[13] However, this catalyst resulted in significantly lower yields of DMF and appeared to be very sensitive to NaCl impurities. Note also that under milder conditions (333 K and 0.1 MPa of H_2), a 36% yield of DMF was achieved using a Pd/C catalyst in 1-propanol (Table 6.1, entry 10).[19] Interestingly, BHMF was obtained as the main product when the hydrogenation of 5-HMF was carried out in 1,4-dioxane under the same conditions. Further hydrogenation of DMF renders DMTHF. This reaction occurred in the presence of Ru- or Pd-based catalysts (Table 6.1, entries 10 and 12).[13,19]

It is noteworthy that the selectivity strongly depends on the adsorption behavior of HMF and its subsequent hydrogenation products at the active catalyst surface. Resasco's group examined the gas-phase HDO of furfural in the presence of mono- and bimetallic catalysts supported on SiO_2 under 0.1 MPa total pressure in the temperature range of 483–563 K in combination with DFT calculations.[27–30] They determined that the adsorption of furfural either parallel to the catalyst surface both with the C atom and the O atom of the aldehyde group bound to the metal surface (*e.g.* on bimetallic Ni–Fe surfaces)[27] or only through the O atom of the aldehyde group (*e.g.* on copper surfaces),[30] leads to 2-methylfuran. However, decarbonylation of furfural to furan is observed as the main reaction when the aldehyde group adsorbs generating surface acyl species, that is, only the C atom of the al-dehyde group is linked to the metal (*e.g.* on Pd[28,29] and Ni[27] surfaces). A strong interaction of the furanic ring with the metal surface (*e.g.* Ni surfaces)[27,29] will lead to hydrogenation of the ring or to acyclic products. The aforementioned adsorption modes are also possible for HMF. Indeed, the formation of 2-methylfuran as a side product with yields of up to 6% in the hydrodeoxygenation of HMF in the presence of Cu–Ru/C catalysts[13] may be explained by the adsorption of the aldehyde group through the C atom, which leads to the decarbonylation of HMF. The adsorption of HMF is also strongly influenced by the solvent[19] and catalyst support.[18,23] In 1-propanol, HMF is converted into DMF, while BHMF is the product in 1,4-dioxane (Table 6.1, entries 10 and 13).[19]

6.3.2 Hydrogen-Transfer Reactions

An alternative HDO route of HMF into DMF is transfer hydrogenation that has been performed using supercritical methanol (*sc*MeOH)[14] or formic acid (FA)[31,32] as hydrogen sources. *sc*MeOH serves a dual purpose both as solvent and as hydrogen source. Molecular hydrogen is formed and supplied to the reaction through methanol reforming in the presence of copper containing porous metal oxides (Cu-PMO). Consequently, the HDO mechanism using *sc*MeOH is similar to that performed under pressure of externally supplied molecular hydrogen. At full conversion of HMF, the highest yield of fuel components, DMF (48%) and DMTHF (10%), was achieved after 3 h at 533 K under a H_2 pressure of 10.7 MPa (Table 6.2 entry 1). In these reactions, 2-hexanol is formed as a side product of DMF.[14]

Molecular hydrogen can also be provided by decomposition of formic acid. By using Pd/C, the hydrodeoxygenation of HMF is highly selective and nearly quantitative yields (> 95%) of DMF have been reported (Table 6.2, entry 3).[31] In the first step, HMF is hydrogenated to BHMF. The HDO of BHMF to DMF only proceeds in the presence of low amounts of sulfuric acid. This observation suggests that the deoxygenation of BHMF is achieved *via* a Pd-catalyzed decarboxylation occurring after the esterification of the hydroxyl groups with formic acid (Scheme 6.2). In this case, sulfuric acid serves as a catalyst for the esterification. Overall, these observations are similar to those concerning the hydrothermal treatment of biomass in the presence of formates (for a detailed discussion on formate chemistry, see Chapter 11).

With formic acid as a hydrogen source, DMF can be obtained from fruc-tose in yields of 30–51% over carbon supported Ru[32] and Pd[31] catalysts (Table 6.2, entries 4 and 5). Moderate to high yields of DMF were also

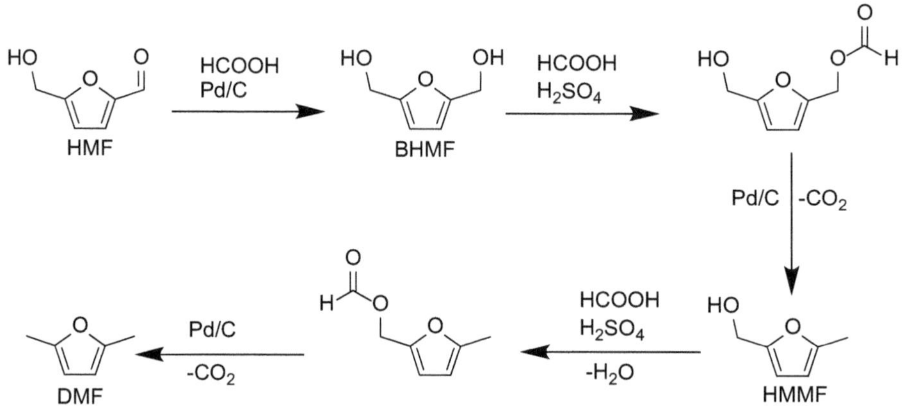

Scheme 6.2   Mechanism of hydrodeoxygenation of HMF *via* hydrogen-transfer reaction in the presence of a Pd/C catalyst using formic acid as hydrogen source. Figure adapted from Thananatthanachon and Rauchfuss.[31]

obtained from the conversion of α-cellulose, sugarcane bagasse and agar (Table 6.2, entries 6–8).[32] However, the use of sulfuric acid as a cocatalyst, the need of a noble-metal catalyst, the prolonged reaction durations and the use of THF as a solvent are the main disadvantages of this approach.

6.3.3   Production of DMF as a Biobased Fuel

Pioneered by Dumesic's group,[13] the production of DMF as a biofuel has gained more interest in recent years. The highest DMF yield of 78% was achieved in the conversion of HMF using 1-hexanol as a solvent in a flow reactor at 493 K and at a H_2 pressure of 0.7 MPa in the presence of a Cu–Ru/C catalyst with a Cu:Ru molar ratio of 3 : 2 (Table 6.1, entry 6).[13] A 72% yield of DMF was obtained in a continuous setup using the HMF formed from dehydration of fructose in a biphasic mixture of water and 1-butanol as the reactor feed (Table 6.1, entry 8). DMF can also be obtained in moderate yields in the presence of a base-metal catalyst (*e.g.* copper chromite). Again, the impurities (*e.g.* NaCl) present in the crude HMF dramatically decrease the catalytic performance of copper chromite, compared to Cu–Ru/C (6% *vs.* 61% DMF yield, Table 6.1, entries 2 and 4).[13]

Pd-based catalysts constitute another class of catalysts for the hydrodeoxygenation of HMF to DMF. A 36% yield of DMF was achieved in the presence of a 10 wt% Pd/C catalyst at 333 K under a H_2 pressure of 0.1 MPa using 1-propanol as a solvent (Table 6.1, entries 10–13).[19] When the reaction was carried out in 1,4-dioxane instead of 1-propanol, the product selectivity shifted from DMF to BHMF.

Binder and Raines[33] used untreated corn stover as feedstock for the production of DMF. Corn stover was processed in a mixture of EMIMCl and *N,N*-dimethylacetamide using LiCl, $CrCl_3$ and HCl as catalysts. The crude HMF was purified by ion-exclusion chromatography. The purified HMF was hydrogenated in 1-butanol under similar conditions reported by Dumesic and coworkers.[13] A 49% yield of DMF (Table 6.1, entry 14), which corresponds to a 9% overall yield relative to the cellulose content of corn stover, was achieved.[33]

Similarly, Chidambaram and Bell[20] presented the production of DMF using a mixture of an ionic liquid and acetonitrile as solvent. In the first acid-catalyzed step crude HMF was produced from glucose in nearly quantitative yield (99% conversion and 98% selectivity). Next, HDO in the same solvent mixture can be performed by merely exchanging the solid acid catalyst with a Pd/C catalyst. After 1 h at 393 K under a H_2 pressure of 6.2 MPa, a 12% yield of DMF at a HMF conversion of 44% was achieved. Interestingly, in this solvent mixture the source of HMF (either neat HMF or crude HMF) did not have a significant effect on the conversion to DMF (Table 6.1, entries 16 and 17). In the pure ionic liquid, however, both the conversion of HMF and the yield of DMF were significantly lower compared to the ionic liquid/acetonitrile mixture (Table 6.1, entry 15) which was attributed to a lower hydrogen solubility in the pure ionic liquid.[20] Nonetheless, HMF separation from the ionic liquid is seemingly elaborate for a commercial consideration on a large scale.

The conversion of sugars and biomass residues to DMF using formic acid as a hydrogen source using microwave heating was successfully demonstrated by Saha and coworkers.[32] Compared to a conventional heating processes, the reaction achieved completion 20 times faster using microwave heating; however, the yield of DMF was comparable to that obtained from experiments performed using conventional heating techniques (16–32% DMF yield, Table 6.2, entries 6–8).[32]

6.4 Hydrodeoxygenation of Levulinic Acid

Levulinic acid (LA) is a C_5-keto acid and one of the top-twelve sugar-based platform molecules.[10] LA is soluble in water and industrially used as a polymer plasticizer. Since LA shows two functional groups, it serves as a platform for the production of a wide range of products. Important chemicals derived from the upgrading of LA are γ-valerolactone, 1,4-pentanediol, levulinic acid esters and diphenolic acid.[10] γ-Valerolactone (GVL) is the most important derivative of LA. It serves as a platform for several specialty chemicals as well as synthetic fuels, as depicted in Scheme 6.3. Due to its herbaceous odor, GVL is used by the perfume and flavor industry. Furthermore, GVL can be converted into α-methylene-γ-valerolactone or dimethyl adipate, which are monomers in the plastic industry. GVL also shows interesting solvent properties, and is hence proposed as a green solvent or even as a precursor for other green solvents.[34] Moreover, GVL holds promise as a synthetic biofuel or fuel additive. Alternatively, GVL is proposed as a platform for the production of jet fuels (C_{8+} alkanes) or diesel fuels (C_9–C_{18} alkanes).

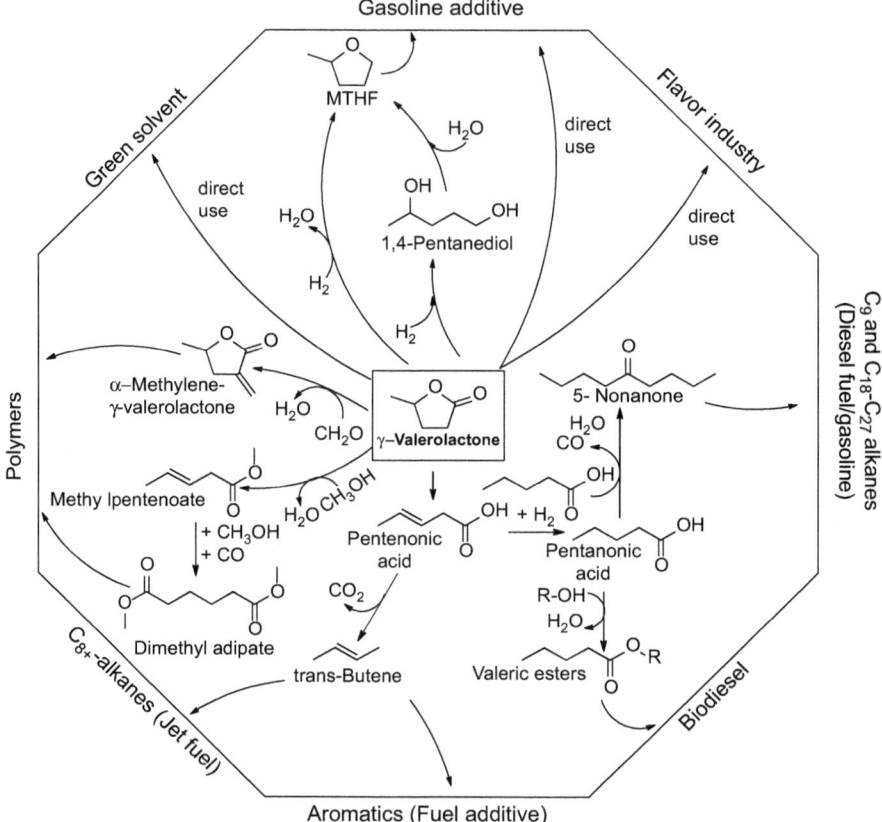

Scheme 6.3 Use of GVL and its derivatives.

Different reaction pathways have been reported to produce GVL from LA (Scheme 6.4). A possible reaction route is the hydrogenation to γ-hydroxyvaleric acid, an unstable intermediate, which undergoes spontaneous lactonization rendering γ-valerolactone. In an alternative pathway, LA is dehydrated to α-angelica lactone (which occurs in equilibrium with β-angelica lactone), and is then hydrogenated. This reaction pathway is limited to systems in which acidic functionalities in the catalyst and water are present. As a result, the yields of GVL are lower due to coke formation during the synthesis of angelica lactones. In both pathways, deoxygenation is achieved by releasing water. A third possibility is the esterification of LA, followed by a hydrogenation and transesterification of the obtained levulinic acid esters to GVL.

Over the past few years, various metal catalysts, both homogeneous and heterogeneous, have been examined in the hydrogenation of LA to GVL. Table 6.3 summarizes a selection of results reported for the conversion of LA into GVL under varying conditions.

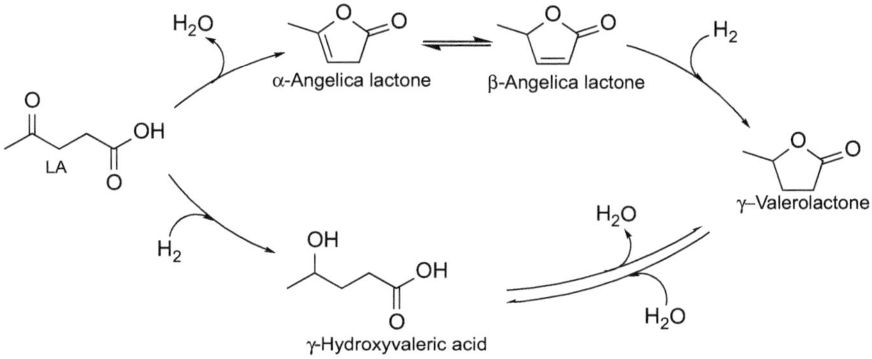

Scheme 6.4 Reaction pathways of the hydrogenation of LA to GVL.

6.4.1 Conversion of Levulinic Acid Using Molecular Hydrogen

Most of the catalysts for the hydrogenation of LA to γ-valerolactone are based on noble metals and the reactions are often carried out in liquid phase. Nonetheless, gas-phase reactions were conducted by the groups of Chang and Hwang[35] and Dunlop and Madden[36] using either noble- or base-metal catalysts (Table 6.3, entries 1, 14, 20 and 33). Quantitative yields of GVL were obtained using a 5% Ru/C[35] or a copper chromite[36] catalyst, whereas Pd and Pt catalysts[35] yielded 90% or only 30% GVL, respectively. Considering liquid-phase reactions, the best results were obtained from experiments using 5 wt% Ru/C at 463 K under a H_2 pressure of 1.2 MPa (Table 6.3, entry 2).[37]

Although a variety of organic solvents was screened, water or even a solvent-free reaction is preferred. Manzer and Hutchenson[38] and Poliakoff's group[39] examined the reaction in supercritical CO_2 using a continuous flow setup as well as a batch reactor. With ruthenium-based catalysts, full conversion of LA and γ-valerolactone yields up to 100% were obtained (Table 6.3, entries 11 and 13).[38,39]

Pd- or Pt-based catalysts often show high catalytic activity for the hydrogenation of LA. However, they can also catalyze the hydrogenation of GVL to 2-methyltetrahydrofuran (MTHF) or 1,4-pentanediol, hence decreasing the selectivity to GVL (Scheme 6.5 and Table 6.3, entry 20).[35]

Base metals were examined in the production of GVL (Table 6.3, entries 29–33). Ni showed the best LA conversion and GVL yields in liquid-phase reactions.[40] The presence of a solvent lowered the catalytic activity of both, Raney Ni and supported Ni catalysts. The use of 15 wt% Ni/Al$_2$O$_3$ as a catalyst gave similar results to those obtained by Ru catalysts for the hydrogenation of LA to GVL.[40] In gas-phase reactions, a quantitative yield of GVL was obtained from experiments in the presence of a CuO/Cr$_2$O$_3$ catalyst.[36]

From an economic perspective, the use of non expensive base metal catalysts is advantageous over their noble-metal counterparts. Unfortunately, the required reaction temperatures for quantitative GVL yields using

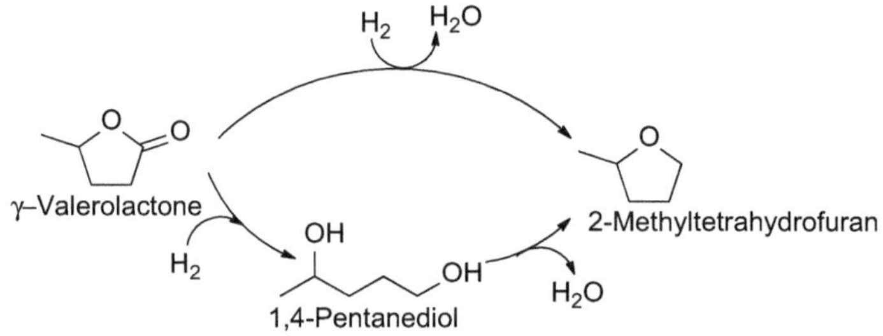

Scheme 6.5 Hydrogenation of GVL as consecutive reaction with Pd or Pt catalysts.

base-metal catalysts are much higher (*e.g.* Ni/Al$_2$O$_3$, $T > 463$ K) than that for noble-metal catalysts (*e.g.* Ru/C, $T < 403$ K). Therefore, the catalytic activity of base-metal catalysts has to be improved at low temperatures.

6.4.2 Conversion of Levulinic Acid Using Formic Acid as a Hydrogen Source

The use of formic acid as a hydrogen source was investigated in the conversion of LA into GVL (Table 6.3, entries 8, 9, 18–19, 22, 25–28). This process option is interesting because formic acid is a stoichiometric side product in the conversion of HMF into LA. Therefore, the use of formic acid as a hydrogen source could hold the key for a more sustainable, hydrogen-efficient production of GVL. Two reaction mechanisms are proposed for the hydrogenation using formic acid as the hydrogen source.[41–43] In the first mechanism, formic acid is catalytically decomposed, on the metal particles, into CO$_2$ and H$_2$.[43] The molecular hydrogen remains adsorbed resulting in two M–H sites, where the hydrogenation of LA takes place following the classical hydrogenation mechanism.[41,43] In the second mechanism, LA is hydrogenated by transfer hydrogenation of formic acid adsorbed on metal surfaces.[42]

Mainly noble metals (*e.g.* AuRu, Pd or Pt) have been studied for the reaction in the presence of formic acid as hydrogen source, since these metals also catalyze the decomposition of formic acid.[43] Quantitative yield of GVL was reported for experiments performed in the presence of a 1 mol% Au/ZrO$_2$ catalyst at 423 K under a N$_2$ pressure of 0.5 MPa and using water as a solvent (Table 6.3, entry 25). Under similar reaction conditions, Pd- and Pt-based catalysts produced only traces of GVL. Du *et al.*[43] studied catalysts impregnated with 1 mol% Au on activated carbon, SiO$_2$ and TiO$_2$. The catalyst Au/TiO$_2$ resulted in 55% GVL yield at about 55% conversion (Table 6.3, entry 28). Au on activated carbon or SiO$_2$ was found to be inactive for the production of GVL. This fact is ascribed to the low stability of these catalysts in the presence of formic acid. In fact, most of the supported metal catalysts often suffer from low stability issues due to leaching of the active phase.[43] Therefore, the improvement of the catalyst stability in the presence

Table 6.3 Selected results for the production of GVL from LA.

Entry	Catalyst	Reactor[e]	Hydrogen source	Solvent	T [K]	p_{H_2} [MPa]	Conv. [%]	Yield GVL [%]	Ref.
1	Ru/C[a]	F	H_2	1,4-dioxane	538[f]	0.1	100	98	35
2	Ru/C[a]	B	H_2	No solvent	463	1.2	100	100	37
3	Ru/C[a]	B	H_2	1,4-dioxane	423	5.5	80	72	52
4	Ru/C[a]	B	H_2	MeOH/H_2O	403	1.2	96	85	37
5	Ru/C[a]	B	H_2	EtOH/H_2O	403	1.2	99	89	37
6	Ru/C[a]	B	H_2	1-BuOH/H_2O	403	1.2	99	75	37
7	Ru/C[a]	B	H_2	H_2O	403	1.2	100	86	37
8	Ru/C[a,d]	B	Formic acid	H_2O	423	4.0	n.a.	67	53
9	Ru/C[a]	F	Formic acid	H_2O	423	n.a.	>80	n.a.	54
10	Ru/SiO$_2$[a]	B	H_2	EtOH/H_2O	403	1.2	98	75	37
11	Ru/SiO$_2$[a]	F	H_2	scCO$_2$/H_2O	473	10	100	100	39
12	Ru/Al$_2$O$_3$[a]	B	H_2	EtOH/H_2O	403	1.2	95	76	37
13	Ru/Al$_2$O$_3$[a]	B	H_2	scCO$_2$	423	25.0[g]	99	99	38
14	Pd/C[a]	F	H_2	1,4-dioxane	538[f]	0.1	100	90	35
15	Pd/C[a]	F	H_2	1,4-dioxane	538[f]	1	100	60	35
16	Pd/C[a]	B	H_2	1,4-dioxane	423	5.5	30	27	52
17	Pd/C[a]	B	H_2	MeOH	403	1.2	17	6	55
18	Pd/C[a]	B	Formic acid	H_2O	423	0.5 (N$_2$)	2	2	43
19	Pd/ZrO$_2$[a]	B	Formic acid	H_2O	423	0.5 (N$_2$)	n.a.	trace	43
20	Pt/C[a]	F	H_2	1,4-dioxane	538[f]	0.1	100	30	35
21	Pt/C[a]	B	H_2	1,4-dioxane	423	5.5	13	10	52
22	Pt/ZrO$_2$[a]	B	Formic acid	H_2O	423	0.5 (N$_2$)	n.a.	trace	43
23	PtO$_2$	B	H_2	Ethyl ether	298	0.3	n.a.	87	56
24	PtO$_2$	B	H_2	EtOH	298	0.3	n.a.	52	56
25	Au/ZrO$_2$[b]	B	Formic acid	H_2O	423	0.5 (N$_2$)	99	99	43
26	Au/C[b]	B	Formic acid	H_2O	423	0.5 (N$_2$)	n.a.	trace	43
27	Au/SiO$_2$[b]	B	Formic acid	H_2O	423	0.5 (N$_2$)	1	1	43
28	Au/TiO$_2$[b]	B	Formic acid	H_2O	423	0.5 (N$_2$)	55	55	43
29	Ni/C[a]	B	H_2	1,4-dioxane	423	5.5	2	0.4	52
30	Raney Ni	B	H_2	No solvent	493	5.0	n.a.	94	57

Table 6.3 (*Continued*)

Entry	Catalyst	Reactor[e]	Hydrogen source	Solvent	T [K]	p_{H_2} [MPa]	Conv. [%]	Yield GVL [%]	Ref.
31	Raney Ni	B	H_2	MeOH	403	1.2	18	5	55
32	Ni/Al$_2$O$_3$[c]	B	H_2	No solvent	473	5.0	100	96	40
33	CuO/Cr$_2$O$_3$	F	H_2	No solvent	473[f]	0.1	100	100	36

Abbreviations: MeOH – methanol, EtOH – ethanol, 1-BuOH – 1-butanol, scCO$_2$ – supercritical CO$_2$, n.a. – not available/not mentioned in the respective article.
[a]5 wt%
[b]1 mol%
[c]15 wt%
[d]decomposition of formic acid by 1.4 wt% Ru-P/SiO$_2$
[e]B = batch, F = continuous flow-reactor
[f]vapor phase
[g]total pressure (105 bar CO$_2$ and 145 bar H$_2$).

of substantial amounts of formic acid is essential for the replacement of externally supplied molecular hydrogen by formic acid.

6.5 Prospects: Mechanistic Understanding of Hydrogenation and Hydrodeoxygenation Reactions Using *In Situ* Spectroscopy

In the previous sections, the conversion of selected model compounds has been discussed including some mechanistic considerations. The understanding of the reaction mechanism(s) involved in the selected examples is still in its infancy and mostly relies upon kinetic data rather than on spectroscopic measurements of the surface structure or of catalytically active species. Recently, DFT calculations have been used[47,48,58,59] to predict the adsorption behavior of several compounds in many catalytic HDO systems in order to shed light on possible reaction mechanisms; however, hardly any *in situ* studies have been performed in order to verify the calculations.

A first step towards the spectroscopic elucidation of reaction pathways has been taken by Zhao *et al.*[44] They uncovered the kinetics not only by online monitoring but also using *in situ* infrared spectroscopy. Indeed, appropriate *in situ* infrared (IR) transmission cells with short path lengths, capable of operating under high pressure, are commercially available.[45,46] Alternatively, extremely small penetration depth is possible through attenuated total reflection infrared spectroscopy (ATR-IR). Furthermore, powdered catalyst layers of several micrometers thickness can be deposited on the reflecting crystal, enabling the assessment of surface intermediates by ATR-IR.[45]

Considering ATR-IR analysis, both commercial and home-built probes exist.[45] An example is shown in Figure 6.6. In this home-built batch reactor, the infrared beam can be directed to two different positions, namely (a) at the bottom of the reactor where the catalyst can be deposited on an ATR crystal and (b) in the middle of the reactor where the fluid phase can be monitored using transmission IR. This setup has previously been used to study the hydrogenation of ethyl pyruvate on Pt/Al$_2$O$_3$.[47] The consumption of ethyl pyruvate and the formation of ethyl lactate during the hydrogenation in carbon dioxide as the solvent were monitored in transmission mode (upper beam position) allowing the kinetic evaluations of the reaction. ATR-IR measurements of Pt/Al$_2$O$_3$ catalyst deposited on a ZnSe crystal (lower beam position) showed only ethyl lactate and no ethyl pyruvate, unraveling some mechanistic aspects of this very quick hydrogenation on Pt particles.[47]

Additionally, *in situ* or *operando* studies are required to monitor the dynamic changes of the surface structure and/or the oxidation state of heterogeneous catalysts, which are often amorphous.[48] Apart from X-ray diffraction, Raman spectroscopy, and UV-vis spectroscopy, *in situ* X-ray absorption spectroscopy (XAS) is a particularly valuable technique since it can monitor both the crystalline and amorphous components for a specific element in a catalyst. Furthermore, questions addressing possible metal

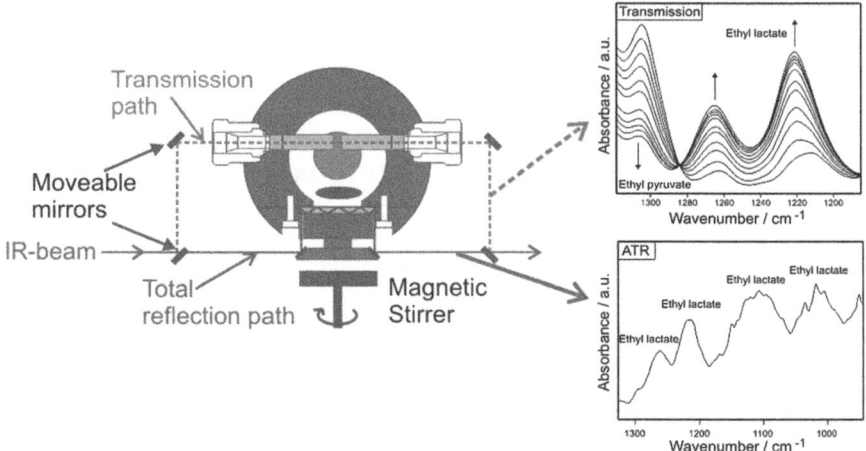

Figure 6.6   Scheme of a high-pressure cell (designed to work up to 150 bar) for combined ATR-IR (bottom of the cell) and transmission IR (middle of the cell); spectra were taken during hydrogenation of ethyl pyruvate over Pt/Al_2O_3 coated on a ZnSe-crystal.
Adapted with permission from ref. 45, Copyright 2005, Royal Society of Chemistry.

leaching are of importance due to the harsh reaction conditions and the presence of acids in the reaction medium (water, polar solvents[49,50]). An approach reported by Pham *et al.*[49] consists in coating the catalysts with a carbon layer. Also flame-made catalysts often show more resistant corrosion behavior.[51] X-ray-based techniques can also be used for mechanistic studies on the stability in addition to structural investigation.

A suitable reaction cell is depicted in Figure 6.7, consisting of an autoclave (designed for pressures up to 18 MPa) that exhibits two pairs of windows for monitoring the catalytic species at two positions. In the present example the hydrogenation and formylation of 3-methoxypropylamine over phosphine-modified Ru/Al_2O_3 catalysts was studied (a) at the bottom of the reactor, where the solid catalyst is placed, and (b) through the liquid phase where the hydrogenation/formylation reaction occurs and soluble ruthenium species may be probed. In fact, the *in situ* formation of soluble Ru phosphine complexes was detected in the presence of phosphino ligands during the hydrogenation and formylation of amines in CO_2, used as a solvent and reactant. This indicated that homogeneous species were, at least partially, responsible for the product formation instead of a truly heterogeneous catalyst. Furthermore, the structure and oxidation state of the catalyst could be uncovered, proving that ruthenium was partially reduced and present as Ru phospino complexes in solution.

In the same way, the structure of catalysts could be investigated both in batch and continuous processes during the HDO of biomass-derived platform molecules, although a special cell design would be required for an *operando* study of continuous processes. A special challenge may be the

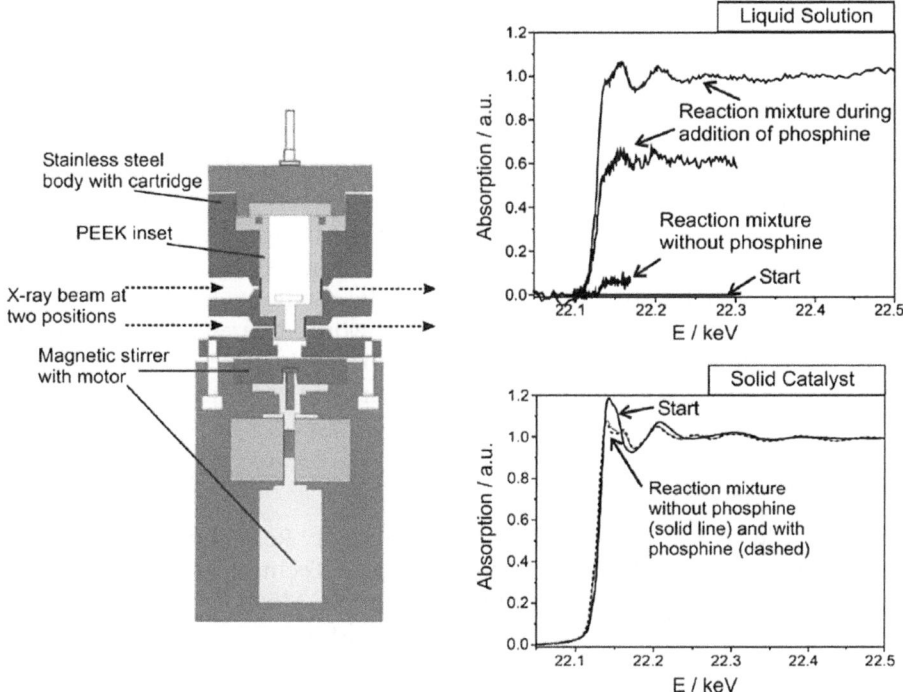

Figure 6.7 High-pressure batch cell for *in situ* X-ray absorption spectroscopy that allows monitoring the structure of the heterogeneous catalyst during hydrogenation (bottom) and the structure of catalytic species in solution (*e.g.* due to leaching) in the middle.
Reprinted with permission from ref. 47, Copyright 2005, AIP publishing LCC and from ref. 45 with permission from the PCCP owner societies.

higher reaction temperatures compared to those applied for typical hydrogenations in fine chemistry.

6.6 Conclusion

Catalytic HDO of biomass-derived chemicals is a very promising pathway for the production of biofuels and chemicals. In analogy to hydrodesulfurization, HDO processes can be investigated by the use of representative platform molecules. In this chapter, HMF and levulinic acid have been chosen as representative examples, since they can be easily obtained from cellulose and hemicellulose, but also regarded as good model examples for organic acids, ketones and alcohols which are typical constituents of pyrolysis oils. Together with the studies on phenolic compounds (*e.g.* guaiacol, as discussed in Chapters 3, 7 and 8) these studies are important for the optimization of the HDO processes performed on the pyrolysis oils.

In the catalytic upgrading of HMF and levulinic acid, the oxygen content is often removed by dehydration and only in certain cases by decarbonylation or decarboxylation. However, hydrogenation is a fundamental component of the HDO processes leading at the same time to a significant increase of the energy content of the chemicals. Therefore, hydrogen should be produced either by gasification of biomass or other sustainable processes (*e.g.* water electrolysis using wind or solar power), to run the process in a fully sustainable way. Finally, some progress in the elucidation of reaction mechanisms and catalyst deactivation has been achieved by *in situ* spectroscopic tools based on infrared or X-ray absorption spectroscopy. However, these studies are still very challenging, since biomass conversion is often performed under harsh conditions (*e.g.* acidic medium, liquid phase, elevated temperature and high pressure) and it would be rewarding to exploit those techniques in more detail to further improve the processes and to gain a full insight into the complex reaction network of HMF and GVL reduction.

Acknowledgements

We acknowledge KIT and the Helmholtz program "Renewable Energies" as well as the European Institute of Innovation and Technology, under the KIC InnoEnergy SYNCON project for financial support. Moreover, we thank Prof. Jörg Sauer and coworkers (IKFT at KIT) for fruitful discussions in this area.

References

1. G. W. Huber, S. Iborra and A. Corma, *Chem. Rev.*, 2006, **106**, 4044–4098.
2. B. Kamm, *Angew. Chem., Int. Ed.*, 2007, **46**, 5056–5058.
3. J. N. Chheda, G. W. Huber and J. A. Dumesic, *Angew. Chem. Int. Ed.*, 2007, **46**, 7164–7183.
4. F. Schüth, *Chem. Ing. Tech.*, 2011, **83**, 1984–1993.
5. (a) www.bioliq.de; (b) www.chemrec.se; (c) www.topsoe.com; (d) www.fuelcenter.rwth-aachen.de; (e) www.catchbio.com; (f) www.case.dtu.dk (15.04.2014).
6. B. Kamm and M. Kamm, *Chem. Ing. Tech.*, 2007, **79**, 592–603.
7. R. Rinaldi, P. Engel, J. Büchs, A. C. Spiess and F. Schüth, *ChemSusChem*, 2010, **3**, 1151–1153.
8. A. V. Bridgwater, S. Czernik, J. Diebold, D. Meier, A. Oasmaa, C. Peakocke, J. Piskorz and D. Radlein, *Fast Pyrolysis of Biomass: A Handbook*, CPL Press, Newbury, 1999.
9. P. M. Mortensen, J. D. Grunwaldt, P. A. Jensen, K. G. Knudsen and A. D. Jensen, *Appl. Catal. A: Gen.*, 2011, **407**, 1–19.
10. G. P. T. Werpy, J. Holladay, J. White, A. Manheim, M. Gerber, K. Ibsen, L. Lumberg, S Kelley, *Top Value Added Chemicals From Biomass. Volume I: Results of Screening for Potential Canditates from Sugars and Synthesis Gas*, U.S. Department of Energy, 2004.
11. F. Vogel, J. Harf, A. Hug and P. R. von Rohr, *Water Res.*, 2000, **34**, 2689–2702.

12. P. N. R. Vennestrøm, C. M. Osmundsen, C. H. Christensen and E. Taarning, *Angew. Chem., Int. Ed.*, 2011, **50**, 10502–10509.
13. Y. Roman-Leshkov, C. J. Barrett, Z. Y. Liu and J. A. Dumesic, *Nature*, 2007, **447**, 982–985.
14. T. S. Hansen, K. Barta, P. T. Anastas, P. C. Ford and A. Riisager, *Green Chem.*, 2012, **14**, 2457–2461.
15. X. L. Tong, Y. Ma and Y. D. Li, *Appl. Catal. A: Gen.*, 2010, **385**, 1–13.
16. Y. Nakagawa and K. Tomishige, *Catal. Commun.*, 2010, **12**, 154–156.
17. V. Schiavo, G. Descotes and J. Mentech, *Bull. Soc. Chim. Fr.*, 1991, 704–711.
18. J. Ohyama, A. Esaki, Y. Yamamoto, S. Arai and A. Satsuma, *RSC Adv.*, 2013, **3**, 1033–1036.
19. G. C. A. Luijkx, N. P. M. Huck, F. van Rantwijk, L. Maat and H. van Bekkum, *Heterocycles*, 2009, **77**, 1037–1044.
20. M. Chidambaram and A. T. Bell, *Green Chem.*, 2010, **12**, 1253–1262.
21. E. J. Ras, M. J. Louwerse and G. Rothenberg, *Catal. Sci. Technol.*, 2012, **2**, 2456–2464.
22. M. Tamura, K. Tokonami, Y. Nakagawa and K. Tomishige, *Chem. Commun.*, 2013, **49**, 7034–7036.
23. R. Alamillo, M. Tucker, M. Chia, Y. Pagan-Torres and J. Dumesic, *Green Chem.*, 2012, **14**, 1413–1419.
24. M. Schlaf, *Dalton Trans.*, 2006, 4645–4653.
25. J. H. Zhang, L. Lin and S. J. Liu, *Energy Fuels*, 2012, **26**, 4560–4567.
26. J. M. R. Gallo, D. M. Alonso, M. A. Mellmer and J. A. Dumesic, *Green Chem.*, 2013, **15**, 85–90.
27. S. Sitthisa, W. An and D. E. Resasco, *J. Catal.*, 2011, **284**, 90–101.
28. S. Sitthisa, T. Pham, T. Prasomsri, T. Sooknoi, R. G. Mallinson and D. E. Resasco, *J. Catal.*, 2011, **280**, 17–27.
29. S. Sitthisa and D. E. Resasco, *Catal. Lett.*, 2011, **141**, 784–791.
30. S. Sitthisa, T. Sooknoi, Y. G. Ma, P. B. Balbuena and D. E. Resasco, *J. Catal.*, 2011, **277**, 1–13.
31. T. Thananatthanachon and T. B. Rauchfuss, *Angew. Chem., Int. Ed.*, 2010, **49**, 6616–6618.
32. S. De, S. Dutta and B. Saha, *ChemSusChem*, 2012, **5**, 1826–1833.
33. J. B. Binder and R. T. Raines, *J. Am. Chem. Soc.*, 2009, **131**, 1979–1985.
34. D. Fegyverneki, L. Orha, G. Lang and I. T. Horvath, *Tetrahedron*, 2010, **66**, 1078–1081.
35. P. P. Upare, J. M. Lee, D. W. Hwang, S. B. Halligudi, Y. K. Hwang and J. S. Chang, *J. Ind. Eng. Chem.*, 2011, **17**, 287–292.
36. A. P. Dunlop, J. W. Madden, Process of Preparing gamma-Valerolactone (Quaker Oats Co.), patent no. US 2786852, 1957.
37. M. G. Al-Shaal, W. R. H. Wright and R. Palkovits, *Green Chem.*, 2012, **14**, 1260–1263.
38. L. E. Manzer and K. W. Huchenson, Production of 5-Methyl-dihydrofuran-2-one from Levulinic Acid in Supercritical Media (E. I. Du Pont De Nemours & Co., USA), US 20040254384, 2004.

39. R. A. Bourne, J. G. Stevens, J. Ke and M. Poliakoff, *Chem. Commun.*, 2007, 4632–4634.
40. K. Hengst, C. Lu, W. Kleist and J. D. Grunwaldt, *Chem. Ing. Tech.*, 2012, **84**, 1247.
41. W. R. H. Wright and R. Palkovits, *ChemSusChem*, 2012, **5**, 1657–1667.
42. J. Horvat, B. Klaić, B. Metelko and V. Šunjić, *Tetrahedron Lett.*, 1985, **26**, 2111–2114.
43. X. L. Du, L. He, S. Zhao, Y. M. Liu, Y. Cao, H. Y. He and K. N. Fan, *Angew. Chem. Int. Ed.*, 2011, **50**, 7815–7819.
44. C. Zhao, S. Kasakov, J. He and J. A. Lercher, *J. Catal.*, 2012, **296**, 12–23.
45. J. D. Grunwaldt and A. Baiker, *Phys. Chem. Chem. Phys.*, 2005, 7, 3526–3539.
46. J.-D. Grunwaldt, R. Wandeler and A. Baiker, *Catal. Rev. Sci. Eng.*, 2003, **45**, 1–96.
47. M. S. Schneider, J.-D. Grunwaldt, T. Bürgi and A. Baiker, *Rev. Sci. Instrum.*, 2003, **74**, 4121.
48. J.-D. Grunwaldt, In *Chemical Energy Storage*, ed. R. Schlögl, Walter de Gruyter GmbH, Berlin/Boston, 2012, p. 311.
49. H. N. Pham, A. E. Anderson, R. L. Johnson, K. Schmidt-Rohr and A. K. Datye, *Angew. Chem., Int. Ed.*, 2012, **51**, 13163–13167.
50. A. R. Ardiyanti, S. A. Khromova, R. H. Venderbosch, V. A. Yakovlev, I. V. Melian-Cabrera and H. J. Heeres, *Appl. Catal. A: Gen.*, 2012, **449**, 121–130.
51. M. J. Beier, B. Schimmoeller, T. W. Hansen, J. E. T. Andersen, S. E. Pratsinis and J. D. Grunwaldt, *J. Mol. Catal. A: Chem.*, 2010, **331**, 40–49.
52. L. E. Manzer, *Appl. Catal. A: Gen.*, 2004, **272**, 249–256.
53. L. Deng, Y. Zhao, J. A. Li, Y. Fu, B. Liao and Q. X. Guo, *ChemSusChem*, 2010, **3**, 1172–1175.
54. D. J. Braden, C. A. Henao, J. Heltzel, C. T. Maravelias and J. A. Dumesic, *Green Chem.*, 2011, **13**, 1755–1765.
55. Z. P. Yan, L. Lin and S. J. Liu, *Energy Fuels*, 2009, **23**, 3853–3858.
56. H. A. Schuette and R. W. Thomas, *J. Am. Chem. Soc.*, 1930, **52**, 3010–3012.
57. R. V. Christian, H. D. Brown and R. M. Hixon, *J. Am. Chem. Soc.*, 1947, **69**, 1961–1963.
58. A. Popov, E. Kondratieva, J.-P. Gilson, L. Mariey, A. Travert and F. Mauge, *Catal. Today*, 2011, **172**, 132–135.
59. A. Popov, E. Kondratieva, L. Mariey, J. M. Goupil, J. El Fallah, J.-P. Gilson, A. Travert and F. Mauge, *J. Catal.*, 2013, **297**, 176–186.

CHAPTER 7

Catalytic Hydrotreatment of Fast Pyrolysis Oils Using Supported Metal Catalysts

AGNES RETNO ARDIYANTI,[a]
ROBERTUS HENDRIKUS VENDERBOSCH,[b] WANG YIN[a] AND
HERO JAN HEERES*[a]

[a] University of Groningen, Groningen, The Netherlands; [b] BTG, Biomass
Technology Group, Enschede, The Netherlands
*Email: h.j.heeres@rug.nl

7.1 Introduction

In fast pyrolysis, biomass is subjected to a rapid thermal treatment (673–773 K), in the absence of O_2. Under these conditions, the decomposition of biomass generates hot vapors and aerosols alongside byproducts: char and non-condensable gases. The products released to the vapor phase are very quickly quenched (*i.e.* 2 s or faster) in order to avoid repolymerization of the components in the vapor phase. At room temperature, the "fast pyrolysis oil" (or bio-oil) is a viscous, dark-brown liquid with a pungent odor. Notably, fast pyrolysis oil shows a higher energy density, is easier to transport, and contains less ash compared to the precursor biomass. In this chapter, the general aspects concerning the conversion of lignocellulosic biomass into liquid products by fast pyrolysis, followed by the catalytic hydrotreatment and refining, and analytics are addressed.

RSC Energy and Environment Series No. 13
Catalytic Hydrogenation for Biomass Valorization
Edited by Roberto Rinaldi
© The Royal Society of Chemistry 2015
Published by the Royal Society of Chemistry, www.rsc.org

7.1.1 The Biorefinery Concept of Fast Pyrolysis Oil

The fast pyrolysis oil can serve as a platform for the production of a wide range of biobased chemicals and transportation fuels in addition to heat and power (Figure 7.1).[1,2] The development of a fast pyrolysis oil biorefinery was the main objective of the European 6th Framework project BIOCOUP. The project focus was on both fast pyrolysis oil upgrading to an intermediate product (suitable as a cofeed in existing oil refineries) as well as recovery/production of valuable chemicals.[3] Indeed, fast pyrolysis oil comprises high levels of various oxygenates. Some of these components are platform chemicals (*e.g.* acetic, formic, and glycolic acids, furans and several others, see also Chapter 6). Other interesting components are phenolics and aldehydes (*e.g.* 2-hydroxyacetaldehyde, a precursor for ethyleneglycol).

7.1.2 Fast Pyrolysis Oil as an Energy Carrier

Fast pyrolysis oil has the same visual appearance of vacuum gas oil (VGO), although its composition differs considerably from VGO. Table 7.1 compares the properties of pyrolysis oil and VGO. Fast pyrolysis oil shows higher moisture and oxygen contents than VGO. In addition, fast pyrolysis oil also contains considerable amounts of organic acids (4.2–6.8 wt%),[5,6] resulting in TAN values (total acid number) typically greater than 100 mg KOH kg^{-1}. For comparison, crude oils have TAN values of lower than 2 mg KOH kg^{-1}.[7] The high oxygen content of the fast pyrolysis oil makes the oil more hydrophilic, and leads to an energy value of about half that found for crude oil. Furthermore, the oils are known to have a limited storage stability and excessive coke formation is noticed when the pyrolysis oil is subjected to elevated temperatures.[2]

   Table 7.2 shows that the direct use of fast pyrolysis oil, as a transportation fuel, is not possible for conventional engines due to its intrinsic properties (*e.g.* high acidity, low stability, *etc.*). Moreover, blending pyrolysis oils with diesel is also a challenge due to the poor miscibility of the two liquids; however, emulsified systems have been proposed to overcome this problem.[9,10] In effect, the upgrading of the pyrolysis oils is required in order to produce drop-in transportation fuels with a reduced oxygen content, allowing for its utilization as fuels. It is worth mentioning that a small amount of oxygenated compounds in the upgraded fuel can be, in some cases, advantageous as they improve the combustion process, *e.g.* leading to reduced soot formation.[11] Another utilization of pyrolysis oils is as a cofeed in conventional oil refineries. As such, substantial deoxygenation is also required, for two main reasons. The first is to reduce the pyrolysis oil polarity, thus improving the miscibility with the conventional stream. The second is to enhance the thermal stability, suppressing the formation of solid products that could block the feed lines and reactors.

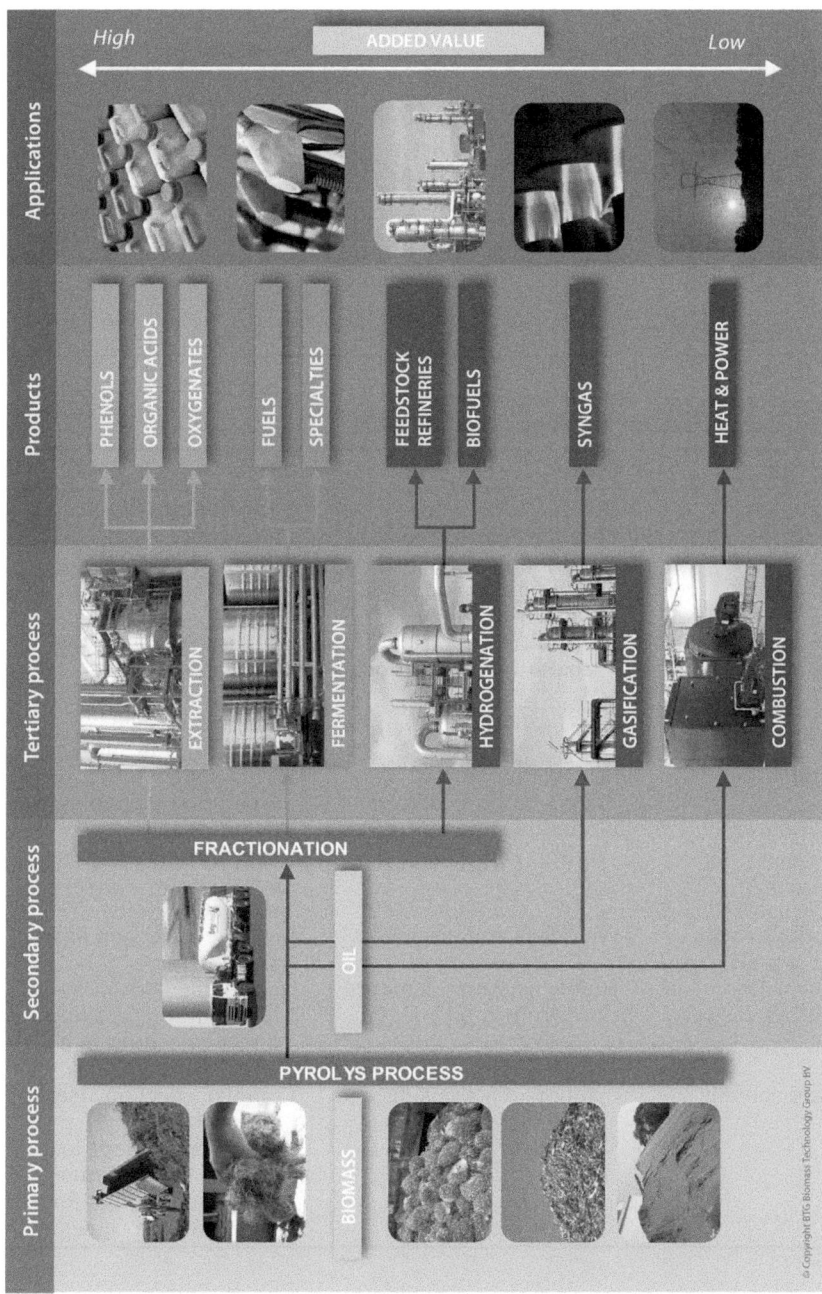

Figure 7.1 Bioliquids refinery scheme using fast pyrolysis oil as the intermediate, suggested by BTG.[4] Reproduced with permission.

Table 7.1 Typical properties of lignocellulosic fast pyrolysis oil.[8]

Physical properties	Pyrolysis oil	Heavy fuel oil
Moisture content, wt%	15–30	0.1
pH	2.5	-
specific gravity	1.2	0.94
elemental composition, wt%		
C	54–58	85
H	5.5–7.0	11
O	35–40	1.0
N	0–0.2	0.3
Ash	0–0.2	0.1
HHV, MJ/kg	16–19	40
Viscosity (at 50 °C, cP)	40–100	180
Solids, wt%	0.2–1	1
Distillation residue, wt%	Up to 50	1

Table 7.2 Characteristics of fast pyrolysis oil.[2]

Characteristic	Cause	Effects
Acidity/low pH	Organic acids from biopolymer degradation	Corrosion of vessels and pipework
Aging	Continuation of secondary reactions, including polymerization	Slow increase in viscosity, solids formation
Presence of small char particles	Incomplete char separation in the process	Enhanced aging of the oil Sedimentation Filter blockage Catalyst blockage Engine injector blockage
Poor distillability	Reactive mixture of degradation products	Limits applicable separation technologies
Low H/C ratio	Biomass feed has low H/C ratio	Upgrading to hydrocarbons requires considerable increase in H/C ratio
Incompatibility with other materials	Phenolics and aromatics	Negative effects on seals and gaskets
Miscibility with hydrocarbons is very low	Highly oxygenated nature of pyrolysis oil	Immiscible with hydrocarbons making direct cofeeding in a refinery difficult
High water content	Pyrolysis reactions, feed water	Complex effect on viscosity and stability. High water content lowers heating value, limits catalyst selections for upgrading processes

Several catalytic and noncatalytic upgrading processes have been proposed.[7,12,13] They can be classified into low- and high-temperature processes. Low-temperature upgrading approaches include reactive-blending

Table 7.3   High temperature processes for pyrolysis oil upgrading.[17,27]

	HPTT	Zeolite cracking	Hydrotreatment/HDO
Temperature (K)	473–623	573–873	523–673
Pressure (MPa)	15–30	No	10–30
Catalyst	No	Yes	Yes
Coke formation	High	High, 26–39 wt%	Low, <5 wt%

with alcohols[14] and solvent addition.[15] High-temperature processes include: i) a noncatalytic high pressure thermal treatment (HPTT),[16,17] ii) upgrading by cracking using zeolites,[18,19] and iii) catalytic hydrotreatment (also known as hydrodeoxygenation or HDO).[20] The main differences of the three high-temperature processes are summarized in Table 7.3. The main advantage of catalytic hydrotreatment over HPTT and zeolite cracking lies in the low levels of coke formation, which reduces carbon losses, and most importantly, allows for a long lifetime of the catalyst.[7,12,13,21–26]

7.2   Fast Pyrolysis Oil Upgrading by Catalytic Hydrotreatment

Catalytic hydrotreatment is a process in which the fast pyrolysis oil is subjected to a catalyst at elevated temperatures under high H_2 pressures. As a result, the reactive oxygenates are largely converted into less reactive components with a lower oxygen content. Consequently, not only is the acidity decreased but also the long-term stability of the oil improved. The catalytic hydrotreatment of fast pyrolysis oil may be carried out either in a slurry batch reactor or in a packed-bed reactor. In batch reactors, the process duration is relatively long (*e.g.* 4 h). In continuous reactors, a low space-velocity (0.2–0.5 kg_{oil} $kg^{-1}_{catalyst}$ h^{-1}) is required for a considerable reduction in the oxygen content.

7.2.1   Catalyst Studies on the Hydrotreatment of Fast Pyrolysis Oils

HDO of fast pyrolysis liquids was initially explored under similar conditions applied to the hydrodesulfurization process (HDS) in use in the petroleum refinery (as described in Chapters 8 and 9). For the upgrade of pyrolytic oil, a simplified reaction equation for the hydrotreatment of fast pyrolysis oil is given by eqn (7.1), where "CH_2" refers to unspecified hydrocarbons.[24]

$$CH_{1.4}O_{0.4} + 0.7\,H_2 \rightarrow 1\,``CH_2" + 0.4\,H_2O \tag{7.1}$$

The identification of highly active and stable catalyst for the catalytic upgrading of pyrolysis oil has been the objective of various studies (Table 7.4). Most of the examples are heterogeneous catalysts, since they can be reused and regenerated more easily than their homogeneous counterparts. HDS catalysts (sulfided NiMo/γ-Al$_2$O$_3$ or sulfided CoMo/γ-Al$_2$O$_3$) and noble-metal catalysts have been extensively explored. Other types of catalysts (Table 7.5)

Table 7.4 Overview of catalyst studies for pyrolysis oil hydrotreatment.

Catalyst	Ref.	Metal content (wt%)	Catalyst (wt%)	Mode	T (K)	P (MPa)	Time (h)[a] or SV (h^{-1})[b]	Oil yield (wt%)	DOD (%)[c]	H/C (product oil)	Remarks	Related studies
CoMoS/γ-Al2O3	25	NA	5	Batch	623	20.0	4	22	83	1.18		29
CoMoS/γ-Al2O3	22	14		Continuous	532/649	13.8	LHSV: 0.10	37	~96	1.68		21,30,31
NiMoS/γ-Al2O3	22	11		Continuous	423 (1st) 653 (2nd)	NA[d]	WHSV: 0.52	42	98.6	NA		21,28,30–33
NiMoS/γ-Al2O3	34	17	10	Batch	673	17.0	4	34.2	50	1.8		25,35
CoMoPS/γ-Al2O3	29	24	NA	Batch	633–663	1.5–3.0	0.1–1	NA	30–90	NA	Tetralin as hydrogen donor	
NiMo/γ-Al2O3 (reduced)	36	NA	5	Batch	373	3.0	2	NA	26	0.17	Feed H/C = 0.7	
Pt/mesoporous ZSM-5	37	0.5	5	Batch	473	4.0	1	NA	53	1.48	n-tridecane as solvent	
Pt/γ-Al2O3	37	0.5	5	Batch	473	4.0	1	NA	34	1.23	n-tridecane as solvent	
Pt/C	25	5	5	Batch	623	20.0	4	22	81	1.34		25,38,39
Pt/Al2O3-SiO2	33	5		Continuous	673	8.5	WHSV: 2		45			
Ru/C	26	5	5	Batch	503–613	29.0	4	50	60	1.58		
Ru/TiO2	25	5	5	Batch	623	20.0	4	59	80	1.31		
Ru/γ-Al2O3	25	5	5	Batch	623	20.0	4	37	81	1.11		
Ru/γ-Al2O3 (1st stage) CoW/γ-Al2O3 (2nd stage)	40	5 (Ru) 16 (Co+W)	5 (1st stage) 3 (2nd stage)	Batch	393 (1st) 623 (2nd)	7.0 (1st), 17.3 (2nd)	4 (total)	37	20	1.14		
Pd/C	25	5	5	Batch	623	20.0	4	83	85	1.27		
Pd/C	31	5	5	Continuous	613	14.0	WHSV: 0.25	45	64	1.53		

[a] for batch hydrotreatment.
[b] for continuous hydrotreatment.
[c] degree of deoxygenation (eqn (7.2)).
[d] NA = not available.

Table 7.5 Relevant examples of nonsulfided catalysts for hydrodeoxygenation of pyrolysis oil model compounds.

Catalyst	Substrate	P (MPa)	T (K)	Yield$_{HDO}$ (mol%)	Conversion (mol%)	Ref.
Ni/HZM-5	Methyl heptanoate	2.0	493	77	92	54
MoP (C support)	4-methylphenol	4.4	623	98	71	60
RuMo/ZrO$_2$	Propanoic acid	6.4	463	49	79	51
Ru/SiO$_2$–Al$_2$O$_3$	Guaiacol		523	60	100	52
Rh/SiO$_2$–Al$_2$O$_3$	Guaiacol		523	57	100	52
Co–Mo–B amorphous	Phenol, benzaldehyde, acetophenone (mixture)	4.0	548	DOD: 100%	100	59
Ni(Co) –W–B amorphous	Cyclopentanone	4.0	573	DOD: 94%	92.7	58
Ru/C + Pt/ZrP	Maple wood carbohydrates	6.2	394 (1st step) 518 (2nd step)	57 (gasoline yield)		61
Raney Ni + Nafion/SiO$_2$	4-n-Propylphenol	4.0	473	98	100	55
Mo$_2$N/γ-Al$_2$O$_3$	Oleic acid	7.2	653	Y$_{liq}$: 84 wt%, DOD: 100%	100	57
WN/ γ-Al$_2$O$_3$	Oleic acid	7.2	653	Y$_{liq}$: 81 wt%, DOD: 100%	100	57
VN/ γ-Al$_2$O$_3$	Oleic acid	7.2	653	Y$_{liq}$: 85 wt%, DOD: 72%	100	57
Pt/ γ-Al$_2$O$_3$	Sorbitol	2.9	518	56	82	62
Pt/ZrO$_2$	Guaiacol	8.0	373		10	50
Pd/ZrO$_2$	Guaiacol	8.0	373		13.7	50
Rh/ZrO$_2$	Guaiacol	8.0	373		98.9	50
Ni/δ-Al$_2$O$_3$	Anisole	1.0	573	76	80	53
NiCu/ δ-Al$_2$O$_3$	Anisole	1.0	573	99	100	53
NiCu/ δ-ZrO$_2$	Anisole	1.0	573	38	64	53
NiCu/ δ-CeO$_2$	Anisole	1.0	573	100	100	53
Ni$_2$P/SiO$_2$	Anisole	0.1	573	48		56
Co$_2$P/SiO$_2$	Anisole	0.1	573	36		56
MoP/SiO$_2$	Anisole	0.1	573	31		56
Pt–Sn/CNF	Guaiacol	0.1	573	60		63
Fe/SiO$_2$	Guaiacol	0.1	573	38		64

have also been reported. However, these catalysts were mainly studied on recalcitrant model compounds for the lignin fraction (*e.g.* guaiacol and anisole).

Typically, the catalytic activity is defined as the amount of a reactant (in mol) converted per gram of catalyst (or per active site) per time. However, for the catalytic hydrotreatment of pyrolysis oil, this definition cannot be directly used, as the exact molecular composition is unknown. Accordingly, Mortensen *et al.*[27] proposed a combination of the degree of deoxygenation (eqn (7.2)) with oil yield (eqn (7.3)) to evaluate the catalyst performance, and select the promising catalysts.

$$\text{Degree of deoxygenation (DOD)} = \left(1 - \frac{\text{wt \% O in product}}{\text{wt \% O in feed}}\right) \qquad (7.2)$$

$$\text{Oil yield} = \left(\frac{m_{\text{product oil}}}{m_{\text{feed}}}\right) \qquad (7.3)$$

Joshi and Lawal[28] measured the hydrogen consumption, and introduced the extent of HDO (eqn (7.4)) and space-time hydrogen consumption (STC, 7.5) in order to assess the catalyst performance. In eqn (7.4), the theoretical hydrogen consumption for the complete oxygen removal is based on the reaction stoichiometry given in eqn (7.1).

Extent of HDO =

$$\frac{\text{amount of hydrogen consumed}}{\text{amount of hydrogen required for complete oxygen removal}} \times 100 \qquad (7.4)$$

$$\text{Space time consumption (STC)} = \frac{\text{amount of } H_2 \text{ consumed}}{\text{amount of catalyst} \times \text{time}} \qquad (7.5)$$

In the next section, the performance of nonsulfided catalysts for model compounds is also discussed. For these catalysts, the performance is represented by the conversion of the model compound (eqn (7.6)) and the HDO yield (eqn (7.7)).

$$\text{Conversion (X)} = \frac{\text{Initial amount of model compound (moles)}}{\text{Final amount of model compound (moles)}} \times 100\% \qquad (7.6)$$

$$\text{HDO yield} = \frac{\text{amount of oxygen free product (moles)}}{\text{total amount of product (moles)}} \times \text{conversion} \qquad (7.7)$$

7.2.2 Active Metals

Early studies on the catalytic hydrotreatment of fast pyrolysis oil were performed with hydrodesulfurization catalysts (sulfided NiMo and CoMo on

γ-Al_2O_3). Deoxygenation levels up to 100% have been reported.[41] The active sites of the catalyst are sulfur vacancies on the edge of MoS_2 slabs.[42] These sites have a Lewis-acid character. As such, they can adsorb molecules containing electron-rich functional groups, and are thus active for both HDS and HDO.[43,44] The catalytic activity is promoted by the addition of Ni or Co. These metals can replace part of the Mo atoms in the slab edge, increasing the adsorption capacity of the vacancy sites,[45–48] as thoroughly discussed in Chapter 8.

Although NiMo and CoMo/γ-Al_2O_3 were demonstrated to be active HDO catalysts, a continuous supply of a sulfur-containing compound is required in order to maintain the high catalytic activity level. Therefore, they are less attractive for the hydrotreatment of sulfur-free biomass streams.[49] To overcome this limitation, noble-metal catalysts (Ru, Pd, Pt) have been explored (see also Chapter 3). They show performances similar or even superior to those of HDS catalysts. Wildschut *et al.*[25] performed hydrotreatment studies at 623 K and 4 h using Ru/C, Pd/C, and Ru/TiO_2 catalysts. Higher oil yields and products with a lower oxygen content were obtained, compared to sulfided catalysts (*e.g.* NiMo/Al_2O_3 or CoMo/Al_2O_3). Improved activity of noble-metal catalysts (Pd/C) was also reported by Elliott *et al.*[31] Notably, noble-metal catalysts were also reported to produce less coke than sulfided catalysts.[50]

Several mechanistic aspects of the hydrodeoxygenation processes in the presence of noble-metal catalysts are still not clear. Until now, detailed reaction pathways have only been established for model compounds.[7,12,13] Chen *et al.*[51] reported that the HDO of propanoic acid in the presence of Ru/ZrO_2 is more selective for the cleavage of C–C bonds compared to the hydrogenation of the C=O bond. Ha and coworkers[52] demonstrated that Rh or Ru were responsible for the saturation of guaiacol, whereas a highly acidic support (SiO_2–Al_2O_3) is indispensable for the deoxygenation through C–O bond cleavage. Yakovlev *et al.*[53] suggested that the first step of the process should involve activation of the oxygenated compounds by the acidic supports, followed by hydrogenation at the active metal site (for a detailed discussion on support effects, see Chapters 3 and 8).

Even though noble-metal catalysts have shown promising activities, they are expensive. Accordingly, several studies on the identification of catalysts based on nonexpensive base metals have been published. Table 7.5 shows relevant studies for HDO in the presence of nonsulfided catalysts, with an emphasis on model compounds for the lignin fraction (*e.g.* guaiacol and anisole).

Ni catalysts have been examined by a significant number of HDO studies due to its high hydrogenation capacity.[53] Shi *et al.*[54] reported that Ni on HZSM-5 is a good catalyst for HDO reactions. However, Ni contents higher than 3 wt% reduced the activity of Ni/HZSM-5 catalysts. They proposed that the acidic sites of HZSM-5 are blocked by Ni species. Zhao *et al.*[55] reported the use of Raney Ni together with a highly acidic Nafion/SiO_2 for the HDO of 4-*n*-propylphenol performed in water. Ni was very effective for hydrodeoxygenation. Surprisingly, the catalytic system prevented the aromatic rings from saturation. Yakovlev *et al.*[53] developed nickel- and nickel/copper-based catalysts. The addition of copper decreases the reduction temperatures for Ni(II) to Ni(0). In the HDO of guaiacol, the nickel/copper-based catalysts inhibited methanation even at high temperatures.

Other transition metals (*e.g.* Co, Mo, and Fe) have also been examined. Zhao *et al.*[56] investigated the catalytic activity of several metal phosphide supported on silica, and found that the activity for HDO of guaiacol follows the order: $Ni_2P > Co_2P > Fe_2P$, WP, MoP Monnier *et al.*[57] developed various metal-nitride catalysts, and reported $Mo_2N/\gamma\text{-}Al_2O_3$ to show higher HDO activity than $WN/\gamma\text{-}Al_2O_3$ and $VN/\gamma\text{-}Al_2O_3$, as the last two catalysts favor decarboxylation and decarbonylation over HDO. Wang *et al.*[58,59] developed amorphous Co–Mo–B and Ni(Co)–W–B catalysts for the HDO of various oxygenates. Reportedly, these catalysts have a high number of coordinatively unsaturated sites resulting in a high hydrogen-adsorption capacity. The presence of Brønsted acid sites was shown to be important for the deoxygenation reaction, as they activate the C=O bonds.

7.2.3   Characterization of Hydrotreatment Products

Characterization of fast pyrolysis oil and the hydrotreatment products is a challenge due to the high number of individual compounds present in the product mixtures. The most relevant analytical techniques and protocols are briefly discussed in this section. Overall, these techniques can be classified into two groups. The first provides insights into the type and amounts of individual components in the product mixtures. The second assesses macroscopic properties (*e.g.* elemental composition) and physical properties (*e.g.* viscosity and solubility in hydrocarbons).

The molecular composition of the feed and product after the hydrotreatment reaction, lends important insights into the reactivity of several compound classes, and is very useful to elucidate the reaction pathways occurring throughout the upgrading process. A major complication arises due to the presence of both low and high molecular weight (M_w) compounds in the product mixtures. Accordingly, the proper analysis of the products requires several analytical techniques in order to describe the population of compounds in the different ranges of M_w.

Indeed, due to the presence of high-M_w components, analyses by GC-MS/FID and 2D GC×GC-MS/FID provide information only regarding the volatile fraction of the oil.[65] Moreover, due to the high number of individual components in both the feed and product, the compounds are often classified based on the functional organic groups (acids, phenolics, *etc.*).[66] In contrast to GC analytical protocols, gel permeation chromatography (GPC) provides valuable information regarding the overall distribution of M_w. However, it should be used only for comparative purposes. Since there is no appropriate calibration standard that could account for the complexity of the fast pyrolysis oil, the determination of the absolute molecular weight is not possible.[40,67]

^{1}H- and ^{13}C-NMR have also been used to probe the various functional organic groups in fast pyrolysis oils.[68,69] It also allows for quantification of monomeric and oligomeric sugar fragments that are difficult to identify and quantify with conventional GC techniques.

An alternative protocol for the characterization of pyrolysis oils and product oils involves solvent fractionation (Figure 7.2).[5,70] This method separates the oils into a number of fractions: extractives, sugars, ether-solubles, LMM (low M_w) lignin, and HMM (high M_w) lignin. Each fraction can be analyzed further by GC-MS/FID, 2D-GC, GPC, or other analytical techniques. An example of the use of this protocol for the characterization of pyrolysis oil and a typical hydrotreatment product is given in Figure 7.3. After hydrotreatment, the water content increased due to HDO reactions. In agreement with this observation, the sugar content decreased, indicating the high reactivity of this fraction. In turn, the LMM lignin/ extractives content increased considerably due to the formation of nonpolar components.

Acidity is a critical property of both fast pyrolysis oil and hydrotreated products. A useful measure for acidity is the TAN (total acidity number), which is used in order to determine the acidity of fossil fuels.[7,35] It involves the titration of a sample with KOH. TAN analysis should be applied with caution for pyrolysis oil due to the presence of compounds other than organic acids (*e.g.* alcohols and phenolic compounds) that can also be titrated with KOH.[71] Capillary electrophoresis (CE) is very useful in the quantification of organic acids in aqueous samples (*e.g.* the aqueous phase after the solvent fractionation). GC-MS/FID can also be applied to determine the acid content, although derivatization is required for quantification.[72]

The elemental composition (C, H, O, N) of the upgraded products and, particularly, its comparison with the composition of the initial feed, lends insights into the changes in composition occurring throughout the fast pyrolysis oil upgrading. The changes in composition are conveniently

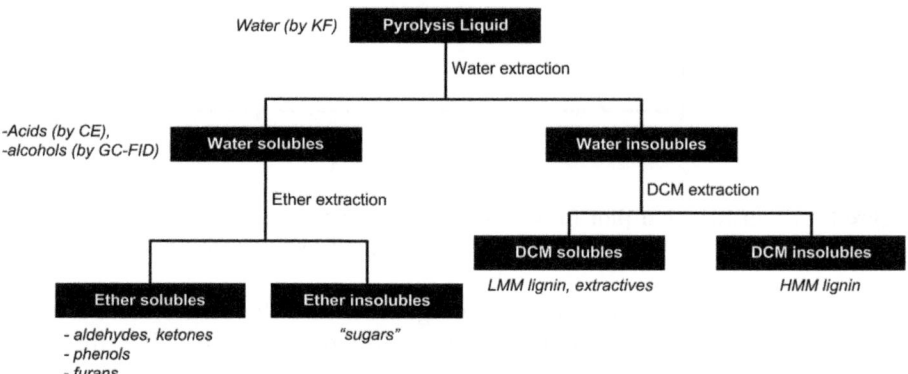

Figure 7.2   Solvent fractionation scheme developed by VTT.[5,70] KF stands for Karl–Fischer analysis; CE, capillary electrophoresis; LMM, low molecular weight; and HMM, high molecular weight.

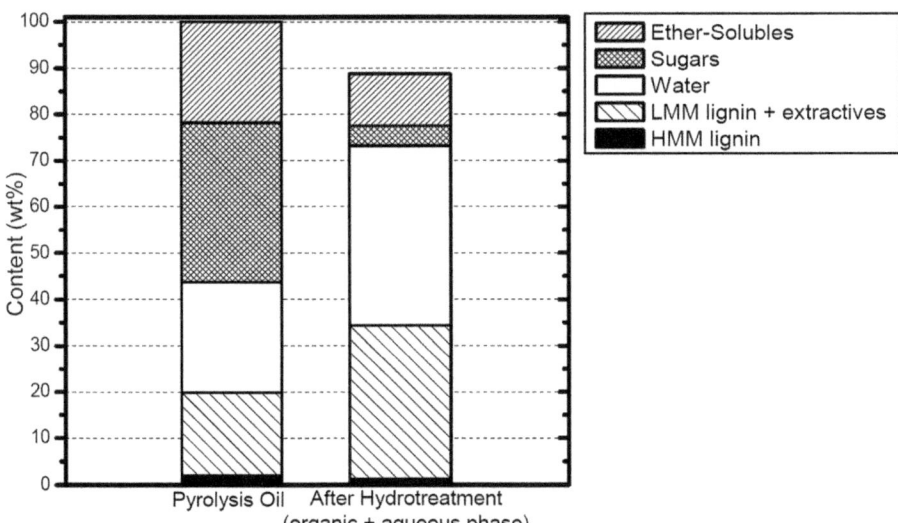

Figure 7.3   Composition of pinewood pyrolysis oil according to the solvent fractionation protocol developed by VTT.[5,70] Hydrotreatment was performed
with Ru/C catalyst in fixed-bed reactors-in-series with gradual increase
of temperature (from 448 K, in the inlet, to 598 K near the outlet).

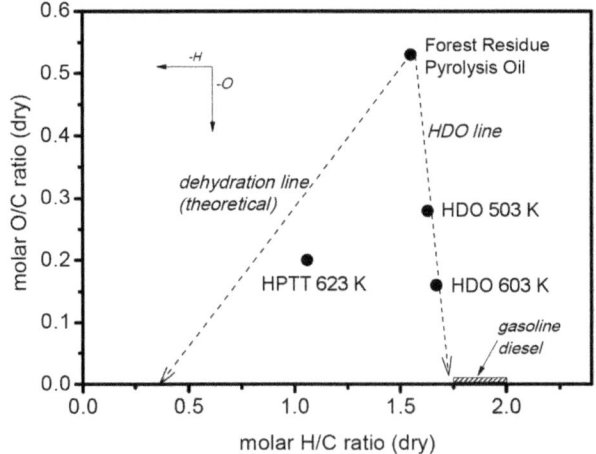

Figure 7.4   Van Krevelen plot for catalytic and noncatalytic, thermal processing of
pyrolysis oil. The data are from de Miguel Mercader *et al.*[17,26]

displayed in a Van Krevelen plot,[73] a diagram in which the molar O/C ratio
is plotted on the Y-axis and molar H/C ratio on the X-axis (Figure 7.4).
Hydrogenation, for example, increases the H/C ratio, while dehydration
decreases both the H/C and O/C ratios.

Viscosity and thermal stability are relevant product properties in order to
assess the potential and performance of the product oils for cofeeding in oil

refineries. Thus, a low viscosity is desirable to facilitate the transportation of the product oil. In turn, thermal stability is of utmost importance for the cofeeding in petroleum refineries. To determine the thermal stability of the hydrotreatment product and its mixtures with fossil feeds (*e.g.* long residue), the Conradson carbon residue, CCR (or MCR, microcarbon residue) analyses are usually performed.[26]

7.3 Hydrotreatment Pathways and Severity Effects

The catalytic hydrotreatment was initially assumed as the hydrodeoxygenation described by oversimplified eqn (7.1). Actually, the upgrading of pyrolysis oil involves several reactions, namely cracking, hydrocracking, decarbonylation, decarboxylation, hydrodeoxygenation, hydrogenation (Scheme 7.1) in addition to repolymerization and methanation (not shown in Scheme 7.1).

To gain more understanding about the major reaction pathways, the hydrotreatment under varying process conditions was performed. In a seminal study, Gagnon and Kaliguine[40] performed a two-stage hydrotreatment of vacuum pyrolysis oil using a batch-reactor setup. The first stage involved the hydrogenation at low temperature range (313–393 K) with Ru/γ-Al$_2$O$_3$. In the second-stage, the hydrodeoxygenation was carried out at 598 K in the presence of a NiO–WO$_3$/γ-Al$_2$O$_3$ catalyst. This study revealed that the hydrogenation of the mono- and oligosaccharides started at a temperature as low as 353 K, and was completed within 1 h. It is not recommended to perform the first stage at a higher temperature as it promotes polymerization, exerting a negative effect on the deoxygenation reactions in the second step of the process. Actually, even after the stabilization of the pyrolysis oil through hydrogenation at the first stage of the process, the distribution of M_w rapidly increased at 598 K in the second stage. However, the distribution of M_w decreased upon prolonged duration of the

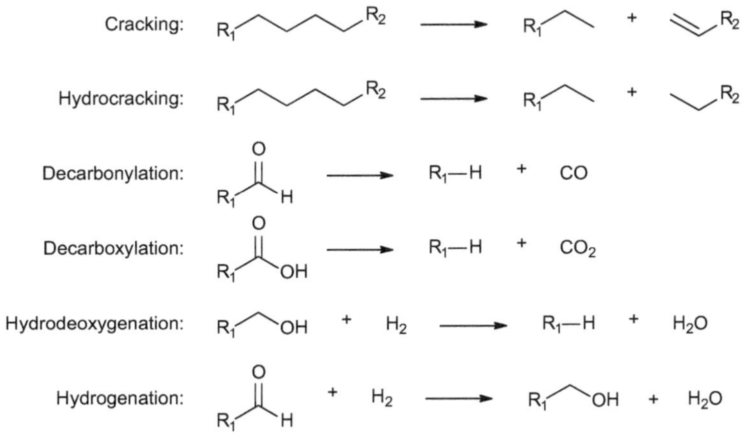

Scheme 7.1 Examples of reactions occurring in the catalytic pyrolysis-oil upgrading.

hydrotreatment, suggesting the occurrence of hydrogenolysis and (hydro)-cracking reactions.

Wildschut *et al.*[74] studied the hydrotreatment of pyrolysis oil in the presence of a Ru/C catalyst (5 wt% Ru) at 623 K and 20.0 MPa at varying process durations (1–6 h). Key findings concerning the process evolution based on ^{1}H-NMR, solvent fractionation, and elemental analyses were provided. In the initial phase of the process, dehydration is evident from the low O/C and H/C ratios of the product oil (Figure 7.5). This finding can be explained either by the thermal repolymerization or by other chemical reactions (*e.g.* alcohol dehydration).

Solvent fractionation showed that most of the sugar fraction was converted within 4 h, most probably into polyols, and subsequently into hydrocarbons.[74] This result confirms the observations by Gagnon on the complete hydrogenation of sugars occurring within a process duration of 1 h. Prolonged process duration increased the H/C ratio of the product oil, indicating that the hydrogenations proceed at slow rates. Surprisingly, the oxygen content also increased. This result suggests that components with high O/C ratios were transferred from the aqueous phase to the organic phase, upon their partial deoxygenation. At durations of 4–6 h, hydrocracking became significant, reducing the product oil yield due to the formation of gaseous products (*e.g.* methane).

Based on studies using Ru/C catalysts, a reaction network for the catalytic hydrotreatment of fast pyrolysis oil was proposed by Venderbosch *et al.* (Scheme 7.2).[75] In the initial steps of the hydrotreatment process, catalytic hydrogenation and thermal (noncatalytic) polymerization occur as parallel reactions. Polymerization leads to the formation of soluble high-M_w fragments, which undergo further condensation forming char. The optimum

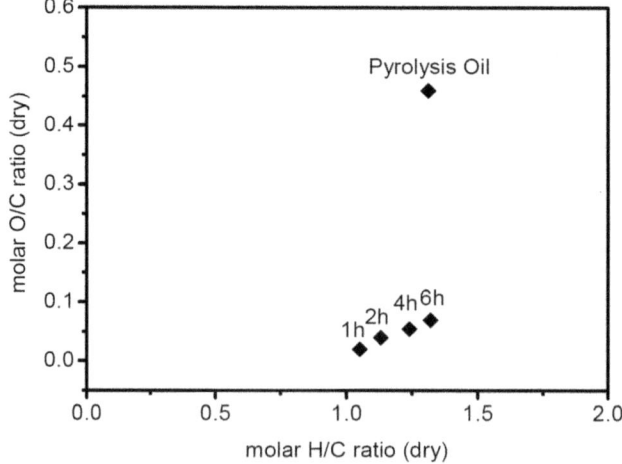

Figure 7.5 Van Krevelen plot for fast pyrolysis oil hydrotreatment with a Ru/C catalyst at 623 K for varying process duration.[74]

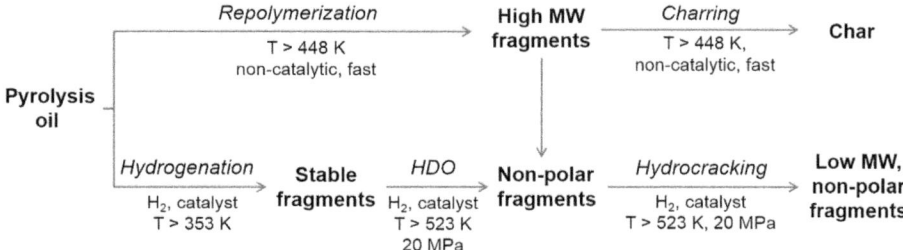

Scheme 7.2   Proposed reaction pathway for the catalytic hydrotreatment of pyrolysis oil.[75]

upgrading pathway involves the hydrogenation of thermally labile components to stable molecules, which are not prone to polymerization. Subsequent reactions (*i.e.* hydrogenations and hydrocracking), on a time scale of hours, render an oil product with reduced oxygen content and ultimately to a high H/C ratio.

Overall, the following conclusions on relevant hydrotreatment pathways can be drawn, i) thermal polymerization/HPTT is a fast reaction and becomes important at about 393 K, ii) hydrogenation/stabilization at a lower temperature is thus recommended to prevent excessive polymerization, iii) therefore, a highly active catalyst is required to ensure that polymerization does not occur to a significant extent in the initial stage.

7.4   Recent Studies Using Bimetallic Ni–Cu Catalysts

Recently, we reported the use of Ni-based catalysts as inexpensive alternatives to noble-metal-based catalysts for the hydrotreatment of fast pyrolysis oil. Bimetallic NiCu catalysts supported on several materials (δ-Al$_2$O$_3$, CeO$_2$–ZrO$_2$, ZrO$_2$, SiO$_2$, TiO$_2$, rice-husk carbon, and sibunite) were examined. Improved performances were obtained, regarding both activity and product properties of the upgraded oils.[76,77] However, in the initial stage of the process, the slow rate of hydrogenation resulted in undesirable char products through thermal polymerization.

In a subsequent study, we have explored the use of a new bimetallic NiCu catalyst, referred to as Picula™ catalyst D. The catalyst is characterized by a high active metal loading (*i.e.* 57.9 wt% Ni, 7 wt% Cu, and 35.1% SiO$_2$). The catalytic hydrotreatment of fast pyrolysis oil was performed in a continuous reactor configuration with Picula™ catalyst D in four fixed beds connected in series (Figure 7.6).

The visual appearance of the liquid phase after reaction is a function of the process severity (*i.e.* temperature and residence time). In the low temperature range (398–448 K, experiments 1 and 2, Table 7.6), a dark brown liquid was obtained. For the experiments 3, 4 and 5 (Table 7.6), two liquid layers were formed as products: a dark-brown organic layer (bottom), and a slightly brown aqueous layer (upper). Considering the entire set of

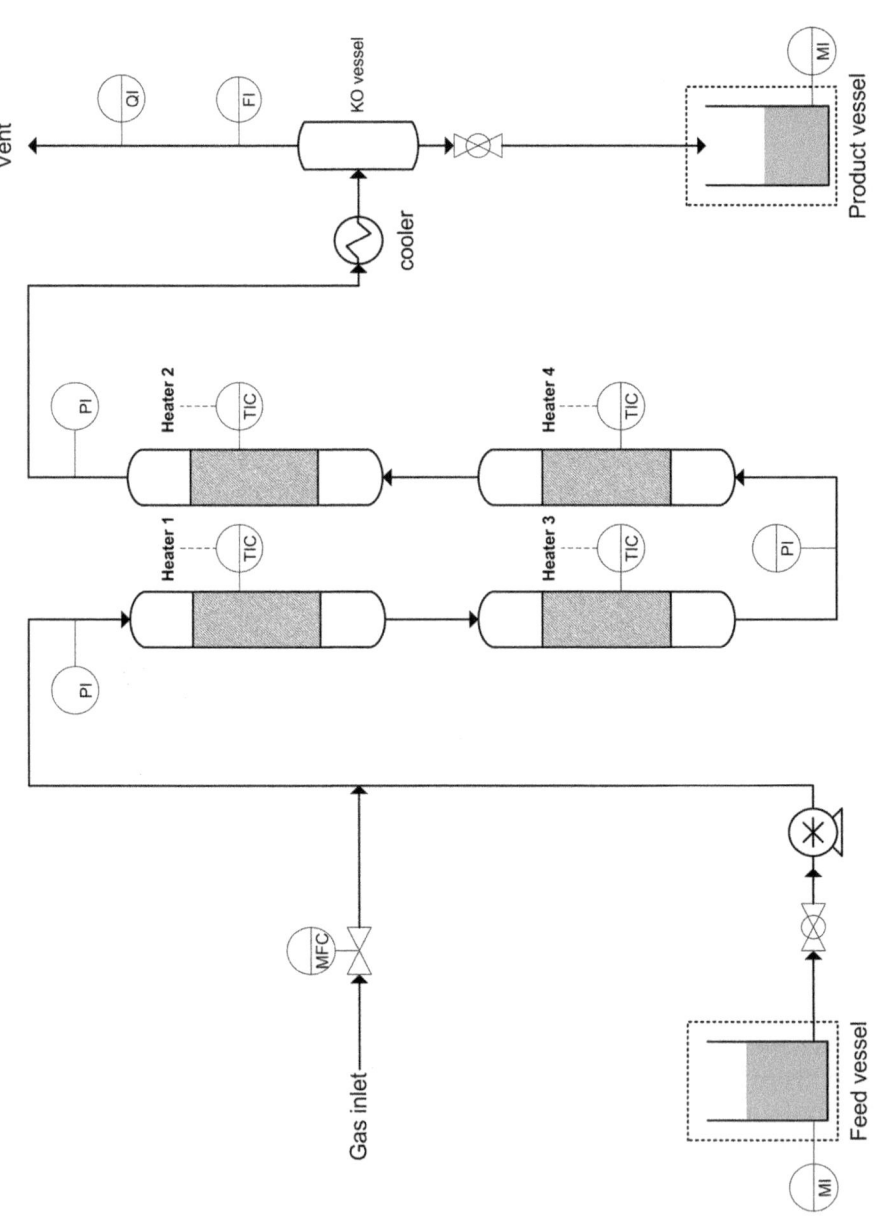

Figure 7.6 Schematic representation of the continuous setup for the hydrotreatment of fast pyrolysis oil in the presence of Picula™ catalyst D.

Table 7.6 Reaction conditions and mass balances for the continuous hydrotreatment of pyrolysis oil in the presence of Picula™ catalyst D.

	Unit	Experiment				
		1	2	3	4	5
T (Bed 1/2/3/4)	K	398/398	398/398	398/448	398/448	398/448
		398/398	448/448	498/548	548/598	548/598
P	MPa	18.9	19.2	21.1	21.1	20.8
WHSV	$(g_{PO}\ h^{-1})/g_{cat}$	0.5	0.6	0.7	1.0	0.6
H_2 consumption	NL/kg	214	249	250	254	345
Liquid yield	wt% feed	97.4	98.6	79.6	94.4	93.5
Organic-fraction yield	wt% feed	75.6	76.4	40.7	51.6	48.1
Aqueous-fraction yield	wt% feed	24.5	23.6	59.3	48.4	51.9
Oxygen content of organic fraction		37.7	38.8	26.6	21.7	20.0
Water content of organic fraction	wt%	20.8	20.3	13.4	9.4	9.7
Carbon content of aqueous fraction	wt%			31.1	20.9	19.4
Carbon distribution						
Organic phase	wt%	97.4	98.6	68.5	66.8	71.2
Aqueous phase	wt%			55.1	20.5	23.5

experiments, the hydrogen consumption increased (from 215 to 345 NL/kg$_{PO}$ for experiments 1 to 5), and the oxygen content of the organic phase decreases gradually (from 37.8 to 20.0 wt%) with the process severity. Carbon recovery in the liquid product ranged from 88 to 100 wt% relative to the pyrolysis oil feed (*i.e.* the formation of gaseous products other than hydrogen was suppressed).

Figure 7.7 shows the H_2 consumption as a function of the time-on-stream for experiments 1–5. Clearly, the highest H_2 uptake was found for experiment 5. For experiments 1, 3–5, the catalyst activity maintained a constant level over the duration of the runtime. For experiment 2, the initial H_2 uptake increased within the first 1–2 h of operation. Deactivation of the catalyst over the time-on-stream was not noticed, even after an experiment duration of 8 h.

For experiment 1, the H_2 uptake with the Picula™ catalyst D (210 Nl/kg$_{PO}$) was significantly higher than that reported for the process with Ru/C (<100 NL/kg$_{PO}$) under comparably low-severity conditions.[66] This result indicates that the Picula™ catalyst D shows a higher activity for hydrogenation than a Ru/C catalyst under the same process conditions. This finding is important because a high hydrogenation rate limits the extent of polymerization that generates high M_w fragments and char. Indeed, Figure 7.8 indicates that the weight-average M_w is nearly constant for the product oils obtained by the hydrotreatment at 448–598 K, showing that the

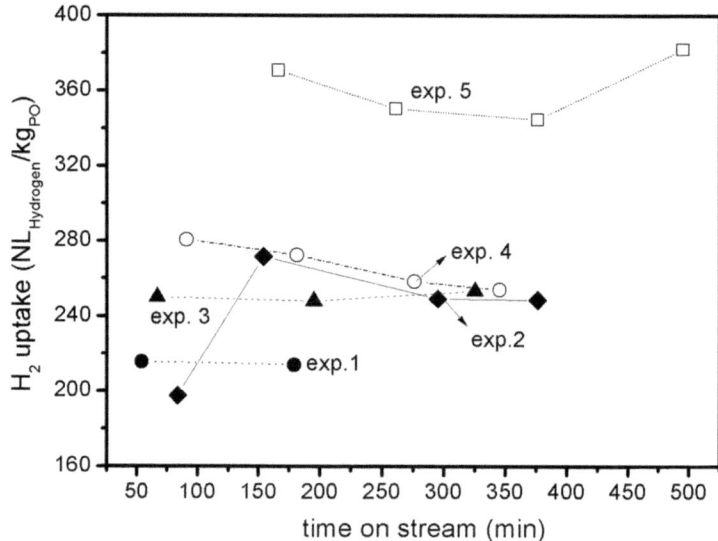

Figure 7.7 Hydrogen uptake versus time-on-stream for continuous hydrotreatment with Picula Cat D.

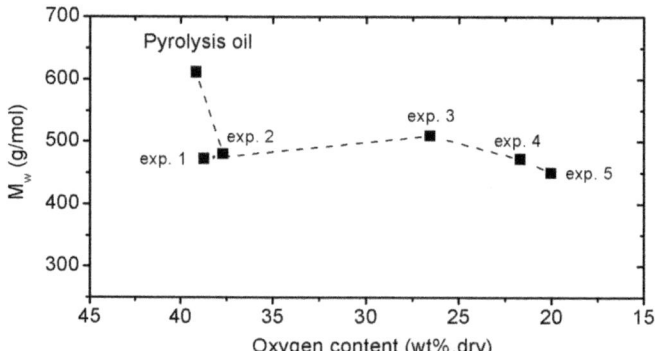

Figure 7.8 Weight-average M_w of the product oils obtained from the continuous hydrotreatment with Picula Cat D. Reaction conditions are given in Table 7.5. The line is drawn for illustrative purposes only.

polymerization pathway does not occur to a significant extent in the process with the Picula™ catalyst D. In contrast, a sharp increase in the weight-average M_w was found for hydrotreatment in the presence of a Ru/C catalyst under similar severity conditions, again indicating that polymerization proceeded faster than the hydrogenation processes.[75]

The van Krevelen plot of the product oils (Figure 7.9) lends further insights into the processes occurring in the catalytic upgrade of the pyrolytic oil. Under low-severity conditions, the H/C values increase significantly, while the O/C ratios remain about constant. These results confirm the hydrogenation processes as the predominant ones at low reaction temperatures and

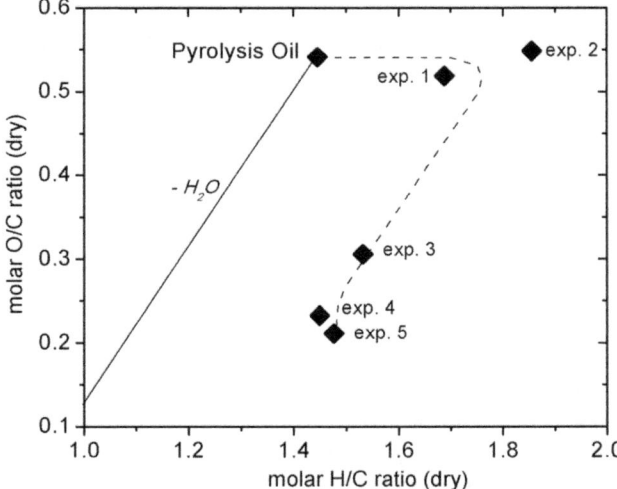

Figure 7.9   Van Krevelen plot of the upgraded oil obtained from hydrotreatment of pyrolysis oil with Picula Cat D. Reaction conditions are given in Table 7.5. The line is drawn for illustrative purposes only.

short residence time. In contrast, the H/C and O/C ratios of the upgraded oils decreased, most likely due to the release of water by dehydration occurring under more severe conditions (498–548 K). At this temperature range, the H_2 uptake is constant, implying that hydrogenation and dehydration occur as parallel reactions.[75] Under higher severity conditions, the H/C ratio slightly increased. This result confirms that the hydrogenation rates are faster than the dehydration rates.

7.5   Conclusions

In the last decade, considerable progress has been made on the catalytic hydrotreatment of fast pyrolysis oils. Improved catalysts are reported for both the upgrading of fast pyrolysis oils and model compounds in batch and continuous reactors. The catalyst stability has been addressed, and significant improvements made, allowing for time-on-stream as long as 400 h without serious deactivation. One of the earlier issues regarding the catalyst performance, *i.e.* repolymerization of reactive molecules leading to negative effects on product properties (high coking tendencies at elevated temperatures) in addition to reactor clogging, has been effectively solved by the development and utilization of active catalysts at low temperature (473 K). By this approach, reactive molecules (*e.g.* aldehydes and ketones) that are prone to thermal polymerization are converted into less-reactive ones (*e.g.* alcohols). Cofeeding in existing refineries has been identified as a very attractive application of hydrotreated pyrolysis oils. Product oils with oxygen contents up to 10 wt% show good performance in coprocessing in the oil refinery. In addition, novel advanced techniques have been developed to characterize the complex nature of the product oils, lending indepth insights into their molecular composition. An example is the use of 2D-GC×GC, allowing for the quantification of the main compound classes present in the upgraded products.

References

1. R. Venderbosch and W. Prins, Biofuels, *Bioprod. Biorefin.*, 2010, **4**, 178–208.
2. A. V. Bridgwater, *Biomass Bioenergy*, 2012, **38**, 68–94.
3. *Biocoup at a glance*. http://www.biocoup.com/index.php?id=18.
4. http://www.btgworld.com/.
5. A. Oasmaa, E. Kuoppala, A. Ardiyanti, R. Venderbosch and H. J. Heeres, *Energy Fuels*, 2010, **24**, 5264–5272.
6. A. Oasmaa and D. J. Meier, *J. Anal. Appl. Pyrol.*, 2005, **73**, 323–334.
7. A. H. Zacher, M. V. Olarte, D. M. Santosa, D. C. Elliott and S. B. Jones, *Green Chem.*, 2014, **16**, 491–515.
8. S. Czernik and A. Bridgwater, *Energy Fuels*, 2004, **18**, 590–598.
9. D. Chiaramonti, M. Bonini, E. Fratini, G. Tondi, K. Gartner, A. Bridgwater, H. Grimm, I. Soldaini, A. Webster and P. Baglioni, *Biomass Bioenergy*, 2003, **25**, 85–99.
10. D. Chiaramonti, M. Bonini, E. Fratini, G. Tondi, K. Gartner, A. Bridgwater, H. Grimm, I. Soldaini, A. Webster and P. Baglioni, *Biomass Bioenergy*, 2003, **25**, 101–111.
11. J. Lange, *Biofuels, Bioprod. Biorefin.*, 2007, **1**, 39–48.
12. H. Wang, J. Male and Y. Wang, *ACS Catal.*, 2013, **3**, 1047–1070.
13. Z. He and X. Wang, *Catal. Sustain. Energy*, 2012, **1**, 28–52.
14. F. Mahfud, I. Melin-Cabrera, R. Manurung and H. Heeres, *Proc. Saf. Environ. Prot.*, 2007, **85**, 466–472.
15. A. Oasmaa, E. Kuoppala, J. Selin, S. Gust and Y. Solantausta, *Energy Fuels*, 2004, **18**, 1578–1583.
16. M. Rep, R. H. Venderbosch, D. Assink, W. Tromp, S. R. A. Kersten and W. Prins, in *Science in Thermal and Chemical Biomass Conversion*, ed. A. V. Bridgwater and D. G. B. Boocock, *CLP Press*, 2006, pp. 1526–1535.
17. F. de Miguel Mercader, M. Groeneveld, S. Kersten, R. Venderbosch and J. Hogendoorn, *Fuel*, 2010, **89**, 2829–2837.
18. S. Vitolo, B. Bresci, M. Seggiani and M. Gallo, *Fuel*, 2001, **80**, 17–26.
19. P. A. Horne and P. T. Williams, *Renew. Energy*, 1996, 7, 131–144.
20. E. Furimsky, *Catal. Rev. - Sci. Eng.*, 1983, **25**, 421–458.
21. D. C. E. G. G. Neuenschwander, in *Developments in Thermochemical Biomass Conversion*, ed. A. V. Bridgwater and D. G. B. Boocock, Blackie Academic & Professional, London, 1996, vol. 1, pp. 611–621.
22. D. C. Elliott, *Energy Fuels*, 2007, **21**, 1792–1815.
23. D. C. Elliott and T. R. Hart, *Energy Fuels*, 2008, **23**, 631–637.
24. J. Wildschut, PhD thesis, University of Groningen, 2009.

25. J. Wildschut, F. H. Mahfud, R. H. Venderbosch and H. J. Heeres, *Ind. Eng. Chem. Res.*, 2009, **48**, 10324–10334.

26. F. de Miguel Mercader, M. Groeneveld, S. Kersten, N. Way, C. Schaverien and J. Hogendoorn, *Appl. Catal. B: Environ.*, 2010, **96**, 57–66.

27. P. M. Mortensen, J. Grunwaldt, P. A. Jensen, K. Knudsen and A. D. Jensen, *Appl. Catal. A: Gen.*, 2011, **407**, 1–19.

28. N. Joshi and A. Lawal, *Chem. Eng. Sci.*, 2012, **74**, 1–8.

29. Z. Su-Ping, *Energy Sources*, 2003, **25**, 57–65.

30. M. Samolada, W. Baldauf and I. Vasalos, *Fuel*, 1998, 77, 1667–1675.

31. D. C. Elliott, T. R. Hart, G. G. Neuenschwander, L. J. Rotness and A. H. Zacher, *Environ. Prog. Sustain. Energy*, 2009, **28**, 441–449.

32. J. Rocha, C. Luengo and C. Snape, *Renew. Energy*, 1996, **9**, 950–953.

33. Y. E. Sheu, R. G. Anthony and E. J. Soltes, *Fuel Proc. Technol.*, 1988, **19**, 31–50.

34. R. J. French, J. Stunkel and R. M. Baldwin, *Energy Fuels*, 2011, **25**, 3266–3274.

35. R. J. French, J. Hrdlicka, R. Baldwin and R. Environ, *Prog. Sustain. Energy*, 2010, **29**, 142–150.

36. Y. Xu, T. Wang, L. Ma, Q. Zhang and L. Wang, *Biomass Bioenergy*, 2009, **33**, 1030–1036.

37. Y. Wang, T. He, K. Liu, J. Wu and Y. Fang, *Bioresour. Technol.*, 2012, **108**, 280–284.

38. J. Wildschut, I. Melian-Cabrera and H. Heeres, *H. Appl. Catal. B: Environ*, 2010, **99**, 298–306.

39. F. de Miguel Mercader, M. J. Groeneveld, S. R. Kersten, C. Geantet, G. Toussaint, N. W. Way, C. J. Schaverien and K. J. Hogendoorn, *Energy Environ. Sci.*, 2011, **4**, 985–997.

40. J. Gagnon and S. Kaliaguine, *Ind. Eng.- Chem. Res.*, 1988, **27**, 1783–1788.

41. W. Baldauf, U. Balfanz and M. Rupp, *Biomass Bioenergy*, 1994, 7, 237–244.

42. E. Furimsky and F. E. Massoth, *Catal. Today*, 1999, **52**, 381–495.

43. E. Ryymin, M. L. Honkela, T. Viljava and A. O. I. Krause, *Appl. Catal. A: Gen*, 2010, **389**, 114–121.

44. C. Dupont, R. Lemeur, A. Daudin and P. Raybaud, *J. Catal.*, 2011, **279**, 276–286.

45. H. Topsøe and B. S. Clausen, *Catal. Rev. Sci. Eng.*, 1984, **26**, 395–420.

46. H. Topsøe and B. S. Clausen, *Appl. Catal.*, 1986, **25**, 273–293.

47. C. Wivel, B. S. Clausen, R. Candia, S. Mørup and H. Topsøe, *J. Catal.*, 1984, **87**, 497–513.

48. C. Wivel, R. Candia, B. S. Clausen, S. Mørup and H. Topsøe, *J. Catal.*, 1981, **68**, 453–463.

49. O. Şenol, E. Ryymin, T. Viljava and A. Krause, *J. Mol. Catal. A: Chem.*, 2007, **277**, 107–112.

50. A. Gutierrez, R. Kaila, M. Honkela, R. Slioor and A. Krause, *Catal. Today*, 2009, **147**, 239–246.

51. L. Chen, Y. Zhu, H. Zheng, C. Zhang and Y. Li, *Appl. Catal. A: Gen.*, 2012, **411**, 95–104.
52. C. R. Lee, J. S. Yoon, Y. Suh, J. Choi, J. Ha, D. J. Suh and Y. Park, *Catal. Commun.*, 2012, **17**, 54–58.
53. V. Yakovlev, S. Khromova, O. Sherstyuk, V. Dundich, D. Y. Ermakov, V. Novopashina, M. Y. Lebedev, O. Bulavchenko and V. Parmon, *Catal. Today*, 2009, **144**, 362–366.
54. N. Shi, Q. Liu, T. Jiang, T. Wang, L. Ma, Q. Zhang and X. Zhang, *Catal. Commun.*, 2012, **20**, 80–84.
55. C. Zhao, Y. Kou, A. A. Lemonidou, X. Li and J. A. Lercher, *Chem. Commun.*, 2010, **46**, 412–414.
56. H. Zhao, D. Li, P. Bui and S. Oyama, *Appl. Catal. A: Gen.*, 2011, **391**, 305–310.
57. J. Monnier, H. Sulimma, A. Dalai and G. Caravaggio, *Appl. Catal. A: Gen*, 2010, **382**, 176–180.
58. W. Wang, Y. Yang, H. Luo, H. Peng, B. He and W. Liu, *Catal. Commun.*, 2011, **12**, 1275–1279.
59. W. Wang, Y. Yang, H. Luo, T. Hu and W. Liu, *Catal. Commun.*, 2011, **12**, 436–440.
60. V. M. Whiffen, K. J. Smith and S. K. Straus, *Appl. Catal. A: Gen.*, 2012, **419**, 111–125.
61. N. Li, G. A. Tompsett, T. Zhang, J. Shi, C. E. Wyman and G. W. Huber, *Green Chem.*, 2011, **13**, 91–101.
62. N. Li and G. W. Huber, *J. Catal.*, 2010, **270**, 48–59.
63. M. Á. González-Borja and D. E. Resasco, *Energy Fuels*, 2011, **25**, 4155–4162.
64. R. Olcese, M. Bettahar, D. Petitjean, B. Malaman, F. Giovanella and A. Dufour, *Appl. Catal. B: Environ.*, 2012, **115**, 63–73.
65. J. Marsman, J. Wildschut, F. Mahfud and H. Heeres, *J. Chrom. A*, 2007, **1150**, 21–27.
66. M. Windt, D. Meier, J. H. Marsman, H. J. Heeres and S. de Koning, *J. Anal. Appl. Pyrol.*, 2009, **85**, 38–46.
67. E. Hoekstra, S. R. Kersten, A. Tudos, D. Meier and K. J. Hogendoorn, *J. Anal. Appl. Pyrol.*, 2011, **91**, 76–88.
68. L. Ingram, D. Mohan, M. Bricka, P. Steele, D. Strobel, D. Crocker, B. Mitchell, J. Mohammad, K. Cantrell and C. U. Pittman, *Energy Fuels*, 2007, **22**, 614–625.
69. C. A. Mullen, G. D. Strahan and A. A. Boateng, *Energy Fuels*, 2009, **23**, 2707–2718.
70. K. Sipilä, E. Kuoppala, L. Fagernäs and A. Oasmaa, *Biomass Bioenergy*, 1998, **14**, 103–113.
71. A. Oasmaa, D. C. Elliott and J. Korhonen, *Energy Fuels*, 2010, **24**, 6548–6554.
72. A. Oasmaa, E. Kuoppala and D. C. Elliott, *Energy Fuels*, 2012, **26**, 2454–2460.
73. D. W. van Krevelen, *Fuel*, 1950, **29**, 269–283.

74. J. Wildschut, M. Iqbal, F. H. Mahfud, I. M. Cabrera, R. H. Venderbosch and H. J. Heeres, *Energy Environ. Sci.*, 2010, **3**, 962–970.
75. R. Venderbosch, A. Ardiyanti, J. Wildschut, A. Oasmaa and H. Heeres, *J. Chem. Technol. Biotechnol.*, 2010, **85**, 674–686.
76. A. Ardiyanti, S. Khromova, R. Venderbosch, V. Yakovlev and H. Heeres, *Appl. Catal. B: Environ.*, 2012, **117**, 105–117.
77. A. Ardiyanti, S. Khromova, R. Venderbosch, V. Yakovlev, I. Melián-Cabrera and H. Heeres, *Appl. Catal. A: Gen.*, 2012, **449**, 121–130.

Hydrodeoxygenation of Biomass-Derived Liquids over Transition-Metal-Sulfide Catalysts

BARBARA PAWELEC AND JOSE LUIS GARCIA FIERRO*

Energy and Sustainable Chemistry Group, Institute of Catalysis and Petrochemistry, CSIC, Marie Curie 2, 28049 Madrid, Spain
*Email: jlgfierro@icp.csic.es

8.1 Introduction

Due to limited oil reserves and the high demand for fuel and petrochemical feedstock, it has become necessary to find new, efficient and economical sources of hydrocarbons with minimal impact on the environment.[1] Thus, the partial replacement of oil by renewable sources, such as biomass, to produce high-quality transportation fuels is a topic of great environmental, chemical and technological relevance.

Several biomass feedstocks can be used for the production of biofuels, namely sugars, vegetable oils ("first-generation" biofuels) and ligno-cellulosic biomass, as residues from agriculture and forestry ("second-generation" biofuels).[2] First-generation biofuels (bioethanol and biodiesel) have increased their market share, mainly because of their blending with oil-derived fuels. However, it should be mentioned that lignocellulosic bio-mass can be grown in combination with food or on nonagricultural lands,

RSC Energy and Environment Series No. 13
Catalytic Hydrogenation for Biomass Valorization
Edited by Roberto Rinaldi

therefore not compromising the sustainable utilization of fertile land, and thus not competing with food supply.[3] Lignocellulosic biomass consists of an abundant renewable source of carbon[4] that could be valorized in existing downstream processes, thus increasing the production of commercial fuels in a sustainable manner.

The main difference between oil and biomass is the higher oxygen content of the latter. Biomass typically contains between 35 and 50 wt% oxygen. Lignocellulosic biomass comprises three types of biopolymers, cellulose (a glucose polymer), hemicellulose (polymers made of pentoses and hexoses), and lignin (a propylphenol polymer).[2,5] These three components form an insoluble three-dimensional composite with a very complex structure, which makes the lignocellulosic biomass difficult to process.[2] The main processes developed to convert biomass into liquid products (bio-oils) are pyrolysis and liquefaction.[6,7] The bio-oils produced by these processes cannot be directly used as transportation fuels. They are complex mixtures of more than 400 different oxygenated molecules, *e.g.* carboxylic acids, aldehydes, alcohols, ketones, esters, ethers, phenols, and carbohydrates. Mainly due to their oxygen-rich composition, bio-oils have some undesired properties for fuel applications: low heating value (less than 50% of those values found for conventional fuel oils), immiscibility with hydrocarbon fuels, thermal and chemical instability, high density and corrosiveness.[8,9] Further details on the properties of pyrolytic oils are described in Section 7.1.2. Overall, in order to become commercially useful, pyrolysis oils must be upgraded to increase their stability, volatility, and to reduce their viscosity through oxygen removal.[10]

Many other studies have been conducted in the production of ester-based biodiesel, which is produced by esterification of vegetable oils with methanol (see, *e.g.*, refs. 11–18). Apart from the importance of adding biodiesel to the oil-derived transportation fuels pool, the utilization of biodiesel still faces serious problems, such as its low oxidative stability, low energy content and cold-flow properties. On this basis, a different processing concept must be developed to transform pyrolysis oils and nonedible oils into a high quality gasoline-kerosene-diesel fuel or diesel blend stock that is fully compatible with oil-derived diesel fuel. Thermal and catalytic routes have been explored to obtain deoxygenated biofuel from triglycerides,[19,20] but these processes provide low C-atom economy, and poor selectivity due to uncontrollable side reactions, such as cracking or polymerization of the hydrocarbons in addition to the formation of undesired hydrocarbon gases, which decreases the yield of liquid fuel products. Thus, the emerging technology for the conversion of triglycerides and other biomass-derived oxygenates into O-free hydrocarbons is catalytic hydrotreating (HDT) or hydrodeoxygenation (HDO). Several reports show that the HDT process utilized worldwide in oil refineries can be used to obtain hydrocarbons in the range of gasoline–kerosene–diesel from pure vegetable oils or from mixtures of vegetable oil and diesel using the conventional catalytic HDT technology and commercially available catalysts.[21–27]

One of the major concerns for the application of the hydrotreatment is the extent of the biomass-derived liquid formed under conventional hydro-processing conditions. Blending of the hydrodesulfurized diesel with vegetable oils, fats and O-containing biomass derived liquids, and a subsequent hydrotreatment of the blend would offer flexibility in the production scheme in addition to a very good alternative to produce a mixture of oil-derived diesel with renewable diesel. Hydrodeoxygenation (HDO) is a promising methodology to remove oxygen from the biomass derived liquids. HDO processes are typically performed in batch[28] or flow reactors[29,30] in the presence of catalysts. In HDO molecular hydrogen is consumed in the reduction of C–O to C–H bonds, decreasing the O content through the formation of water. Since the upgraded products contain less oxygen, they show higher heating values, and are chemically more stable than in their unrefined state.[31]

HDO reactions are performed at high temperatures, under high pressures of H_2, in the presence of a heterogeneous catalyst.[32] The reactions occurring are elimination of oxygen as water, elimination of nitrogen as ammonia, and hydrogenation–hydrocracking of large molecules, leading to the production of hydrocarbons in the diesel range. The reaction conditions and the catalysts first explored were similar to those used in the petroleum-refining HDT processes (*e.g.*, sulfide NiMo, CoMo catalysts). The extent of HDO can be modulated from elimination of more reactive functions, such as carbonyl, olefins, and carboxyls, to complete refining with a maximal hydrocarbon yield that will be a function of the extent and how the oxygen atoms are removed. Lastly, the residual amount of oxygenates is a function of the HDO severity (*i.e.* temperature, H_2 pressure, and residence time).[33]

Several recent reviews reported the boom of research activity concerning the upgrade of biomass-derived oils.[34,35] This chapter focuses on the use of metal-sulfide catalysts in hydrotreatment processes of renewable bio-oils and nonedible feedstocks for the production of drop-in liquid fuels (second-generation biofuels).

8.2   Catalytic Hydrodeoxygenation (HDO)

Hydrodeoxygenation (HDO) of pyrolysis oils, low-range vegetal oils and animal fats uses conventional technology existing in refineries. In the HDO process, oxygen is removed by catalytic hydrogenation according to the general equation:

$$C_nH_mO + H_2 \rightarrow C_nH_m + H_2O \qquad (8.1)$$

In this equation C_nH_m represents a general hydrocarbon product. The reaction is exothermic and calculations have shown an average heat of reaction in the order of 2.4 MJ kg^{-1} when using pyrolysis oil. The formation of water leads to phase separation into an organic and an aqueous layer. Regarding the operating conditions, high pressures of H_2 (7.5–30 MPa) are generally used.[36,37] The high pressure contributes to a high solubility of H_2

in the oil, and thereby a high availability of hydrogen in the vicinity of the catalyst. As a result, the reaction rate is increased and problems related to coking are alleviated.[38]

It should be emphasized that most of the HDO studies have been conducted on model molecules representative of the complex mixtures of pyrolysis oils derived precursors (or fats). While many studies have been focused on catalyst developments, recent studies have been undertaken to demonstrate scale processing. Since the first studies in the early 1980s, the growing concern about the upgrade of low-quality coal and biomass-derived liquids led to many catalytic, kinetic and mechanistic studies.[31,39] HDO has been performed at high pressures of H_2, in the presence of catalysts based on those typically relevant to conventional hydrotreating (*e.g.* sulfided CoMo and NiMo). The extensive work that has been undertaken over the past two decades in the field of catalytic hydrotreating of biomass-derived liquids is documented in recent reviews.[34,35,40]

8.2.1 Sulfide Catalysts

Metal-sulfide catalysts have been studied extensively because they display good hydrodeoxygenation activity. These catalysts require sulfidation at temperatures slightly above 573 K in the presence of a sulfiding agent. Moreover, an additional sulfiding agent is required to be added into the feed stream to maintain the sulfidation level of the metal-sulfide catalyst because of the negative effect of the water on the catalytic activity.[41] Hence, it is very important to develop a new kind of nonsulfur-based, nonexpensive and highly active catalyst for the HDO of vegetable oils. However, compared to sulfided CoMo and NiMo on Al_2O_3, relatively few studies have been attempted with noble-metal-based catalysts.[42–44]

8.2.1.1 The Active Phase and its Nature

The nature of the active phase in unpromoted and promoted MoS_2 catalysts has been widely investigated.[45,46] For unpromoted MoS_2 catalysts, it has been proposed that coordinatively unsaturated sites (CUS) or exposed Mo ions with sulfur vacancies at the edges and corners of MoS_2 crystallites are active in hydrogenation and hydrogenolysis reactions of C=C, C–N and C–S bonds. The basal planes of MoS_2 layers are inactive for the adsorption of molecules, and thus probably less important for hydrotreating reactions. For Co- or Ni-promoted catalysts, several different structural models (*e.g.* monolayer model, intercalation model, contact synergy model, Co(Ni)–Mo–S phase model, and catalytic Co-site model) have been proposed to explain the role of the promoter and its location in the catalyst. Among these, the Co(Ni)–Mo–S model proposed by Topsøe[45] is now widely accepted.

The Co(Ni)–Mo–S structures are small MoS_2 nanocrystals with the Co or Ni atoms located at the edges of the MoS_2 layers (Figure 8.1).[45] The relative number of Co atoms present as Co–Mo–S phase was found to correlate

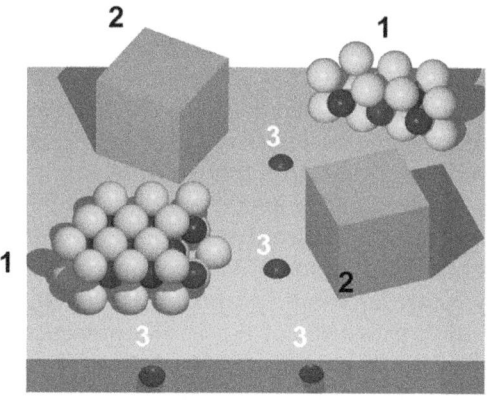

Figure 8.1 Schematic representation of the CoMoS model under reaction conditions. Co is present in three different phases: (1) Active CoMoS nanoparticles; (2) A thermodynamically stable cobalt sulfide (Co_9S_8); (3) Cobalt dissolved in the Al_2O_3 support. Only the CoMoS nanoparticles are catalytically active.
Reproduced with permission from ref. 45.

linearly with the HDS activity. Further studies[47] on the structure–activity correlation of these catalyst systems identified two types of Co–Mo–S structures: type I and type II with the later much more active than the former.[48] Co–Mo–S structures of type I were found to be incompletely sulfided and have some remaining Mo–O–Al linkages to the support.[48] The presence of such linkages was related to the interaction developed during calcination between Mo and surface alumina OH groups. Co(Ni)–Mo–S structures of type II show weaker and fully sulfided support interactions. The underlying MoS_2 in type II Co–Mo–S phases is less dispersed, consisting of multiple slabs not linked to the support. Further studies[49] showed that the degree of stacking in MoS_2 layers and Co–Mo–S structures is controlled by support properties and preparation parameters.[49,50] These Co–Mo–S will have a high MoS_2 edge dispersion that can accommodate more Co atoms to form higher active single slab type II Co–Mo–S phase.[47]

8.2.1.2 The Origin of Synergy between Co(Ni) and Mo

A model correlating stacking degree and selectivity properties for nonpromoted and supported MoS_2 catalysts was proposed by Daage and Chianelli.[51] Two types of sites were distinguished based on their location on the layers: the "rim" sites on the top and bottom layers capable of hydrogenating and cleaving C–S bonds, and the "edge" sites located at the outer edges of the interior layers that can only cleave C–S bonds. The specific location of "rim" sites on exterior layers is responsible for their hydrogenating character by facilitating the adsorption of dibenzothiophene (DBT) parallel to the catalytic surface – a requirement for the hydrogenation step.[51]

Regarding the hydrogenation (HYD) and direct desulfurization (DDS) pathways of the HDS of DBT, it was shown that cobalt promotion sharply enhances the quality of the sites involved in C–S bond cleavage. Following the identification of Co–Mo–S phase structures as the catalytically active phase, many theories were proposed to explain the electronic origin of the synergy between Co and Mo in the Co–Mo–S phase in enhancing the HDS reaction. Norskov *et al.*[52] approached the problem using *ab initio* calculations. Their basic idea was that the binding energy of a sulfur atom was the key variable in order to explain the periodic trends and promoted systems. They found a monotonic relationship between sulfur bond strength per sulfur atom and the specific catalytic activity of the sulfides using literature data. The weaker the metal–sulfur bond strength, the higher the HDS activity. Consequently, the HDS rate would be proportional to the number of active sites (*i.e.* sulfur vacancies) on the catalyst surface, which in turn will be inversely proportional to the metal–sulfur bond strength.

Another view to explain the specific and unique activity of the Co(Ni)MoS phase is related to an electron donation from Co to Mo decreasing the Mo–S bond strength to an optimum range.[53] The electron donation will increase the electron density in the antibonding orbitals related to the Mo–S bond, thus weakening the Mo–S bond strength. Concurrently, this electron transfer will decrease the occupancy of antibonding orbitals for the Co–S, strengthening the Co–S bond. The S atom shared between Co and Mo will then present an intermediate metal–S bond strength. This weakening effect of Co(Ni) promotion on the Mo–S bond strength is now well accepted and was confirmed by theoretical calculations as well as experimental techniques.[54,55]

Further insight into atomic structure and morphology of MoS_2 and Co(Ni)–Mo–S phases was revealed by using a combination of scanning tunneling microscopy (STM), high-angle annular dark-field scanning transmission electron microscopy (HAADF-STEM) and density functional theory (DFT) calculations. HAADF-STEM showed atom-resolved images of the catalytically active edges of MoS_2 and Co(Ni)–Mo–S nanoclusters. The Mo-edge was found to have special electronic edge states identified as brim sites.[56] DFT calculations shed light on the metallic nature of the brim sites. The metallic nature of the brim sites means that they may bind S-containing molecules. When hydrogen is available at the neighboring edge sites in the form of SH groups, hydrogen transfer and hydrogenation reactions can take place. Therefore, these brim sites are catalytically active for hydrogenation reactions, whereas direct sulfur removal can take place at both edges.

8.2.1.3 Catalyst Supports

The selection of an appropriate support is an important aspect of catalyst formulation for HDO. In the presence of water, alumina has been shown to be an unsuitable support, since transitional aluminas convert either into boehmite, AlO(OH),[57,58] or into trihydroxide phase (*e.g.* bayerite).[160]

For sulfide NiMo catalysts supported on alumina, it has been shown that not only is alumina transformed into boehmite but also nickel sulfide becomes oxidized.[57] These nickel oxides were inactive with respect to HDO. In addition, they could further block other Mo or Ni sites on the catalyst. By treating the catalyst in a mixture of *n*-dodecane and water for 60 h, a decrease by two thirds of the activity was seen relative to the catalyst treated in *n*-dodecane alone.[57] Furthermore, it was found that about 65% of alumina was covered with phenolic species when saturating it at 673 K in a flow of phenol in argon. The detected surface species were believed to be potential carbon precursors, indicating that a high affinity for carbon formation exists on this substrate. The high surface coverage was linked to the relative high acidity of the alumina support.

As an alternative to alumina, carbon has been found to be a more promising support.[28,59] The neutral nature of carbon is advantageous, since it is less prone to form carbon, compared to alumina. In addition, SiO_2 has been indicated as a prospective support for HDO as it has a general neutral nature; therefore, it shows a relatively low affinity towards intermediates for carbon formation.[60] Similarly, mesoporous silicas show great potential as catalyst supports.[61,62] MCM-41 and SBA-15 mesoporous silicas offer several advantages over alumina as a catalyst support. In contrast to standard alumina supports and similarly to microporous molecular sieves (zeolites), organized mesoporous silicas have very large specific areas that can be used to achieve very high dispersions of the supported active phase[62] and high loadings of active phase. Moreover, their larger pore diameters, as compared to zeolites, make these mesoporous supports excellent candidates for applications in which large organic molecules have access to the well-dispersed active sites located inside the pores. Mesoporous silicas are conducive to performing catalytic transformations on large oxygenated molecules (*e.g.* triglycerides). Nava *et al.*[63] recently investigated the effect of mesoporous silica supports (SBA-15, HMS, SBA-16, DMS-1) on the hydro-processing of low range, nonedible oils. The catalysts supported on SBA-15, SBA-16 and DMS-1 were reported to be more active in HDO reactions than the HMS-supported catalyst.[63]

8.3   HDO of Model Compounds

Different oxygenate model compounds have been used to elucidate catalyst performances, contributing to understanding HDO mechanisms. Most of these studies were performed in fixed-bed reactors under hydrogen pressure within the range of 2–5 MPa and at temperatures in the range of 523–723 K. The most relevant oxygenates and catalysts employed in the HDO reaction are compiled in Table 8.1.[64–83] In most of the studies on hydrodeoxygenation, phenolic-based compounds were used as model molecules.[74,84–89] It is generally accepted that phenolic compounds react on Co(Ni)–Mo(W)–S catalysts through two main pathways: (i) involving direct C–O bond scission (direct deoxygenation, DDO) yielding aromatic products, and (ii), *via*

Table 8.1 Catalysts and main products obtained from the HDO conversion of model compounds.

Molecule	Catalysts	Products	Refs.
Guaiacol	CoMo	phenol, benzene	64–71
Phenol	Mo, CoMo, NiW	cyclohexane, benzene	72–75
Cresol	CoMo	cyclohexane, benzene	76, 77
Esters	CoMo, NiMo	alkanes, alkenes, methanol, carboxylic acids	78, 79
Acids	CoMo, NiMo	alkanes	80, 81
Furans	Mo, NiMo	cyclohexane, C_3–C_4 alkanes	82, 83

hydrogenation of the aromatic ring leading to cyclohexenes and cyclohexanes (hydrogenation, HYD).[74,84–89] When cobalt is used as a promoter, the DDO was found to be the main route of hydrodeoxygenation of phenolic compounds,[74] whereas in the presence of nickel as a promoter, the HYD was always the main reaction pathway.[74,87,88]

8.3.1 Guaiacol

Guaiacyl species constitute one of the primary structures occurring in either native or technical lignins in addition to pyrolysis oils.[93,94] Indeed, several guaiacyl species, *e.g.* guaiacol ($C_7H_8O_2$), vanillin ($C_8H_8O_3$), and eugenol ($C_{10}H_{12}O_2$), can be obtained by pyrolysis[90] and liquefaction of lignin.[91,92] Therefore, it appears that an effective upgrading process for guaiacyl species should facilitate the design of a lignin-conversion platform. The most common catalysts employed to remove oxygen from guaiacyl species are sulfided CoMo phases supported on alumina.[64] The extensive work developed by Delmon and coworkers[65–67] demonstrated that HDO reactivity of supported CoMo and NiMo catalysts can be controlled by an appropriate catalyst design. More recently, Pinheiro *et al.*[68] examined the coprocessing of straight run gas oil (SRGO) with pyrolysis model compounds, including guaiacol in a hydrotreating pilot plant, in the presence of sulfided CoMo/Al$_2$O$_3$ catalysts. These results demonstrated that it may be possible to simultaneously treat oxygenates and hydrocarbons mixtures.

Lin *et al.*[69] reported activity data on the HDO of guaiacol as a function of reaction temperature and time on sulfided CoMo and NiMo catalysts. Within the temperature interval of 573–673 K, the guaiacol conversion was rather low (below 15%), yielding an insignificant amount of products.[70,71] Increasing temperature improved the product yields and decreased coke formation. The main products included: (i) mono-oxygenated species (*e.g.* methoxybenzene, methylphenol, and phenol) and (ii) hydrocarbons (*e.g.* benzene, cyclohexene, and cyclohexane). In addition, minor unidentified compounds might include methane, methanol, and heavier products (*e.g.* cyclohexylbenzene and methylcyclopentane).[71] In the presence of a NiMo catalyst, the selectivity to cyclohexane was markedly high (*ca.* 80%) at 673 K for 40 min and 623 K for 60 min, whereas phenol and methylphenol were the main products obtained from the experiment using a CoMo catalyst

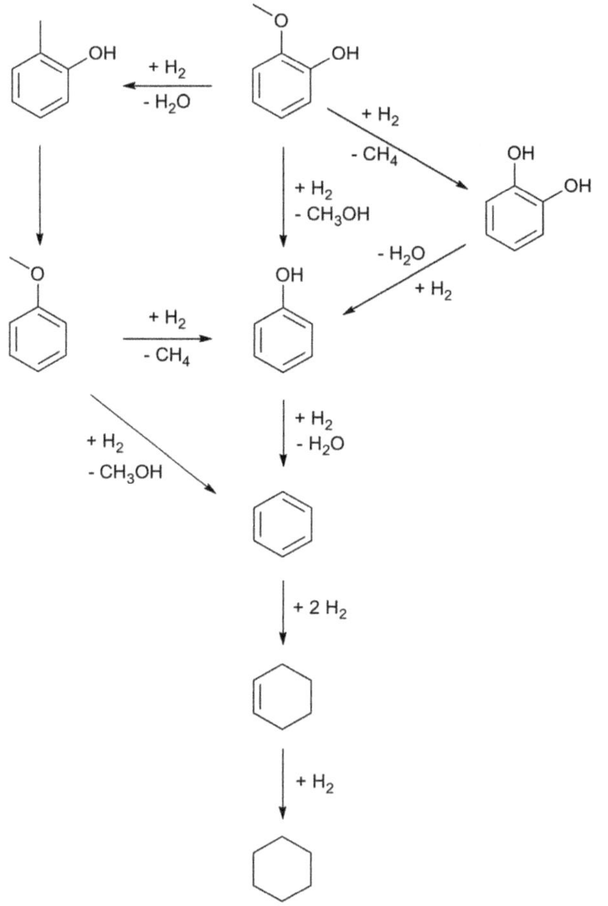

Figure 8.2 Simplified mechanism of the hydrodeoxygenation of guaiacol. Adapted and reproduced with permission from ref. 69.

(Figure 8.2). This finding implies that sulfided NiMo had a stronger influence on the benzene saturation, compared to sulfided CoMo. Both catalysts produced some coke, but the amount was larger on the NiMo catalyst. Sulfur stripping from catalyst surface was observed for all trials using sulfided catalysts. With increasing experimental severity, the sulfur contents in liquid samples increased from 0.03 to 0.44 wt% for CoMo and 0.03 to 0.15 wt% for NiMo. The sulfur-containing species formed by the HDO of guaiacol may be methanethiol, dimethyl sulfide, cyclohexanethiol, and methylthiocyclohexane.[69]

The mechanism of guaiacol HDO on sulfided CoMo and NiMo catalysts proceeds through demethylation, demethoxylation, and deoxygenation. The simplified mechanism of the HDO of guaiacol is shown in Figure 8.3. Early studies suggested that demethylation, which produces catechol through hydrogenolysis of the methyl–oxygen bond, was the onset of guaiacol HDO.[95]

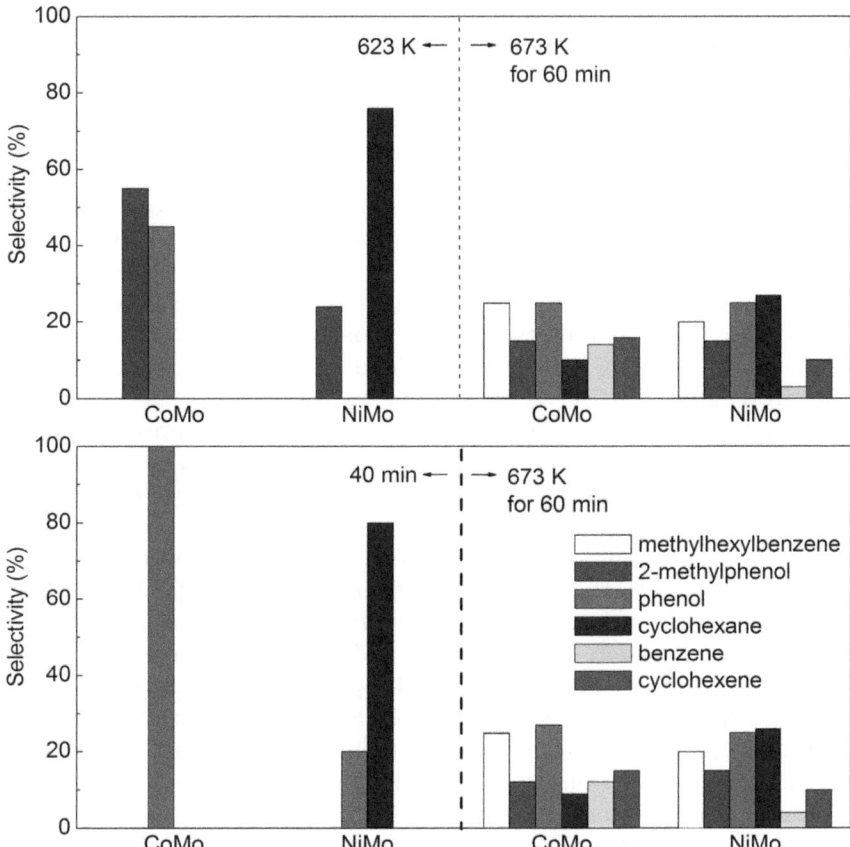

Figure 8.3 Product selectivity of sulfided CoMo and NiMo catalysts. Reproduced with permission from ref. 69.

However, as a substantial concentration of phenol was detected at the initial stage of the reaction,[96] hydrogenolysis of the aromatic carbon–oxygen bond *via* demethoxylation may be operative. This hypothesis was confirmed by Bui *et al.*[70] They identified both phenol and methanol at the initial stage of the reaction in the presence of sulfided CoMo catalysts. Therefore, it is likely that demethylation, demethoxylation, and deoxygenation proceed concurrently in the proposed system.

The mono-oxygenates (*e.g.* phenol, methoxybenzene, and methylphenol) can be subsequently converted into benzene. In a hydrogen-rich environment, benzene can be partially or fully saturated to form cyclohexene and cyclohexane, respectively. Notably, both of these compounds were not detected. A possible explanation might be that catechol is converted during the sampling or handling period. This proposition was confirmed by the significant amount of coke formed under the initial conditions (573 K, 60 min and 673 K, 20 min).[86] Zhao *et al.*[97] reported similar observations using transition metal phosphides in the gas-phase HDO of guaiacol performed in a continuous fixed-bed system. Indeed, no catechol could be detected in experiments carried out using long contact times.

8.3.2   Phenol

Phenol has been extensively used as a model compound for the HDO of pyrolysis oils,[98–102] since it is the main final product of other phenol transformations (*e.g.* alkoxy and methoxyphenols). It presents the most difficult C–O bond to cleave, and is particularly resistant to HDO.

Unsupported MoS_2 catalysts have been used for HDO of phenol.[99] The amorphous MoS_2 with a bent and folded multilayered structure has a significantly greater HDO activity than the well-crystalline and highly multilayered MoS_2. HDO of phenol in the presence of an amorphous MoS_2 catalyst achieved conversion (71 mol%) almost 2.4-fold higher than that with a crystalline MoS_2 catalyst. The products of phenol HDO over the two different MoS_2 catalysts were the same (*i.e.* cyclohexanone, benzene, cyclohexene and cyclohexane) but varied in their relative proportions. Benzene, which is formed *via* the direct deoxygenating route was 3.2-fold more prevalent using amorphous MoS_2 than with crystalline MoS_2, whilst cyclohexene and cyclohexane were 1.6-fold and 3.7-fold more prevalent with crystalline MoS_2 than with amorphous MoS_2 catalyst, respectively. Thus, the product distribution obtained from phenol HDO depends strongly upon the structure of catalyst. Benzene is the main product of phenol HDO by amorphous MoS_2 catalysis in which the product selectivity increases in the order: cyclohexanone < cyclohexene < cyclohexane < benzene. However, the catalysis with crystalline MoS_2 shows product selectivity increasing in the order: cyclohexanone < benzene < cyclohexene < cyclohexane. Hydrodeoxygenation (HDO) of phenol and methyl heptanoate separately and as mixtures was carried out over a sulfided NiMo catalyst at 523 K[102] in order to compare the HDO of aromatic and aliphatic reactants. The conversion of phenol was suppressed in the presence of methyl heptanoate, whereas the conversion of methyl heptanoate was practically unaffected by phenol. In addition, distributions of the hydrocarbon products were different for reactants in the mixture and the reactants tested separately. Some experiments were also carried out in the presence of a sulfur additive. The sulfur additive changed the product distribution of the separate components more than that of the mixture. These findings indicate that reduction (including hydrogenation) reactions occur on coordinatively unsaturated sites (CUS) independently of the aromatic or aliphatic character of the component. Nonetheless, the sulfur additive also adsorbs on CUS, and thus competes with other reactants that have an affinity to CUS. Decarbonylation and acid-catalyzed reactions are, instead, proposed to occur on sulfur-saturated sites.[102]

Phenol conversion was also investigated in the presence of Ni, W and NiW catalysts supported on active carbon at temperatures of 423–573 K and a H_2 pressure of 1.5 MPa.[75] Activity of all catalysts increased with reaction temperature, with the exception of the W(Si) and W catalysts. In order to compare the activities in the absence of deactivation, the steady-state phenol conversions at 423 K were compared (Figure 8.4(a)). It is noteworthy that carbon support shows a low activity in the reaction at 423 K, but drops at higher temperatures as no phenol adsorption is expected above 423 K. Regardless of the tungsten precursor, Ni incorporation to the W/C catalyst series leads to a marked synergistic effect in all catalysts. Such an effect took place even when tungstic acid (H_2WO_4) was utilized as the tungsten precursor. The nickel-promotion effect becomes more evident at high reaction temperatures. Furthermore, the effect is more pronounced on catalysts prepared using silicotungstic (W(Si)) and phosphotungstic (W(P)) acids.

Figures 8.4(b) and (c) show the influence of temperature on the phenol conversion and selectivity in the HDO reaction over Ni–W(Si) and Ni–W(P) catalysts. An increase of reaction temperature up to 523 K led to full phenol conversion and high selectivity to cyclohexane. Simultaneously, the selectivity to cyclohexanol decreased. Further increase in temperature from 523 to 573 K enhances the formation of other products (benzene, methylcyclopentane and cyclohexene).

The activity for HDO of 2-ethylphenol, another model molecule representative of oxygenated compounds present in pyrolysis oil, was evaluated on sulfided (Ni or Co)Mo catalysts supported on alumina under conditions close to those of a conventional hydrotreatment process (*e.g.* 7.0 MPa and 613 K).[101] For sulfided catalysts, the overall promoting effect of nickel was slightly higher than that of cobalt. The transformation of 2-ethylphenol was also studied on CoMo and NiMo catalysts and the results were compared to those obtained from experiments with the unpromoted Mo catalyst under the same experimental conditions. The initial conversion over CoMo was 39 mol% and decreased to 36 mol% after 35 h on stream, indicating a good stability regardless of the carbon content deposited during the reaction. The NiMo catalyst was a slightly more active than CoMo, reaching an initial conversion of 45 mol%. Seemingly, the NiMo active phase was less stable than the CoMo phase during the HDO of 2-ethylphenol. Indeed, after 35 h on stream, a conversion of 38 mol% was achieved in the presence of NiMo. Factors accounting for the deactivation of NiMo may be coke deposition and desulfurization of the NiMo active phase.

By examining the DDO/HYD ratios obtained by Mo catalysts,[99] the HDO of phenol was found to mainly occur *via* two different pathways. The DDO pathway is the primary route of HDO taking place in the presence of an amorphous MoS_2 catalyst, while the HYD pathway is the favored route of HDO occurring on a crystalline MoS_2 catalyst. Upon adding the Co promoter into the amorphous MoS_2 catalyst, the activity and selectivity of the catalysts markedly changed. Phenol conversion increased up to 98 mol%, while the selectivity to benzene increased to 80 mol% for Co–MoS_2. Clearly, Co–MoS_2 catalysts enhanced the DDO pathway, with benzene as the main product. Overall, the reaction scheme for the phenol HDO on these amorphous catalysts is likely to include both the DDO and HYD routes. The improved

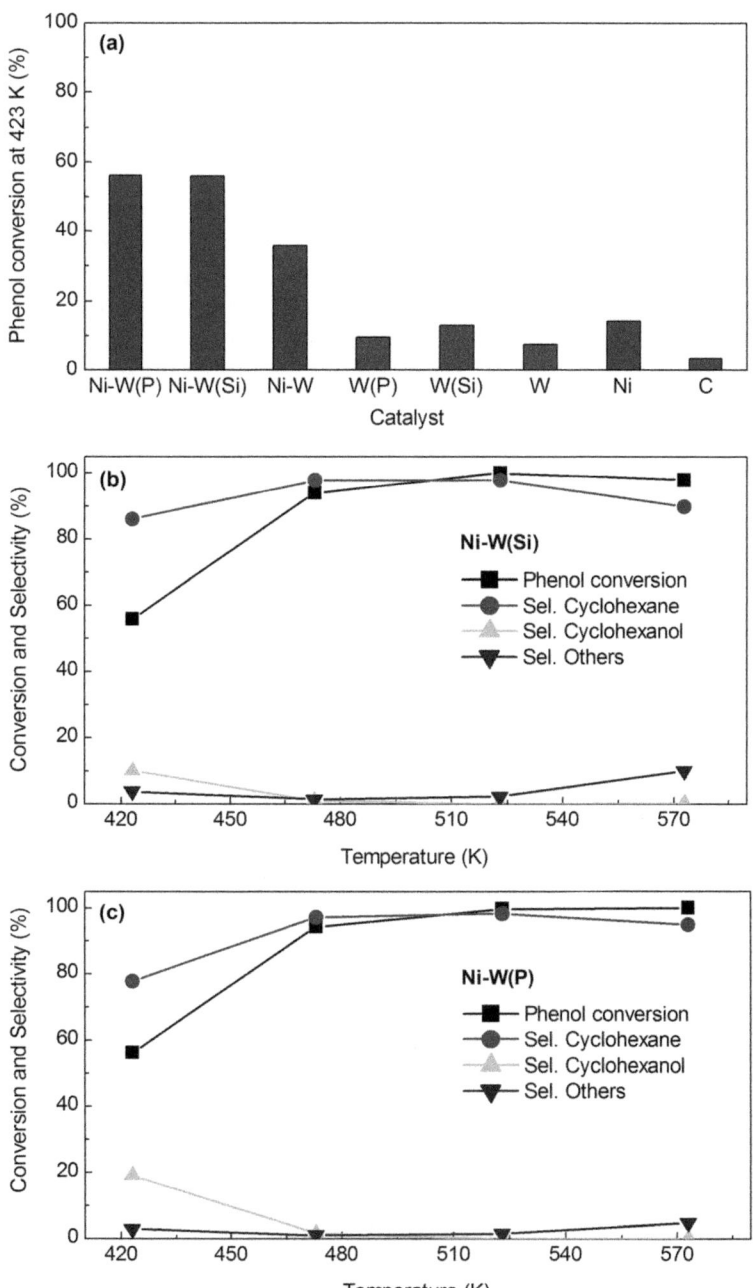

Figure 8.4   (a) Comparison of several catalysts in the steady-state phenol conversion (Reaction conditions: $T = 423$ K, $P(H_2) = 1.5$ MPa and WHSV $= 0.5$ h^{-1}. Influence of temperature on the phenol conversion and selectivity in HDO of phenol (WHSV $= 0.5$ h^{-1}) over (b) Ni–W(Si); and (c) Ni–W(P) catalysts.
Reproduced with permission from ref. 75.

catalytic activity seen in the presence of amorphous Co–MoS$_2$ catalysts was essentially due to the enhanced rate of DDO route. The reaction temperature plays an important role in the product selectivity of the phenol HDO reactions. The pseudo-first-order rate constants for the HDO of phenol increased with temperature.[99] However, they decreased at temperatures above 623 K. In the HDO of phenol, high reaction temperatures (633–653 K) favored the DDO route, while low temperatures (563–593 K), the hydrogenation route.

It is generally accepted that the active sites in sulfided catalysts are sulfur vacancies, *i.e.* CUS, present on the edges of the MoS$_2$ phase.[103] DFT calculations predicted the creation of sulfur vacancies by removal of H$_2$S to be more favorable on the metallic edge than on the sulfur edge over the unpromoted catalyst.[104] In addition, taking into account the predictions by *ab initio* calculations,[105,106] the shape of MoS$_2$ slabs is a bulk-truncated hexagon exposing more metallic edges than sulfur edges. As a first approach, the active sites of the unpromoted catalyst for 2-ethylphenol deoxygenation could be S vacancies, most likely located on the metallic edge of the MoS$_2$ slabs. These active sites would be similar to those accounting for both HYD and DDO pathways. The main difference between these two routes is probably the adsorption mode of the oxygenated molecule on the active site. A DDO site would adsorb 2-ethylphenol through its oxygen atom, whereas the HYD sites would adsorb 2-ethylphenol parallel to the catalyst surface through interactions with the aromatic ring.

Figure 8.5 represents the proposed mechanism for the DDO pathway.[99] Upon heterolytic dissociation of H$_2$, an S–H and a Mo–H group are formed. Accordingly, 2-ethylphenol may adsorb through its oxygen atom on the sulfur vacancy. In the same way, using DFT calculations, it has recently been reported that the η^1 adsorption of furan by the oxygen atom on the metallic edge is more favored than on the sulfur edge.[107] The addition of a proton

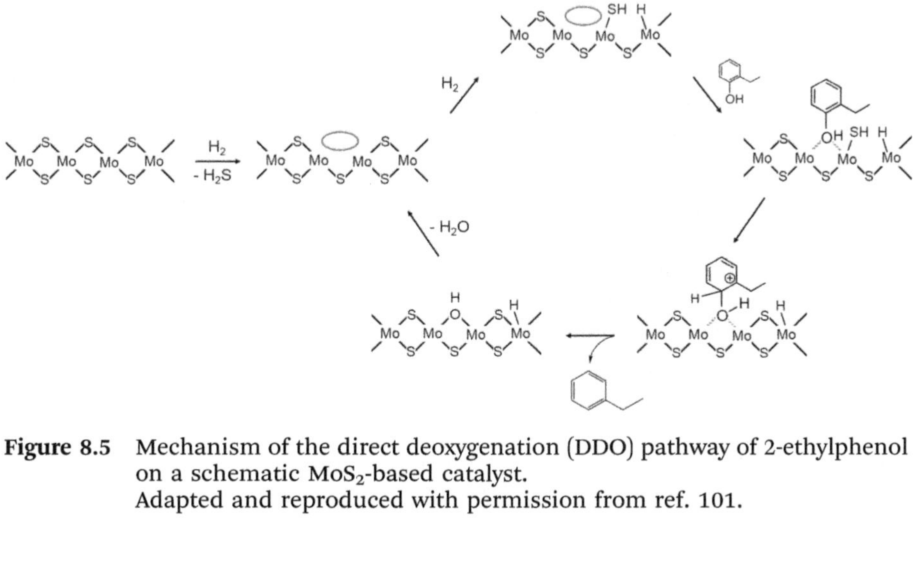

Figure 8.5 Mechanism of the direct deoxygenation (DDO) pathway of 2-ethylphenol on a schematic MoS$_2$-based catalyst.
Adapted and reproduced with permission from ref. 101.

species to the adsorbed oxygenated molecule leads to an adsorbed carbocation. This intermediate can directly undergo C–O bond cleavage, and the aromatic ring is regenerated leading to ethylbenzene. The vacancy is recovered by removal of water. As for phenol, the HDO of 2-ethylphenol takes place by two parallel routes: the HYD pathway yielding mainly ethylcyclohexane, and the DDO pathway yielding only ethylbenzene. For all the studied catalysts, the HYD pathway was the main route of the deoxygenation.

Compared to sulfided Mo, promotion by Ni markedly favors hydrogenation activity of the catalyst by a promoting factor of 3.4, whereas Co promotion mainly enhances the direct C–O bond scission activity by a promoting factor of 3.8.[101] Therefore, the highest DDO/HYD selectivity was obtained by the CoMo catalyst. The promoting effects were explained by an increase of the number of active sites, some of them being new sulfur vacancies. The HYD pathway implies the adsorption of the aromatic ring occurring parallel to the catalytic surface, whereas the DDO pathway requires the adsorption through an oxygen atom. Both adsorption modes could occur on S vacancies, which are located on metallic and sulfur edges of promoted catalysts, and likely only on metallic edges of an unpromoted catalyst. As the sulfur vacancies are probably sparse over the latter, fully sulfided metallic edges (brim sites) present on the basal plane of the MoS_2 slabs could also participate as active sites in the HYD pathway of the unpromoted catalyst. Over promoted catalysts, metallic edges could be active in the HYD pathway, while sulfur edges are more likely responsible for the DDO pathway.

The acidity of the alumina plays a role in the transformation of 2-ethylphenol leading to disproportionation and isomerization products. The disproportionation supposedly takes place through a bimolecular cationic mechanism. In turn, the isomerization occurs by an intermolecular mechanism involving transalkylations. Over the sulfided phases, the products are also deoxygenated *via* both the HYD and DDO pathways.

8.3.3 Other Oxygenated Compounds

Aryl ethers are of great importance because of their high stability, given by a much greater bond strength of the C_{AR}–O bond than that of the C_{AL}–O bond (C_{AR} = aromatic carbon, C_{AL} = aliphatic carbon).[108–110] Artok *et al.*[111] compared dinaphthyl ether with diphenyl ether HDO in the presence of a MoS_2 catalyst at temperatures between 598 and 648 K under a H_2 pressure of 6.9 MPa. HDO of dinaphthyl ether involved HYD of the substrate and O-containing intermediates, followed by direct dehydroxylation. However, the HDO of diphenylether initially led to benzene and phenol. The HDO was completed by the conversion of phenol into benzene and cyclohexane. Cyclohexane underwent isomerization to methylcyclopentane. Similarly, Petrocelli and Klein[112] also reported phenol and benzene as primary products of the diphenyl ether HDO in the presence of sulfided CoMo catalyst under 7.0 MPa of H_2. At higher conversions (above 573 K), phenol was converted into benzene and cyclohexane.

Esters can be fully hydrodeoxygenated on NiMo and CoMo catalysts supported on alumina at temperatures in the range of 523–573 K and a H_2 pressure of 1.5 MPa,[79] rendering C_5–C_7 alkanes and alkenes, methanol and carboxylic acids. Moreover, the deoxygenation of heptanoic acid as well as of methyl heptanoate was also investigated on sulfide NiMo and CoMo catalysts; the NiMo catalysts displayed higher HDO activity when compared with CoMo counterparts.[81]

Laurent and Delmon[33,113] used 4-methylacetophenone (4MA) and diethylsebacate (DES) to study the HDO of ketones and esters.[113] For these substrates, coke formation was less severe than that for guaiacol.[114] The carbonyl group of the 4MA could be readily hydrogenated to a methylene group at temperatures as low as 473 K in the presence of sulfided CoMo and NiMo catalysts supported on alumina. 4-Methyl-1-ethylbenzene was formed as the major product. While the decarboxylation and HYD of the carboxylic group to a methyl group occurred simultaneously, a temperature of at least 573 K was required for HYD of the carboxylic group. At reaction temperatures of about 553 K, the HDO of decanoic acid (DEC) and ethyldecanoate (EDEC) mainly rendered *n*-nonane and *n*-decane with selectivity ratios nonane/decane of 1.5 and 1.1, respectively. In addition, Afonso *et al.*[115] showed that under severe hydrogenation conditions (673 K and 12.5 MPa), the conversion of carboxylic groups into methyl groups prevails over decarboxylation.

8.4   HDO of Pyrolytic Oils

While removal of sulfur (HDS) and nitrogen (HDN) from of oil-derived fractions is a well-established refinery process, oxygen removal from pyrolysis oils is at its early stages. The optimal performance of current industrial hydrotreating catalysts for petroleum feedstocks cannot be reproduced over HDO of pyrolysis oils, thus making the coprocessing in the existing hydrotreating units difficult. High HDO conversion of pyrolysis oils can be achieved in a single step under high-pressure conditions.[116] This approach is well exemplified by an oxygen removal higher than 95% from wood liquefaction oil or pyrolysis oils at 573 K reached in the presence of a sulfided CoMo catalyst.[117]

Pyrolysis oils are prone to polymerization. To improve the stability upon storage, a first stabilization step was proposed in order to remove an important fraction of the oxygen content associated with highly reactive groups (*e.g.* aldehydes, ketones, alcohols, *etc.*).[118] Hence, the HDO of the stabilized pyrolysis oils should become similar to the phenol HDO. This two-stage process was developed at the Pacific Northwest National Laboratory (PNNL).[10] In this improved process, the oil is stabilized in a first reactor that operates at temperatures lower than 573 K and under high pressures of hydrogen. In sequence, the products are fed to a second reactor in which a deeper deoxygenation is achieved at temperatures in the range of 623–673 K by using typical petroleum hydrotreating (alumina-supported NiMo or CoMo) catalysts.[119]

Other examples of two-stage treatments have been reported for the upgrade of bio-oils originated from low-range edible oils[120] or cellulose.[121] In the case of cellulose hydropyrolysis, single- and two-stage treatments have been developed using a fixed-bed reactor. In the single-stage concept, a FeS catalyst was used, whereas for the two-stage test a commercial $NiMo/Al_2O_3$ supported catalyst (Criterion 424) was employed.[121]

Although the two-stage process is a promising option, it is uneconomical due to several factors, *e.g.* the high costs associated with the large amount of hydrogen consumed, the low product yields, the need of further upgrade of the products in a refinery, and the corrosiveness of the raw oil.[119] However, some analyses suggest that lower costs are achieved by exporting partially deoxygenated liquids to the refineries instead. Indeed, hydrogen consumption and hydrotreating capital costs are markedly reduced by limiting the severity of HDO in order to leave about 7 wt% oxygen content and avoid the hydrogenation of aromatics in the pyrolysis oil. In Europe, a large research project "BioCoup" has been developed aimed at evaluating the potential of the coprocessing of pyrolysis-oil-derived products in petroleum refineries.

8.5   HDO of Vegetable Oils

Obviously, the reaction mechanism and rate of HDO depend on the type of compound to be hydroprocessed. Reactions using model compounds or directly performed on different fatty acids and their esters or triglycerides have been studied.[23,24,62,122–136] Efforts have been focused on the HDO of vegetable oils/fats in the presence of transition-metal-sulfide catalysts. A list of most catalysts utilized for this purpose are summarized in Table 8.2. Conventional hydrotreating catalysts (*e.g.* CoMo or NiMo) are suitable for hydrodeoxygenation of triglycerides. Unlike reduced Ni catalysts, sulfided Ni–Mo catalysts require high temperatures (503–553 K) to be active for the hydrogenation of C=C bonds. The primary product of hydroprocessing of the low-rank vegetable oils is linear paraffins, which can be hydroisomerized to high-quality diesel or hydrocracked to kerosene. A variety of vegetable oils (*e.g.* edible, nonedible oils and animal fats) can be processed to yield the same high-quality diesel product. In addition, coprocessing these renewable oils with petroleum-derived middle distillates is possible in conventional refinery units. This is an emerging research field with great potential.

Several authors have been working on the elucidation of the mechanism of the hydroprocesing of vegetable oils into diesel.[137] The main products are aliphatic hydrocarbons that are formed either by full HDO (*i.e.* they contain the same carbon number as the parent acids) or hydrodecarboxylation (HDC, *i.e.* the products show a C-atom number smaller by one unit compared to the corresponding parent carboxylic acid). If the reaction conditions are severe enough, all carboxylic acids (reaction intermediates) are converted

Table 8.2 Feeds, active phases and supports of the sulfided catalysts applied to HDO of edible, nonedible vegetable oils and animal fats.

Feed precursor	Active phase	Support	Ref.
Waste cooking oil	CoMo, NiMo, NiW	Al_2O_3	24
	NiMo	Al_2O_3	122
	CHTC*	-	130
Jatropha oil	NiMo	Al_2O_3	115
Rapeseed oil	NiMo	Al_2O_3	23,124
	CoMo	Al_2O_3	126
	CoMo	MCM-41	62
	CoMo	mesoporous Al_2O_3	126
Rapeseed + diesel	NiMo	Al_2O_3	130
Waste soybean oil	NiMo, NiW	Al_2O_3, SiO_2-Al_2O_3	125
Cotton seed	CoMo	Al_2O_3	24
Palm oil	NiMo	Al_2O_3	127
Sunflower oil	CHC**	-	128
Sunflower oil + HVO	NiMo	Al_2O_3	131
Vegetable oil	Ni, NiMo	Al_2O_3	135
Vegetable oil + diesel	NiMo, NiW	Al_2O_3	133
Vegetable oil + VGO	CHC**	-	134
Olive oil	CoMo	SBA-15, SBA-16, HMS, DMS-1	136

CHTC* = commercial hydrotreating catalyst; CHC** = commercial hydrocracking catalyst

into the hydrocarbons and negligible cracking of the hydrocarbon chains is observed. Metal-sulfide catalysts are extensively studied because of the good HDO activity, but the sulfidation process of these catalysts requires high temperature and a sulfiding agent.

Many hydroprocessing catalysts, both conventional and unconventional formulations have been explored on the conversion of triglycerides present in vegetable oils into hydrocarbons within the boiling range of diesel. Conventional hydroprocessing catalysts are appropriate because they give high yields of diesel fractions. The vast majority of studies dealing with the processing of vegetable oils by hydrotreament have utilized NiMo sulfide supported on alumina.[22,138–141] However, studies exploring the use of CoMo/ Al_2O_3,[24,123,142] commercial hydrocracking catalysts,[124,143] NiMoW[144,145] as well as the use of CoMo and NiMo sulfide phases on supports, such as SiO_2,[124,126] MCM-41,[22,146] and zeolites[24,126,139,147] have also been reported.

Interestingly, CoMo catalysts supported on different mesoporous silicas (DMS-1, SBA-15, SBA-16, HMS) display very high activity in the HDO reaction of low-range olive oil.[63] The influence of the substrate of CoMo catalysts on the total oxygen content after HDO reaction at 553 K, overall pressure of 3 MPa, and time on stream of 5 h is plotted in Figure 8.6(a) (normalized to BET surface area) and Figure 8.6(b) (normalized to mmol of Co and Mo).[63] The results indicate that the CoMo catalysts supported on SBA-16, SBA-15 and DMS-1 are more effective for oxygen removal and more selective towards desirable products than their CoMo/HMS counterpart. The HDO activity of these catalysts is influenced by the support acidity and much less so by the sulfidation degree of surface cobalt species. The selectivity to desired

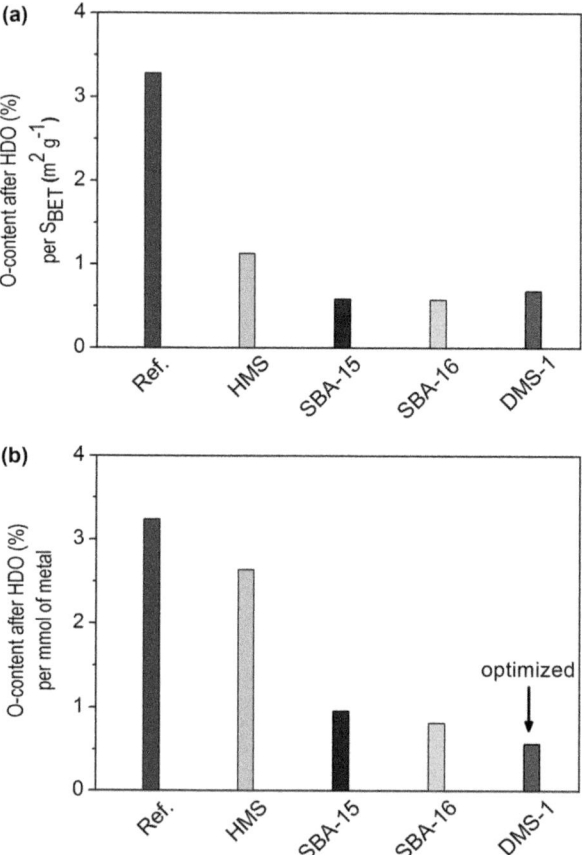

Figure 8.6 Influence of support on the total oxygen content after HDO of bioliquid over sulfided CoMo catalysts ($T = 523$ K, $P = 3$ MPa, WHSV $= 2.7$ h^{-1}, TOS $= 5$ h) normalized to (a) specific BET surface area and (b) mmol of Co and Mo.
Reproduced with permission from ref. 63.

products is influenced by support morphology. The CoMo/SBA-16 catalyst is the most appropriate considering the formation of desirable products (paraffins and alcohols) and the removal of oxygen-containing products and polyaromatics (PAH).

It is generally accepted that the triglycerides are first saturated on their side chain, followed by scission of the C–O bond, leading to the formation of diglycerides, monoglycerides, carboxylic acids and waxes. These intermediates are then transformed into hydrocarbons by hydrotreating. The predominant hydroconverted products are *n*-heptadecane and *n*-octadecane in addition to byproducts, carbon monoxide, propane, carbon dioxide, and water. Due to the acid functionality of the catalyst, isomerization and cyclization of the olefin intermediates can occur, leading to the formation of isoparaffins and naphthenes. Aromatics may also be produced when the process is performed under low-severity conditions.

Under the HDO conditions, the hydrogenation equilibrium may be established rapidly. Then, the ratio of unsaturated-to-saturated free fatty acids is directly proportional to temperature, while being inversely proportional to the H_2 pressure. HDO and decarboxylation reactions determine the overall rate of the triglycerides conversion into hydrocarbons. The reactions (8.2)–(8.5) below were proposed to account for the catalytic transformation, *i.e.* hydrodecarboxylation (8.2), HDO (8.3), CO formation (8.4) and reversed steam reforming of methane (8.5):

$$(CH_2)_2CH[(CH_2)_{16}]_3(COO)_3(CH_3)_3 + 3H_2 \rightarrow 3C_{17}H_{36} + C_3H_8 + 3CO_2 \quad (8.2)$$

$$(CH_2)_2CH[(CH_2)_{16}]_3(COO)_3(CH_3)_3 + 12H_2 \rightarrow 3C_{18}H_{38} + C_3H_8 + 6H_2O \quad (8.3)$$

$$CO_2 + H_2 \rightleftharpoons CO + H_2O \quad (8.4)$$

$$CO_2 + 4H_2 \rightleftharpoons CH_4 + 2H_2O \quad (8.5)$$

The predictions of this thermodynamic approach were compared with the experimental observations assuming a total hydrogenation of tristearate to hydrocarbons. The predictions suggested that C_{18} hydrocarbons were the main products and that their concentration was influenced by temperature and by H_2 pressure. On the basis of product distribution achieved in the presence of a sulfided NiMo catalyst between 573 and 723 K, 5 MPa and a liquid hourly space-velocity (LHSV) of 5 h^{-1}, a mechanism of HDO of triglyceride was proposed.[131] Straight-chain hydrocarbons, as primary products, may continue to react *via* isomerization and hydrocracking. The extent of these reactions depends on the structure of catalyst (*e.g.*, they are favored by bifunctional catalysts). Boda *et al.*[41] demonstrated that the type of catalyst may influence the mechanism of conversion of triglycerides into hydrocarbons in experiments with tricaprylin and caprylic acid and NiMo catalysts. The reactions proceed stepwise beginning with hydrogenolysis to carboxylic acid and propane, followed by the HDO of the carboxylic acid. The HDO route includes both decarbonylation and carboxyl group reduction: the HDO involved the consecutive H_2 addition and dehydration giving predominantly C_8 alkenes, alkanes and water. Decarbonylation proceeded *via* a formic acid intermediate, which rapidly decompose to CO and H_2O.

8.6 Coprocessing of Biofeeds

The coprocessing of biofeeds with petroleum fractions of a similar boiling range under hydroprocessing conditions has been explored with some positive results.[148,149] For these processes, biofeeds of vegetable origin are the most suitable. Model oxygenates (*e.g.* 2-propanol, cyclopentanone, anisole, guaiacol, propanoic acid and ethyldecanoate) have been used as additives (0.5 wt% oxygen) to SRGO of petroleum origin to study their effect on HDS and HDN.[68] Experiments conducted at 603 K and under a H_2 pressure of 5 MPa with a sulfided CoMo catalyst showed little inhibiting effect of water on the HDS activity. The inhibition was mainly caused by the hydrogen depletion due to methanation reaction of CO and CO_2. For coprocessing of guaiacol with SRGO over CoMo catalyst above 593 K, full HDO of guaiacol was achieved and HDS proceeded without inhibition.[150]

A NiMo catalyst was employed for hydroprocesing two mixtures of diesel and rapeseed oils containing 10 and 20 wt% of rapeseed oil.[151] Under the reaction conditions (593–653 K, 3–5 MPa H_2 and LHSV of 2 h^{-1}), high conversion of vegetable oil into diesel-fuel hydrocarbons was obtained. Under similar conditions, the HDO of the mixture of SRGO and rapeseed oil in the presence of a sulfide CoMo catalyst was investigated by Aribert *et al.*[152] Apparently, additional hydroprocessing may be required in order to meet the cold-flow property specifications of commercial diesel fuel. Pretreatment of the pyrolysis feedstock may improve efficiency of the subsequent upgrading step as it was reported by Lappas *et al.*[153] In this study, the primary pyrolysis oil quality was enhanced by thermal treatment without the use of a catalyst in the presence of H_2. In this case, almost 85% oxygen removal from the pyrolysis oil could be achieved with the final liquid containing only 6.5 wt% of oxygen. The fractions of gasoline, diesel and VGO (vacuum gas oil) derived from the upgraded bioliquid were suitable for blending with the corresponding oil fraction for further upgrading. When blending with the fluid catalytic cracking (FCC) feed, a decrease in the oxygen content down to about 20% was adequate for achieving a smooth hydroprocessing operation.[154]

8.6.1 Commercial Developments

Several commercial plants have been designed for the hydrogenation of low-range vegetable oils and animal fats. UOP/Eni Ecofining™ commercialized a process for the conversion of nonedible, second-generation natural oils into Honeywell Green Diesel™, a drop-in diesel fuel for use in any percentage.[155] Petrobras in Brazil built up a plant for coprocessing of vegetable oils with middle distillates (H-Bio process, for a detailed discussion on this process, refer to Chapter 9).[156] Similarly, Neste Oil[157] developed the NExBTLR process by starting with the first commercial plant in Finland and then expanding in Singapore and the Netherlands.

Still considering these technological achievements, there are problems to be solved. For instance, the relatively high levels of water in the hydro-treatment reactor could exert a negative effect on the sulfided catalyst performance. In order to make this an attractive process not only minimum consumption of hydrogen is needed throughout the HDO process but also the catalyst should be robust enough to tolerate high concentrations of water and free fatty acids present in nonedible oils. Renewable diesel is becoming more attractive than conventional biodiesel due to its more flexible feedstock, low cost, high oxidative stability, high blendability, and its drop-in use in existing distribution networks.

For coprocessing of nonedible oils with diesel, the CO formed by hydrogenolysis of the vegetable oil molecules has an inhibiting effect on HDS activity. Under the hydrotreatment conditions, unsaturated fatty chains are hydrogenated. The resulting straight chains, mainly C_{12} to C_{18}, are completely saturated. Such paraffinic structures have excellent cetane index, but generally bad cold-flow properties compared to the corresponding esters.[158] As a result, an additional hydroisomerization step may be required. In the coprocessing, there are no longer any constraints on the chemical composition of the vegetable oils or fats to be employed. If, as it seems nowadays with the average crude oil price about US$110 per barrel, the incentive for producing fuels from renewable resources will become competitive, especially from nonedible vegetable oils/fats feedstocks. The use of biobased feeds will lead to the presence of oxygen atoms in feed molecules and HDO or any other techniques for the removal of oxygen may be needed. Therefore, more basic knowledge of HDO appears to be imperative.

8.7 Deactivation of Sulfide Catalysts in HDO Reactions

Naturally occurring impurities of vegetable oils or triglycerides are in general responsible for the deactivation of sulfide catalysts employed in deoxygenation reactions. Among the different impurities present in triglycerides – sulfur, phosphorus and alkali metals are the major agents responsible for catalyst deactivation. The effects of these impurities on the HDO performance of sulfide catalysts are briefly analyzed below.

For a refined rapeseed oil (RRO) in which the level of impurities is very low, a sulfided CoMo catalyst showed gradual activity decay throughout the HDO processing.[126] This deactivation can be due to two main factors. First, a decrease in catalyst active sites by progressive removal of sulfur from the sulfide phase. Secondly, the formation of carbonaceous deposits that block the active sites. The addition of 0.5 wt% dimethyldisulfide (DMDS) to the RRO feed significantly improved catalyst stability over long times on-stream. By comparing the oxygen removal percentage at 144 h on-stream, the removal of oxygen dropped to *ca.* 75% when neat RRO was used; however, it remained above 95% when DMDS was added to the feedstock. From this result, it can be inferred that the main reason for the gradual deactivation during the HDO of vegetable oils was the removal of sulfur from the active sulfide phase. Previous studies on tetrahydrofuran hydrogenation demonstrated that the activity loss was attributed to the decrease in sites, which sulfur is incompletely coordinated in the sulfided CoMo catalyst.[159] Aiming to assess the reversibility of the deactivation, DMDS was added to the feed once the CoMo catalyst became partially deactivated. Under the presence of DMDS, the HDO activity was only partially recovered. This observation indicates that some of the changes responsible for catalyst deactivation are irreversible. These changes can be attributed to partial loss of sulfur under reaction conditions due to insufficient quantities of sulfur in the feedstock to maintain the catalyst in a sulfided state.[116] As water is a product of the HDO process, it is likely that the Co–Mo–S active phase became oxidized to some extent ($Mo^{4+} \rightarrow Mo^{6+}$ and $S^{2-} \rightarrow S^{6+}$).[136]

Addition of DMDS to the feedstock improved the stability of catalyst and considerably affected the product selectivity. In the absence of DMDS in the feed, HDO was the main reaction pathway, that is, formation of HDO products was 3–4 times faster than formation of hydrodecarboxylation (HDC) products. However, when the feedstock contained DMDS, the hydrocracking (HDC) reaction pathway played a more important role. Incorporation of DMDS in the feed stream also influenced distribution in the group of oxygenates. Only low yields of fatty esters, fatty acids and fatty alcohols were achieved throughout the process duration. In the absence of DMDS, fatty esters and alcohols were the main oxygenated products and their yield increased gradually with reaction time. However, in the presence of DMDS the yield of fatty esters and fatty alcohols immediately dropped to zero. Since fatty esters are tertiary-reaction intermediates (*i.e.* formed by esterification of fatty acids by fatty alcohols), their disappearance can be explained by disappearance of fatty alcohols. Since fatty alcohols are formed by hydrogenation of fatty acids, the sudden decrease in the yield of fatty alcohols in the presence of DMDS can be due to either a decreased hydrogenation rate of fatty acids to fatty alcohols or due to an increased consumption rate of fatty acids by the competing decarboxylation.

In summary, the presence of DMDS in the RRO feed is conducive to the catalyst deoxygenation performance. In fact, the catalytic performance was sustained in the presence of DMDS. The changes in the structure of active sites caused by their desulfurization under a hydrogen atmosphere were found to be partially reversible, since a minor fraction of the active sites was irreversibly deactivated. Besides modification of catalyst activity, addition of DMDS (which generates H_2S during its decomposition) led to significant changes in selectivity in the HDO process, *i.e.* enhancement of (hydro)-decarboxylation products likely originating by an increase in catalyst acidity.

8.8   Conclusions

HDO appears to be one of the most promising routes for production of high-quality fuels through upgrading of pyrolysis oils and low-rank vegetable oils. This process has further been found economically feasible with production costs equivalent to conventional fuels from crude oil, but challenges still exist within the field. So far, hydrotreatment processes have been evaluated on the industrial scale to some extent, elucidating which unit operation should be performed when going from biomass to fuel.

However, research into the several aspects of the catalysts, reaction mechanisms and the high-pressure requirement in the actual HDO reactions is still required to optimize the processes and to bring it closer to industrial realization. A great concern within the field is catalyst formulation. Much effort has focused on Mo(W) catalysts. A significant effect of the catalyst structure on the two main pathways in the HDO system was also observed. Indeed, hydrogenation of the aromatic ring (HYD pathway) is favored in the presence of a crystalline MoS_2 catalyst with a high stacking of MoS_2 slabs, whereas the DDO pathway is also an operating pathway in the presence of amorphous MoS_2 catalysts with low degree of stacking. A general trend in HDO of both model oxygenate compounds and real feeds is the promotion effect of Co or Ni on Mo and W sulfides. The growth of crystallized MoS_2 particles was inhibited when the promoter was incorporated. Moreover, a decrease in the strength of the Co(Ni)–Mo(W)–S bonding was demonstrated to be the factor accounting for the increase in catalytic activity found for the Co- or Ni-promoted Mo or W sulfides.

The upgrading of pyrolysis oils remains the primary area of the current interests in HDO. It is believed that breakthroughs in the catalyst development may be required to make this source of fuels more attractive. Other Mo-, W-based catalysts may be more suitable, as is already indicated by some preliminary information. The main requirement for tailor-made catalysts is the high resistance towards carbon formation, and a sufficiently high activity and stability under the conditions imposed by HDO processes.

Acknowledgements

This research was supported by the Ministry of Science and Competitivity (Spain) and the Autonomous Government of Madrid, Madrid (Spain) under grants ENE2010-21198-C04-01 and S2009ENE-1743, respectively.

References

1. E. Furimsky and F. E. Massoth, *Catal. Rev.-Sci. Eng.*, 2005, **47**, 297.
2. J. P. Lange, *Biofuels Bioprod. Biorefin.*, 2007, **1**, 39.
3. J. C. Escobar, E. S. Lora, O. J. Venturini, E. E. Yáñez, E. F. Castillo and O. Almazam, *Renew. Sustain. Energy Rev.*, 2009, **13**, 1275.
4. O. Stéphane and T. Daniel, *Biochimie*, 2009, **91**, 659.
5. M. Stöcker, *Angew. Chem. Int. Ed.*, 2008, **47**, 9200.
6. A. V. Bridgwater, D. Meier and D. Radlein, *Org. Geochem.*, 1999, **30**, 1479.
7. A. Demirbas, *Energy Conv. Manag.*, 2001, **42**, 1357.
8. D. Mohan, C. U. Pittman, Jr. and P. H. Steele, *Energy Fuels*, 2006, **20**, 848.
9. L. Qiang, L. Wen-Zhi and L. Z. Xi-Feng, *Energy Conv. Manag*, 2009, **50**, 1376.
10. D. Elliott, *Energy Fuels*, 2007, **21**, 1792.
11. E. Ma and M. A. Hanna, *Bioresour. Technol.*, 1999, **70**, 1.

12. L. C. Meher, D. V. Sagar and S. N. Naik, *Ren. Sust. Energy Rev.*, 2006, **10**, 248.
13. R. Aafaqi, A. Rahman, M. Bhatia and S. Bhatia, *J. Chem. Technol. Biotechnol.*, 2004, **79**, 1179.
14. A. A. Kiss, A. C. Dimian and G. Rothenberg, *Adv. Synth. Catal.*, 2006, **348**, 75.
15. M. di Serio, R. Tesser, L. Pengmei and E. Santacesaria, *Energy Fuels*, 2008, **22**, 207.
16. A. Sivasamy, K. Y. Cheah, P. Fornasiero, F. Kemausuor, S. Zinoviev and S. Miertus, *ChemSusChem*, 2009, **2**, 278.
17. J. A. Melero, J. Iglesias and G. Morales, *Green Chem.*, 2009, **11**, 1285.
18. D. Srinivas and J. K. Satyarthi, *Catal. Surv. Asia*, 2011, **15**, 145.
19. Y. S. Ooi and S. Bhatia, *Micropor. Mesopor. Mater.*, 2007, **102**, 310.
20. H. Lappi and R. Alen, *J. Anal. Appl. Pyrol.*, 2011, **91**, 154.
21. B. Donnis, R. G. Egeberg and P. Blom, *Top. Catal.*, 2009, **52**, 229.
22. S. Bezergianni, S. Voutetakis and A. Kalogianni, *Ind. Eng. Chem. Res.*, 2009, **48**, 8402.
23. P. Šimáček, D. Kubička, G. Šebor and M. Pospíšil, *Fuel*, 2009, **88**, 456.
24. I. Sebos, A. Matsoukas, V. Apostolopoulos and N. Papayannakos, *Fuel*, 2009, **88**, 145.
25. P. Šimáček, D. Kubička, G. Šebor and M. Pospíšil, *Fuel*, 2010, **89**, 611.
26. D. Kubiča and L. Kaluţa, *Appl. Catal. A: Gen.*, 2010, **372**, 199.
27. P. Priecel, D. Kubička, L. Čapek, Z. Bastl and P. Ryšánek, *Appl. Catal. A: Gen.*, 2011, **397**, 127.
28. D. C. Elliott and T. R. Hart, *Energy Fuels*, 2009, **23**, 631.
29. J. N. Chheda, G. W. Huber and J. A. Dumesic, *Angew. Chem. Int. Ed.*, 2007, **46**, 7164.
30. N. Li and G. W. Huber, *J. Catal.*, 2010, **270**, 48.
31. E. Furimsky, *Appl. Catal. A: Gen.*, 2000, **199**, 147.
32. P. de Wild, R. van der Laan, A. Kloekhorst and E. Heeres, *Environ. Progress Sust. Energy*, 2009, **28**, 461.
33. E. Laurent and B. Delmon, *Appl. Catal. A: Gen.*, 1994, **109**, 77.
34. P. M. Mortensen, J. D. Grunwaldt, P. A. Jensen, K. G. Knudsen and A. D. Jensen, *Appl. Catal. A: Gen.*, 2011, **407**, 1.
35. T. V. Choudhary and C. B. Phillips, *Appl. Catal. A: Gen.*, 2011, **397**, 1.
36. R. Maggi and B. Delmon, *Stud. Surf. Sci. Catal.*, 1997, **106**, 99.
37. A. V. Bridgwater, *Biomass Bioenergy*, 2012, **38**, 68.
38. K. C. Kwon, H. Mayfield, T. Marolla, B. Nichols and M. Mashburn, *Renew. Energy*, 2011, **36**, 907.
39. F. de Miguel, M. J. Groeneveld, S. R. A. Kersten, N. W. J. Way, C. J. Schaverien and J. A. Hogendoorn, *Appl. Catal. B: Environ.*, 2010, **96**, 57.
40. I. Graca, J. M. Lopes, H. S. Cerqueira and M. F. Ribeiro, *Ind. Eng. Chem. Res.*, 2013, **52**, 275.
41. Boda, G. Onyestyak, H. Solt, F. Lonyi, J. Valyon and A. Thernesz, *Appl. Catal. A: Gen.*, 2010, **374**, 158.

42. M. Snare, I. Kubickova, P. Maki-Arvela, K. Eranen and D. Yu Murzin, *Ind. Eng. Chem. Res.*, 2006, **45**, 5708.
43. S. Lestari, P. Maki-Arvela, J. Beltramini, C. Q. Max Lu and D. Yu Murzin, *ChemSusChem*, 2009, **2**, 1109.
44. K. Murata, Y. Liu, M. Inaba and I. Takahara, *Energy Fuels*, 2010, **24**, 2404.
45. H. Topsøe, *Appl. Catal. A: Gen.*, 2007, **322**, 3.
46. S. Eijsbouts, S. W. Mayo and K. Fujita, *Appl. Catal. A: Gen.*, 2007, **322**, 58.
47. S. M. A. M. Bouwens, F. B. M. van Zon, M. P. van Dijk, A. M. van der Kraan, V. H. J. de Beer, J. A. R. van Veen and D. C. Koningsberger, *J. Catal.*, 1994, **146**, 375.
48. R. Candia, O. Sorensen, J. Villadsen, N. Y. Topsøe, B. S. Clausen and H. Topsøe, *Bull. Soc. Chim. Belg.*, 1984, **93**, 763.
49. E. J. M. Hensen, P. J. Kooyman, Y. Meer, A. M. van der Kraan, V. H. J. de Beer, J. A. R. van Veen and R. A. van Santen, *J. Catal.*, 2001, **199**, 224.
50. G. Alonso, M. D. Valle, J. Cruz, A. Licea-Claverie, V. Petranovskii and S. Fuentes, *Catal. Lett.*, 1998, **52**, 55.
51. M. Daage and R. R. Chianelli, *J. Catal.*, 1994, **194**, 414.
52. J. K. Nørskov, B. S. Clausen and H. Topsøe, *Catal. Lett.*, 1992, **13**, 1.
53. S. Harris and R. R. Chianelli, *J. Catal.*, 1986, **98**, 17.
54. P. Raybaud, J. Hafner, G. Kresse, S. Kasztelan and H. Toulhoat, *J. Catal.*, 2000, **190**, 128.
55. R. R. Chianelli, *Oil Gas Sci. Technol. Rev. IFP*, 2006, **61**, 503.
56. J. Kibsgaard, J. V. Lauritsen, B. S. Clausen, H. Topsøe and F. Besenbacher, *J. Am. Chem. Soc.*, 2006, **128**, 13950.
57. E. Laurent and B. Delmon, *J. Catal.*, 1994, **146**, 281.
58. R. H. Venderbosch, A. R. Ardiyanti, J. Wildschut, A. Oasmaa and H. J. Heeres, *J. Chem. Technol. Biotechnol.*, 2010, **85**, 674.
59. K. V. R. Charry, H. Ramakrishna and G. Murali Dhar, *J. Mol. Catal.*, 1991, **68**, L25.
60. H. Y. Zhao, D. Li, P. Bui and S. T. Oyama, *Appl. Catal. A: Gen.*, 2010, **391**, 305.
61. S. Lestari, P. M. Arvela, H. Bernas, O. Simakova, R. Sjoholm, J. Beltramini, G. Q. M. Lu, J. Myllyoja, I. Simakova and D. Y. Murzin, *Energy Fuels*, 2009, **23**, 3842.
62. D. Kubička, M. Bejblova and J. Vlk, *Top. Catal.*, 2010, **53**, 168.
63. R. Nava, B. Pawelec, P. Castano, M. C. Alvarez-Galvan, C. V. Loricera and J. L. G. Fierro, *Appl. Catal. B: Environ.*, 2009, **92**, 154.
64. S. B. Alpert and S. C. Shuman, Canadian Patent 851709, 1970.
65. M. Ferrari, R. Maggi, B. Delmon and P. Grange, *J. Catal.*, 2001, **198**, 47.
66. M. Ferrari, B. Delmon and P. Grange, *Carbon*, 2002, **40**, 497.
67. M. Ferrari, B. Delmon and P. Grange, *Micropor. Mesopor. Mater.*, 2002, **56**, 279.
68. A. Pinheiro, D. Hudebine, N. Dupassieux and C. Geantet, *Energy Fuels*, 2009, **23**, 1007.

69. Y. C. Lin, C. L. Li, H. P. Wan, H. T. Lee and C. F. Liu, *Energy Fuels*, 2011, **25**, 890.

70. V. N. Bui, D. Laurenti, P. Afanasiev and C. Geantet, *Appl. Catal. B: Environ.*, 2010, **101**, 239.

71. V. N. Bui, D. Laurenti, P. Delichere and C. Geantet, *Appl. Catal. B: Environ.*, 2010, **101**, 246.

72. Y. Yang, A. Gilbert and C. Xu, *Appl. Catal. A: Gen.*, 2009, **360**, 242.

73. Y. Q. Yang, C. T. Tye and K. J. Smith, *Catal. Commun.*, 2008, **9**, 1364.

74. E. Laurent and B. Delmon, *Ind. Eng. Chem. Res.*, 1993, **32**, 2516.

75. S. Echeandia, P. L. Arias, V. L. Barrio, B. Pawelec and J. L. G. Fierro, *Appl. Catal. B: Environ.*, 2010, **101**, 1.

76. E. J. Shin and M. A. Keane, *Ind. Eng. Chem. Res.*, 2000, **39**, 883.

77. H. Weigold, *Fuel*, 1982, **61**, 1021.

78. O. I. Şenol, T. R. Viljava and A. O. I. Krause, *Catal. Today*, 2005, **100**, 331.

79. O. I. Şenol, T. R. Viljava and A. O. I. Krause, *Appl. Catal. A: Gen.*, 2007, **326**, 236.

80. M. Snare, I. Kubickova, P. Maki-Arvela, K. Eränen and D. Y. Murzin, *Ind. Eng. Chem. Res.*, 2006, **45**, 5708.

81. O. I. Şenol, E. M. Ryymin, T. R. Viljava and A. O. I. Krause, *J. Mol. Catal. A: Chem.*, 2007, **268**, 1.

82. K. V. R. Chary, K. S. Rama Rao, G. Muralidhar and P. Kanta Rao, *Carbon*, 1991, **29**, 478.

83. C. Kordulis, A. Gouromihou, A. Lycourghiotis, C. Papadopoulou and H. K. Matralis, *Appl. Catal.*, 1990, **67**, 39.

84. B. S. Gevert, M. Eriksson, P. Eriksson and F. E. Massoth, *Appl. Catal. A: Gen.*, 1994, **117**, 151.

85. T. R. Viljava, R. S. Komulainem and A. O. I. Krause, *Catal. Today*, 2000, **60**, 83.

86. F. E. Massoth, P. Politzer, M. C. Concha, J. S. Murray, J. Jakowski and J. Simons, *J. Phys. Chem. B*, 2006, **110**, 14283.

87. O. I. Senol, E. M. Ryymin, T. R. Viljava and A. O. I. Krause, *J. Mol. Catal. A: Chem.*, 2007, **277**, 107.

88. I. Gandarias, V. L. Barrio, J. Requies, P. L. Arias, J. F. Cambra and M. B. Güemez, *Int. J. Hydrogen Energy*, 2008, **33**, 3485.

89. Y. Romero, F. Richard, Y. Renème and S. Brunet, *Appl. Catal. A: Gen.*, 2009, **353**, 46.

90. A. Pattiya, J. O. Titiloye and A. V. Bridgwater, *J. Anal. Appl. Pyrol.*, 2008, **81**, 72.

91. J. Zakzeski, P. C. A. Bruijnincx, A. L. Jongerius and B. M. Weckhuysen, *Chem. Rev.*, 2010, **110**, 3552.

92. M. Kleinert and T. Barth, *Energy Fuels*, 2008, **22**, 1371.

93. Z. Hou, C. A. Bennett, M. T. Klein and P. S. Virk, *Energy Fuels*, 2010, **24**, 58.

94. J. B. Binder, M. J. Gray, J. F. White, Z. C. Zhang and J. E. Holladay, *Biomass Bioenergy*, 2009, **33**, 1122.

95. E. Furimsky, *Appl. Catal. A: Gen.*, 2000, **199**, 147.
96. G. de la Puente, A. Gil, J. J. Pis and P. Grange, *Langmuir*, 1999, **15**, 5800.
97. H. Y. Zhao, D. Li, P. Bui and S. T. Oyama, *Appl. Catal. A: Gen.*, 2010, **391**, 305.
98. Y. Yang, A. Gilbert and C. Xu, *Appl. Catal. A: Gen.*, 2009, **360**, 242.
99. B. Yoosuk, D. Tumnantong and P. Prasassarakich, *Chem. Eng. Sci.*, 2012, **79**, 1.
100. M. M. Ahmad, M. F. R. Nordin and M. T. Azizan, *Am. J. Appl. Sci.*, 2010, 7, 746.
101. Y. Romero, F. Richard and S. Brunet, *Appl. Catal. B: Environ.*, 2010, **98**, 213.
102. E. M. Ryymin, M. L. Honkela, T. R. Viljava and A. O. I. Krause, *Appl. Catal. A: Gen.*, 2010, **389**, 114.
103. H. Topsøe, B. S. Clausen and F. E. Massoth, *Hydrotreating Catalysis, Science and Technology*, Vol. 11, Springer Verlag, 1996.
104. J. F. Paul and E. Payen, *J. Phys. Chem. B*, 2003, **107**, 4057.
105. J. V. Lauritsen, J. Kibsgaard, G. H. Olesen, P. G. Moses, B. Hinnemann, S. Helveg, J. K. Nørskov, B. S. Clausen, H. Topsøe, E. Laegsgaard and F. Besenbacher, *J. Catal.*, 2007, **249**, 220.
106. P. Raybaud, *Appl. Catal. A: Gen.*, 2007, **322**, 76.
107. M. Badawi, S. Cristol, J. F. Paul and E. Payen, *C.R. Chim.*, 2009, **12**, 754.
108. R. Parthasarathi, R. A. Romero, A. Redondo and S. Gnanakaran, *J. Phys. Chem. Lett.*, 2011, **2**, 2660.
109. J. M. Younker, A. Beste and A. C. Buchanan, *ChemPhysChem*, 2011, **12**, 3556.
110. E. Dorrestijn, L. J. J. Laarhoven, I. W. C. E. Arends and P. Mulder, *J. Anal. Appl. Pyrol.*, 2000, **54**, 153.
111. L. Artok, O. Erbatur and H. H. Schobert, *Fuel Proc. Technol.*, 1996, **47**, 153.
112. F. P. Petrocelli and M. T. Klein, *Fuel Sci. Technol. Int.*, 1987, **5**, 25.
113. E. Laurent and B. Delmon, *Appl. Catal.*, 1994, **109**, 97.
114. E. Laurent, A. Centeno and B. Delmon, *Stud. Surf. Sci. Catal.*, 1994, **88**, 573.
115. J. C. Afonso, M. Schmal and J. N. Cardoso, *Ind. Eng. Chem. Res.*, 1992, **31**, 1045.
116. F. Goudriaan and D. G. R. Peferoen, *Chem. Eng. Sci.*, 1990, **45**, 2729.
117. E. G. Baker and D. C. Elliot, In *Research in Thermochemical Biomass Conversion*, A. V. Bridgwater, J. L. Kuster, ed., Elsevier, London, 1988, p. 883.
118. D. Mohan, C. U. Pittman, Jr. and P. H. Steele, *Energy Fuels*, 2006, **20**, 848.
119. R. J. French, J. Hrdlicka and J. R. Baldwin, *Environ. Progress Sustain. Energy*, 2010, **29**, 142.
120. E. Churin, R. Maggi, P. Grange and B. Delmon, In *Research in Thermochemical Biomass Conversion*; A. V. Bridgwater and J. L. Kuster, ed.; Elsevier, London, 1988, p. 896.
121. J. D. Rocha, C. A. Luengo and C. E. Snape, *Renew. Energy*, 1996, **9**, 950.

122. Y. Liu, R. Sotelo-Boyas, K. Murata, T. Minowa and K. Sakanishi, *Chem. Lett.*, 2009, **38**, 552.
123. S. Bezergianni, A. Dimitriadis, A. Kalogianni and K. G. Knudsen, *Ind. Eng. Chem. Res.*, 2011, **50**, 3874.
124. R. Sotelo-Boyas, Y. Liu and T. Minowa, *Ind. Eng. Chem. Res.*, 2011, **50**, 2791.
125. R. Tiwari, B. S. Rana, R. Kumar, D. Verma, R. Kumar, R. K. Joshi, M. O. Garg and A. K. Sinha, *Catal. Commun.*, 2011, **12**, 559.
126. D. Kubička and J. Horaček, *Appl. Catal A: Gen.*, 2011, **394**, 9.
127. C. Templis, A. Vonortas, I. Sebos and N. Papayannakos, *Appl. Catal. B: Environ.*, 2011, **104**, 324.
128. D. Kubička, P. Šimaček and N. Žilkova, *Top. Catal.*, 2009, **52**, 161.
129. P. Šimaček, D. Kubička, I. Kubičkova, F. Homol, M. Pospišil and J. Chudoba, *Fuel*, 2011, **90**, 2473.
130. S. Bezergianni, A. Dimitriadis, A. Kalogianni and P. A. Pilavachi, *Bioresour. Technol.*, 2010, **101**, 6651.
131. G. W. Huber, P. O'Connor and A. Corma, *Appl. Catal A: Gen.*, 2007, **329**, 120.
132. J. Walendziewski, M. Stolarski, R. Łusny and B. Klimek, *Fuel Process. Technol.*, 2009, 3036.
133. J. Mikulec, J. Cvengroš, L. Jorikova, M. Banič and A. Kleinova, *J. Cleaner Prod.*, 2010, **18**, 917.
134. S. Bezergianni, A. Kalogianni and I. A. Vasalos, *Bioresour. Technol.*, 2009, **100**, 3036.
135. J. Gusmao, D. Brodzki, G. Djega-Mariadassou and R. Frety, *Catal. Today*, 1989, **5**, 533.
136. R. Nava, B. Pawelec, P. Castano, M. C. Alvarez-Galvan, C. V. Loricera and J. L. G. Fierro, *Appl. Catal. B: Environ.*, 2009, **92**, 154.
137. S. Lestari, P. M. Arvela, J. Beltramini, G. Q. M. Lu and D. Y. Murzin, *ChemSusChem*, 2009, **2**, 1109.
138. B. Donnis and R. G. Egeberg, *P. Blom, Top. Catal.*, 2009, **52**, 229.
139. P. Šimáček and D. Kubička, *Fuel*, 2010, **89**, 1508.
140. A. Guzman, J. E. Torres, L. P. Prada and M. L. Nuñez, *Catal. Today*, 2010, **156**, 38.
141. P. Priecel, D. Kubička, L. Čapek, Z. Bastl and P. Ryšánek, *Appl. Catal. A: Gen.*, 2011, **397**, 127.
142. V. N. Bui, G. Toussaint, D. Laurenti, C. Mirodatos and C. Geantet, *Catal. Today*, 2009, **143**, 172.
143. K. Murata, Y. Liu, M. Inaba and I. Takahara, *Energy Fuels*, 2010, **24**, 2404.
144. R. Kumar, B. S. Rana, R. Tiwari, D. Verma, R. Kumar, R. K. Joshi, M. O. Garg and A. K. Sinha, *Green Chem.*, 2010, **12**, 2232.
145. S. Kovács, L. Boda, L. Leveles, A. Thernesz and J. Hancsók, *Chem. Eng. Trans.*, 2010, **21**, 1321.
146. T. R. Viljava, E. R. M. Saari and A. O. I. Krause, *Appl. Catal. A: Gen.*, 2001, **209**, 33.

147. M. Krár, S. Kovács, D. Kalló and J. Hancsók, *Biores. Technol.*, 2010, **101**, 9287.
148. P. Bielansky, A. Weinert, C. Schönberger and A. Reichhold, *Fuel Process. Technol.*, 2011, **92**, 2305.
149. J. A. Melero, M. M. Clavero, G. Calleja, A. Garcia, R. Miravalles and T. Galindo, *Energy Fuels*, 2010, **24**, 707.
150. V. N. Bui, G. Toussaint, D. Laurenti, C. Mirodatos and C. Geantet, *Catal. Today*, 2009, **142**, 172.
151. J. Walendziewski, M. Stolarski, R. Luzny and B. Klimek, *Fuel Process. Technol.*, 2009, **90**, 686.
152. N. Aribert, A. Daudin and T. Chapus, *Prep. Pap. ACS Div. Fuel Chem.*, 2012, **57**, 775.
153. A. A. Lappas, S. Bezergianni and I. A. Vasalos, *Catal. Today*, 2009, **145**, 55.
154. R. Kumar, B. S. Rana, R. Tiwari, D. Verma, R. Kumar, R. K. Joshi, M. O. Garg and A. K. Sinha, *Green Chem.*, 2012, **12**, 2232.
155. http://www.uop.com.
156. http://www.petrobras.com.br.
157. http://www.nesteoil.com.
158. D. Casanave, J. L. Duplan and E. Freund, *Pure Appl. Chem.*, 2007, **79**, 2071.
159. E. Furimsky, *Ind. Eng. Chem. Prod. Res. Dev.*, 1983, **22**, 3.
160. R. Rinaldi, F. Y. Fujiwara and U. Schuchardt, *Appl. Catal. A: Gen.*, 2006, **315**, 44.

CHAPTER 9

Biofuels Generation via Hydroconversion of Vegetable Oils and Animal Fats

JEFFERSON ROBERTO GOMES,[a] STELLA BEZERGIANNI,[c] JOSÉ LUIZ ZOTIN[a] AND EDUARDO FALABELLA SOUSA-AGUIAR*[a,b]

[a] Petrobras Research Center (CENPES), Ilha do Fundão, Q7, Cidade Universitária, CEP 21949-900, Rio de Janeiro, Brazil; [b] Federal University of Rio de Janeiro (UFRJ), School of Chemistry, Department of Organic Processes, Centro de Tecnologia, Bloco E, Ilha do Fundão, Rio de Janeiro, Brazil; [c] Center for Research & Technology Hellas/CERTH, Chemical Process & Energy Resources Institute/CPERI, Laboratory of Environmental Fuels & Hydrocarbons, 6th km Harilaou-Thermi Rd, GR-57001, Thermi, Thessaloniki, Greece
*Email: efalabella@petrobras.com.br

9.1 Introduction

Production of biofuels from renewable sources has been the object of several studies during the last decade. In this context, the production of first-generation biodiesel (*i.e.* fatty acid methyl esters, FAME) from vegetable oils and animal fats has received sustained attention.[1–4] However, first-generation biodiesel presents several inconveniences regarding its fuel quality compared to that of conventional diesel. In fact, biodiesel shows a lower weight calorific value (39 MJ kg^{-1} vs. 42.6 MJ kg^{-1}),[5,6] generates higher

RSC Energy and Environment Series No. 13
Catalytic Hydrogenation for Biomass Valorization
Edited by Roberto Rinaldi

NO_x emissions,[7–9] and displays lower storage stability[9–11] than conventional diesel. In addition to these issues, pipeline contamination and hygroscopicity pose problems to the commercialization of FAME.[12] Hence, research into alternative chemical processes (*i.e.* hydrotreatment) aiming at conversion of biomass triglycerides[13,14] into a second-generation biofuel (also referred to as "green diesel") have been carried out.

Hydroconversion processes of vegetable oils and animal fats usually involve the saturation of the C=C bonds occurring in the paraffinic chain in addition to hydrodeoxygenation of the acid or ester groups of the molecule taking place in a catalytic hydrotreating (HDT) unit. As a result, linear paraffins (C_{12}–C_{18}, depending on the vegetable oil and animal fat sources), propane and water are formed. Although the hydroconversion process of gasoil is well known, the development of a new process for the production of second-generation biodiesel from vegetable oils still presents a major challenge. This is mainly due to deactivation issues that sulfide catalysts face. As discussed in Chapter 8, the main causes for such deactivation are (1) catalyst poisoning; (2) coking; (3) fouling; and (4) active-phase sintering.

In petroleum refineries, the HDT unit is dedicated to improve the quality of refinery streams (diesel, kerosene, *etc.*) upon the removal of contaminants, such as nitrogen, sulfur, oxygen, metals, and unsaturated hydrocarbons. In the HDT unit, the stream undergoes hydrogenolysis of the carbon–heteroatom bond, leading to hydrodenitrogenation (HDN), hydrodesulfurization (HDS), hydrodeoxygenation (HDO), and hydrodemetallization (HDM) in addition to hydrogenation of olefin (HO) and aromatics (HAr).

A conventional HDT process comprises the flow of petroleum hydrocarbons mixed with a hydrogen stream through a fixed catalytic bed reactor, under pressures in the range of 2–12 MPa and average temperatures from 553–673 K. Since the reactor is operating under adiabatic conditions, and exothermic reactions are involved, the temperature increases along the fixed catalytic bed. However, to guarantee a minimum catalyst lifetime of 1 to 2 years, the process imposes a limit on the increase in the catalytic bed temperature of 40–50 K per catalytic bed. When the increase in temperature is excessive due to highly exothermic reactions, the reactor can be designed to comprise more than one fixed catalytic bed. In addition, a recycled gas stream can also be injected between the beds for quenching. In the case of multiple catalytic beds, the thermal release is greater in the first bed because of a high concentration of reactive compounds. The more refractory compounds continue reacting at lower reaction rates through the subsequent catalytic beds within the reactor. The operating conditions are set to maximize the product quality for the required specifications. However, the product distillation range is essentially the same as the feedstock submitted to the HDT unit. Light secondary products may be produced through hydrocracking in the presence of acidic catalysts. In this case, the hydrogenolysis of C–C bonds is responsible for converting heavy feeds into more valuable product fractions.

This chapter provides a review of the most significant industrial progress to produce biofuels from vegetable oils *via* hydroprocessing. An indepth industrial analysis of the reaction steps and processes in current development by the major global oil companies is given.

9.2 Processes of Hydroconversion of Vegetable Oils and Animal Fats

9.2.1 Coprocessing Triglycerides in a Conventional HDT Unit

The conversion of triglycerides into hydrocarbons has been demonstrated to be a feasible route for second-generation biodiesel through the coprocessing of vegetable oils and animal fats in a petroleum refinery.[15–17] The use of an established refinery scheme with minor modifications is conducive to the coprocessing of triglycerides with petroleum. A potential benefit in terms of production costs and distribution of bioproducts also exists on a small scale. However, the economic benefits are limited by technological barriers, such as the limit of renewable feedstock volume admissible in the coprocessing, the hydrogen consumption for triglyceride hydroconversion, and side reactions (*e.g.* decarboxylation and decarbonylation of fatty acids) resulting in the undesired generation of CO and CO_2.

Figure 9.1 shows the general coprocessing flow sheet of the H-Bio process by Petrobras.[17] During a normal operation of conventional diesel production, the vegetable oil or animal fat stream is blended with petroleum prior to the inlet into the unit pump (indicated by "1" in Figure 9.1). The stream can comprise up to 20% renewable feedstock. Before the reactor entrance, it is necessary to mix a H_2 stream with the feedstock, to obtain a first reaction stream, which is heated and injected in the first catalytic bed (5) at temperatures in the range of 493 to 623 K. As already mentioned, to reduce the temperature before the next catalytic bed, a hydrogen stream must be injected between the catalyst beds in order to "quench" the reactor. After the reactor, the stream is cooled (2), the liquid hydrocarbons and the water are separated (3) and the unused H_2 stream is pressurized (6) and reinjected into the process flow. The hydrocarbons are further rectified (7) and the dissolved gases and light fractions, *e.g.* C_3 and naphtha recovered. Finally, a hydrotreated liquid product that contains *n*-paraffins from the hydroconversion of triglycerides is obtained.

Under the high-severity operational conditions, triglycerides are prone to undergo side reactions (*e.g.* decarbonylation and decarboxylation),[18] forming CO and CO_2 as byproducts. The carbon monoxide generated in the process decreases the hydrogenation activity of the catalyst, but its effect is completely reversible. The CO poisoning effect is higher for sulfided CoMo catalysts, compared to sulfided NiMo catalysts.[19–21] Considering these facts, the coprocessing of vegetable oil diluted in a less-saturated stream of mineral hydrocarbons would be the most appropriate choice for four main reasons. The first is related to the high exothermicity of the hydroconversion

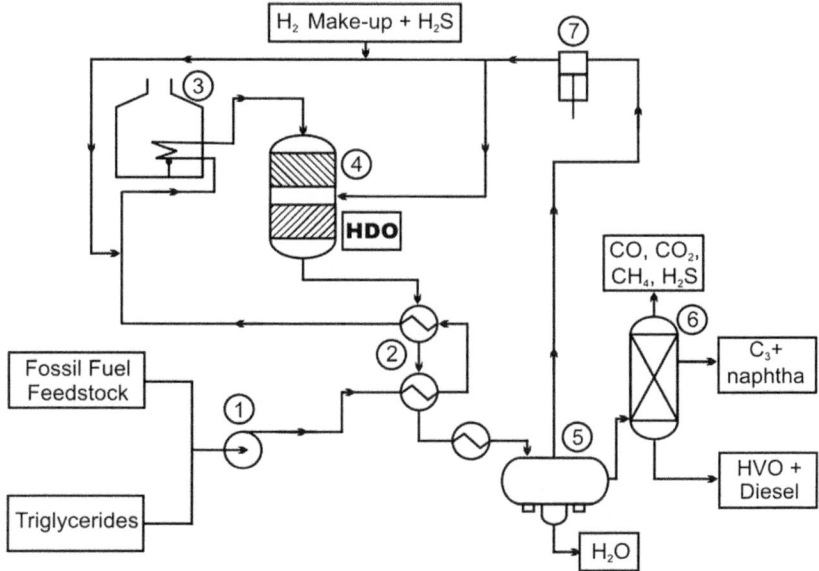

Figure 9.1   Flow sheet of the H-Bio process by Petrobras. Operation units: (1) Feedstock pump, (2) Heat exchanger, (3) Furnace, (4) Reactor, (5) High-pressure separator (6) Distillation column (7) Recycle compressor. Adapted from ref. 17.

of vegetable oils and animal fats. The dilution with the mineral stream helps to keep the catalyst at appropriate working temperatures. The second reason is the minimization of coke formation, and thus catalyst deactivation, through the dilution of highly reactive intermediates (acids, alcohols, *etc.*). The third concerns the formation of low boiling point products, which may vaporize under the processing conditions for the hydroconversion of trigly-cerides. Consequently, the catalyst bed could potentially dry out, disturbing the reactor fluid dynamics. This obstacle combined with the presence of highly exothermic reactions could result in "hot-spot" formation along the catalyst bed. Therefore, the dilution contributes to the flow stability and act as a heat sink, preventing the formation of hot spots. Last but not least, the fourth reason is that the dilution contributes to a low level of CO_2 in the recycled gas, thus increasing the partial pressure of H_2.

The impact of the soya-oil coprocessing in two industrial units with a NiMo catalyst is depicted in Table 9.1. One may notice an increase in the concentration of gases (CO and CO_2) in the reaction medium. The reduction of the H_2 partial pressure decreases the catalytic activity with regard to HDS, HDN and hydrogenation of unsaturated hydrocarbons in refinery streams. Hence, to obtain a hydrotreated liquid product, it is necessary to increase the reaction temperature.[26]

The hydroconversion of triglycerides is proposed to initiate through the hydrogenation of C=C, followed by thermal cracking reactions on the aliphatic chains.[22] The formation of acrolein and carboxylic acids also takes

Table 9.1 Impact of soya oil processing on the hydrogen concentration in the recycle gas.[26]

Refinery unit	Total pressure (MPa)	WABT[a] (K)	H_2 purity (mol/mol)	CO (mol/mol soya oil)	CO_2 (mol/mol soya oil)
I	5.7	595	0.88	1.14	0.91
II	7.8	591	0.92	0.41	0.11

[a]WABT stands for "weighted average bed temperature."

place.[23] Further reactions occurring in the hydroconversion process, as well as the role of the catalyst and the mechanisms involved, are still the subject of discussion. Decarboxylation, decarbonylation and hydrodeoxygenation are undoubtedly important reactions taking place in the process, as the product distribution contains CO_2, CO, H_2O, CH_4, C_3 and linear paraffins. However, the reverse water-gas shift reaction $(CO_2 + H_2 \rightleftharpoons CO + H_2O)$ and methanation $(CO + 3H_2 \rightleftharpoons CH_4 + H_2O)$ may also be significant under high-severity processes. Therefore, it is difficult to establish whether the detected CO and CO_2 are only produced through decarboxylation or decarbonylation.[52–56]

In a fixed catalytic bed HDT reactor, bimetallic catalysts are generally made of metal oxides (Ni–Mo, Co–Mo, Ni–W and Co–W), supported on materials with a high specific surface area (*e.g.* γ-alumina, silica, *etc.*). These catalysts are usually sulfided in order to obtain the highest activity for the catalytic bed in the process. As thoroughly discussed in Section 8.2.1.2 by Pawelec and Fierro, there is a highly important synergism between the metal sulfides from group 6 (Mo and W) and those from group 9 (Co and Ni). The activity of a catalyst containing both metals is much higher than the activity of each one alone. The selectivity found for HDT coprocessing of vegetable oils is affected by catalyst composition. The use of a bimetallic catalyst containing metals from groups 6 and 8 results in a high gas yield, mainly CO_2.[24,25] However, catalysts containing only group 6 metals show very low selectivity to gas, resulting in the production of *n*-paraffins, with the same carbon number as the fatty acids originally present in the vegetable oil or animal fat.[26,27]

The coprocessing of triglycerides in a petroleum refinery HDT improves the quality of the diesel oil, as summarized in Table 9.2. The coprocessing of a stream comprising a small fraction of triglycerides (6–12 wt%) does not affect the HDS of the petroleum content.[16,28] In effect, the sulfur content decreases by dilution with a sulfur-free feedstock. The hydrotreatment of triglycerides renders *n*-paraffins that contribute to a decrease in the specific gravity of the product. Moreover, the high content of *n*-paraffins in the product improves the cetane number. However, the coprocessing exerts a negative effect on the cold flow (CFPP, *i.e.* cold filter plugging point) properties of the final fuel. Hence, an ideal formulation of the HDT stream should target at an appropriate CFPP of the diesel blend in order to determine the maximum natural oil percentage to be coprocessed. An improvement in the CFPP is achievable by hydrorefining–hydrocracking

Table 9.2 Properties of conventional diesel compared to those from second-generation biodiesel fuels obtained by coprocessing in a refinery HDT unit (Petrobras).

Properties	Diesel reference	Diesel reference *plus*	
		6 wt% soybean oil	12 wt% soybean oil
Sulfur (ppm)	12.5	11.8	10.9
Cetane number	38.1	42.3	45.4
CFPP[1] (°C)	−25	− 19	− 11
Viscosity 40 °C (cST)	3.784	3.763	3.725
Specific gravity @ 20/4 °C	0.8662	0.8620	0.8576

Table 9.3 The catalyst effect on the yields of gas products and soya oil hydroconversion.[26]

Catalyst	C_1	CO	CO_2	H_2O	H_2 Consumption	Yield	
	mol/mol of soya oil				NL/L soya[c]	wt%	vol%
100% decarboxylation[a]	0.00	0.00	3.00	0.00	183	79.6	94.4
COM I (CoMo)	0.19	0.79	1.21	2.79	276	80.8	95.8
COM II (NiMo)	0.57	0.75	0.52	4.21	335	81.4	96.4
BRbio1	0.00	0.00	0.07	5.86	390	84.1	99.3
100% HDO[b]	0.00	0.00	0.00	6.00	393	84.3	99.5

"a" and "b" correspond to theoretical values of gas formation and H_2 consumption for selective decarboxylation and hydrodeoxygenation, respectively; "c" NL stands for the volume of H_2 converted to standard air conditions (0.10 MPa, 273.15 K, and 0% relative humidity).

(NiMo) catalysts that promote the partial hydrocracking and isomerization of *n*-paraffins; however, the hydrocracking increases the production of light hydrocarbons, thus reducing the volume yield of diesel.[29]

The industrial unit can also be modified to process high vegetable oil concentrations (up to 30%) in the feedstock.[20,21] In that case, a catalyst that is capable of enduring high temperature and high water concentration at the reactor entrance is required. Furthermore, the stripping of H_2S and CO_2 from the recycled gas through extraction by amines is also needed. In addition, some processes should be designed with special materials in order to cope with the high concentration of CO_2 (due to corrosion issues). Finally, the recycled gas and make-up compressors must be retrofitted due to the high quenching and high hydrogen demand (Figure 9.1, operation unit 6).

Table 9.3 summarizes the results of gas formation obtained from soya oil coprocessing in the presence of different catalysts. Regular hydrotreatment catalysts (NiMo or CoMo sulfides supported on alumina) promote both HDO and decarboxylation reactions, as indicated by the formation of substantial amounts of CH_4, CO and CO_2. In this regard, the CoMo catalyst is slightly less selective to HDO than the NiMo catalyst. However, in a coprocessing approach, highly active NiMo or CoMo sulfided catalysts are of paramount importance to guarantee the extensive conversion of sulfur, nitrogen and aromatic compounds present in the mineral oil. Nevertheless, it is possible to use a more selective catalyst to reduce gas production. Despite its low

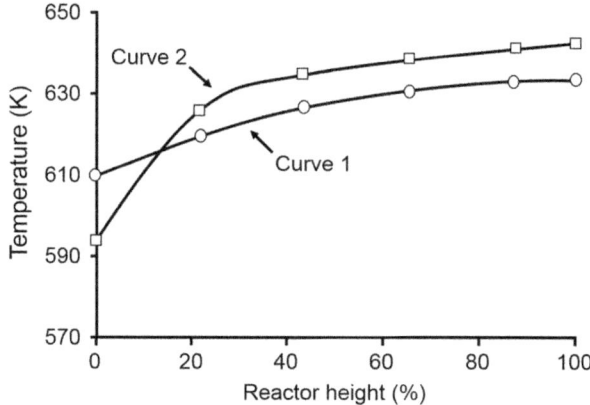

Figure 9.2 Temperature profile of an adiabatic reactor obtained from the hydro-conversion of a feed with 10 wt% soya oil (curve 1) and a feed without addition of soya oil (curve 2).

activity to HDS and HDN, the catalyst BRbio1 listed in Table 9.3 has a very high selectivity. The use of a composite bed reactor with a selectivity grading[27] proved effective in the processing of feeds with up to 20% of vegetable oil (since the unit has enough make-up hydrogen and recycled gas). This approach allows triglycerides to be converted with low selectivity to gas products in the top of the reactor without affecting the conversion of the petroleum stream by the main catalyst.

Again, the hydroconversion of vegetable oils are fairly exothermic in view of the olefin saturation and water generation. For soya oil, the estimated hydroconversion heat is 1.15 MJ/L of vegetable oil. It is noteworthy that the temperature profile of an adiabatic pilot reactor shows that the reactions proceed also quickly (Figure 9.2). The addition of 10% soya oil to the feed (curve 1) causes a fast temperature increase at the beginning of the reactor and then, after 20% of the reactor height, the curve is parallel to that obtained from the feed without soybean oil (curve 2). This observation indicates that the conversion of vegetable oil occurs at the entrance of the catalyst bed. Thus, the effects of the soy oil on the energy release and reaction rate are clear – the presence of a hydrocarbon stream to dilute the vegetable oil stream is of crucial importance to control the reactor temperature. Ultimately, these observations translate into low coke formation, prolonging the catalyst lifetime. In fact, it could be verified that the pilot plant operated for 40 days without the catalyst showing any abnormal deactivation.

Overall, the coprocessing of diluted triglycerides feeds in a refinery stream of petroleum hydrocarbons is feasible, without compromising the integrity of the fixed catalytic bed of a refinery HDT unit. The increase in the hydrotreated liquid yields is achieved by selective hydroconversion of triglycerides into *n*-paraffins. In addition, the levels of hydrogenation of unsaturated hydrocarbons and removal of heteroatom contaminants from the petroleum hydrocarbon stream are maintained.

9.2.2   Standalone Hydroprocessing of Triglycerides

Despite the high investment needed, the standalone unit for processing only triglycerides offers several advantages concerning product quality.[15] First, the unit can be designed to process different feedstocks with high levels of contaminants. Secondly and most importantly, the unit shows considerable degree of flexibility in the production of biojet fuel as well as diesel. Accordingly, many oil companies have directed research into hydrotreated vegetable oil (HVO) technologies.[21,30–33] Indeed, some industrial units are currently in operation.[34] A general process flow sheet is presented in Figure 9.3.

To prolong the catalyst lifetime, the raw biomass oil is first refined removing inorganic contaminants (*e.g.* P, Na, Ca, *etc.*). Similarly to coprocessing, it is necessary to dilute the oil. However, the biomass oil is blended with the recycle stream in an HVO process. The dilution proportion depends upon both catalyst type as well as number of catalyst beds needed for controlling the reactor temperature. The catalyst in the HDO reactor is loaded either in its sulfided form or a sulfiding process is performed at the unit start-up, as a common practice in the operation of HDS units.[15] The sulfur concentration on the catalyst surface is maintained stable by the continuous injection of a sulfiding agent (*e.g.* H_2S, DMDS or other organosulfur compound) in the feed or recycle gas.

Following the HDO reactor, the stream is cooled (Figure 9.3, unit operation 2) and the gaseous products are separated (3). The liquid hydrocarbon stream is recovered and rectified so that the dissolved gases are removed (7). The end product is a hydrotreated vegetable oil stream containing *n*-paraffins. The gas stream is purified in an amine column to remove the CO_2

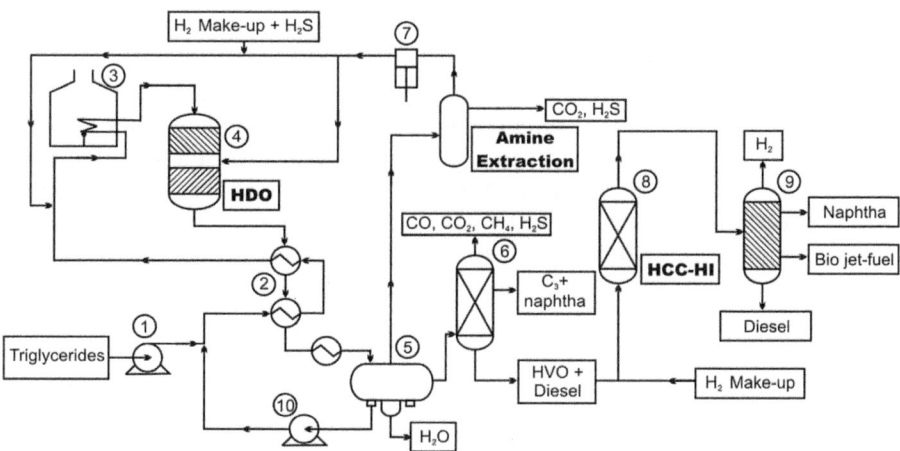

Figure 9.3   General flow sheet of a standalone HVO process. Operation units: (1) Feedstock pump, (2) Heat exchanger, (3) Furnace, (4) HDO reactor, (5) High-pressure separator, (6) Stripping tower (7) Recycle compressor, (8) HCC-HI reactor, and (9) Distillation column.

formed by the hydrotreatment processes, and is thereafter mixed with make-up hydrogen and recycled to the reactor entrance and quench lines (6).

The distribution of the carbon atoms in the end product depends on the fatty acid composition of the vegetable oil or animal fat as well as on the selectivity of the HDO catalyst. Typical feedstocks are palm oil, soya oil and tallow, which are composed of saturated and unsaturated C_{16} and C_{18} fatty acids in different proportions. Should a catalyst showing low decarboxyla-tion selectivity be used, the hydrogen consumption will be high, the CO_2 output will be low, and the product will show a high concentration of C_{16} and C_{18} n-paraffins. However, should a catalyst displaying high decarbox-ylation selectivity be used, the hydrogen consumption will decrease, the CO_2 output will be high, and the product will be composed of C_{15}, C_{16}, C_{17} and C_{18} n-paraffins.

Since n-paraffins are formed in the presence of the HDO catalyst, hydro-isomerization (HI) is needed to improve the cold-flow properties (Figure 9.3, unit operation 8). Different technological approaches are used in order to tackle this problem. Reduced noble-metal catalysts with high HI selectivity may be used; however, the n-paraffins must first be purified to decrease the H_2S concentration to less than 1 ppm. This approach requires the addition of an HI reactor (as well as pumps, compressor, *etc.*). This option is expen-sive and the total energy consumption is high. A more economical alter-native is the use of a catalyst for HI capable of tolerating elevated H_2S concentrations, which allows for the integration of the HI processes within the HDO reactor. Nevertheless, lower selectivity is obtained by this approach compared to that utilizing a noble-metal catalyst. Furthermore, the inte-gration of the HI within an HDO reactor increases the formation of light products (mainly naphtha and gas). In regard to aviation fuels, the design of the industrial unit to provide the maximum yield of biojet fuels is also possible by recycling the nonconverted diesel.[33] Again, the major drawback of this approach is the increase in gas and naphtha yields.[36]

In 2011, biojet fuels have been approved as drop-in fuels. The stream named SPK (Synthetic Paraffin Kerosene) can be blended up to 50% with petroleum-derived kerosene. In practice, the fraction of SPK blended with kerosene is limited in order to meet the standard specifications. Some critical properties are regulated by ASTM D7566; freezing point and distil-lation curve, which are directly affected by the HVO technology, are regulated by ASTM D2386 and ASTM D86, respectively. The respective normal freezing point of C_{16} and C_{18} paraffins are 18.1 and 28.2 °C.[35] Accordingly, deep HI and extensive HDO of the stream are required to meet these fuel specifi-cations. In regard to the distillation curve specifications, the optimal carbon range for a SPK to fit the standards is C_9 to C_{14}. It is therefore crucial to hydrocrack the HDO products.

Due to the standard specifications for freezing point and distillation curve, the application of coprocessing technology to produce biojet fuels is very difficult even for very specific and very expensive vegetable oils (*e.g.* babassu and coconut oils). The products of hydroconversion of such

Table 9.4   Freezing point of biojet fuel blends of interest.[26]

	ASTM D7566	3.3 wt% babassu	4.8 wt% coconut
Freezing point (°C)	−40 (jet A)	−41	−41

vegetable oils are *n*-paraffins with a boiling point compatible with jet fuel specifications. However, as shown in Table 9.4, in order to meet the specification for cold-flow properties, only small amounts of babassu and coconut oils can be coprocessed with kerosene.[26]

Overall, the standalone technology to convert triglycerides into diesel is seemingly one of the most important alternatives to FAME technology. Despite the higher investment compared to the FAME or H-Bio technologies, the better product quality of HVO does not impose percentage volume limits on blending with conventional diesel. Importantly, HVO is currently one of only two drop-in biofuels (the other is SPK from biomass-to-liquid or BTL-technology) that are allowed to be added to aviation fuels (maximum 50 vol%).

9.3   Hydroconversion of Waste Cooking Oil

In order to overcome the increasing costs associated with the cultivation of energy crops as well as the inherent food vs. energy problem, waste cooking oil (WCO) has been explored as an alternative feedstock for the production of biodiesel through transesterification and hydroconversion routes. Several transesterification-based technologies were exploited in the conversion of WCO into FAME.[37–39] However, the variability of WCO quality (*e.g.* free fatty acid and water contents) in addition to its instability under storage conditions pose serious problems to efficient WCO transesterification. Accordingly, catalytic hydroprocessing of WCO has been investigated as an alternative to the transesterification route. The fuel obtained by the catalytic hydrotreating of WCO is often referred to as "white diesel."

Utilizing a commercial hydrocracking catalyst, catalytic hydrocracking of WCO and raw sunflower oil were compared. Equivalent product yields were obtained from both feedstocks; however, slight heteroatom removal superiority was achieved in the case of WCO.[40] The effect of reactor temperature on the catalyst performance was examined on the hydrocracking of WCO.[41] The heteroatom (S, N and O) removal increases with temperature. HDO is particularly the most favored reaction. Nevertheless, the limiting effect of temperature on the saturation of WCO reveals the need for a pretreatment in order to enable saturation of the C=C bonds prior to heteroatom removal.

9.3.1   Operating Parameters, Catalysts, Diesel Yields

Catalytic hydrotreating of WCO was evaluated as a potential catalytic hydroconversion process aimed primarily at diesel production.[42,43] The catalytic hydrotreating technology appears to be an effective conversion technology of WCO to diesel; offering high yields (*ca.* 90%) of a new renewable diesel fuel consisting of mainly linear and some branched paraffins within the C_8–C_{25} range, in the presence of a commercial sulfided NiMo/Al_2O_3 catalyst. The catalytic hydrocracking of WCO could also hold the key for the production of renewable gasoline depending upon the hydrocracking catalyst employed. A study on the conversion of WCO into diesel was performed by examining a wide range of hydrotreating temperatures (603–673 K) while maintaining pressure constant at 8.3 MPa, the liquid hourly space velocity (LHSV) at 1.0 h^{-1}, H_2/oil ratio at 4000 SCFB (standard cubic feet per barrel), liquid feed at 0.33 mL/min, and gas feed at a volumetric flow rate of 0.4 SCFH (standard cubic feet per hour). It was found that low hydrotreating temperatures (603–623 K) maximize the diesel yield, whereas high temperatures (623–673 K) favor the product isomerization, as expected.

Besides the effect of temperature upon the product yield and quality, three additional operating parameters were also explored, *i.e.* pressure, H_2/oil ratio and LHSV.[44] Diesel yield was favored by low pressures (8.2–9 MPa) and low H_2/oil ratios (3000–4000 SCFB), since cracking reactions are not promoted under these low-severity conditions. High LHSV values (1–1.5 h^{-1}) increased the diesel production. In all cases examined, heteroatom-removal effectiveness was also assessed. Excellent results (*i.e.* 99% removal of both sulfur and nitrogen in addition to over 90% removal of oxygen) were achieved.[44]

Table 9.5 compares key properties (density, heteroatom content and distillation curve) of the WCO, unrefined hydrotreated WCO and white diesel. It is clear from Table 9.5 that catalytic hydrotreatment leads to a significant decrease in density and extensive removal of heteroatom contents. In particular, the oxygen content is lowered from 14.57% in WCO to 0.38% in the unrefined hydrotreated WCO. Furthermore, Figure 9.4 shows that the distillation curve of the unrefined hydrotreated WCO (raw product) is markedly shifted to a lower temperature range, compared to that of WCO. The comparison of the data in Table 9.5 and Figure 9.4 shows the white diesel (*i.e.* the refined product) to be a significantly better fuel, compared to the organic phase of the hydrotreated WCO. In fact, the oxygen content of the white diesel is negligible, since the unconverted fatty acids were removed by

Table 9.5 Key properties of WCO, hydrotreated WCO (organic) and white diesel.

		WCO	Hydrotreated WCO (organic)	"White Diesel"
Density (at 60 °C)	g/cm^3	0.8960	0.7562	0.7532
C	wt%	76.74	84.59	86.67
H	wt%	11.61	15.02	14.74
S	ppm	38.00	11.80	1.54
N	ppm	47.42	0.77	1.37
O	wt%	14.57	0.38	0

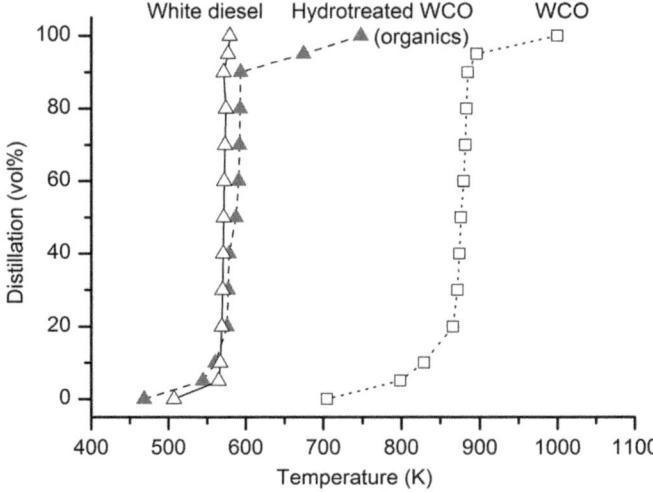

Figure 9.4 Distillation curves for WCO, hydrotreated WCO (organics) and white diesel.

Table 9.6 Properties of the "white diesel" obtained from catalytic hydrotreatment of WCO.

Property	Units	Value
Density (at 60 °C)	g/cm³	0.7532
Sulfur	mg/kg	1.54
Nitrogen	mg/kg	1.37
Cetane index		77.23
Flash point	°C	116
Water	mg/kg	0
Viscosity (at 40 °C)	cSt	3.54
Total acid number (TAN)	mg KOH/g	0
Copper strip corrosion (3 h in 50 °C)		1b
Induction time/oxidation stability (at 110 °C)	h	>22
Distillation of 90%	°C	302.6
Net heating value	MJ/kg	49
Cold filter plugging point (CFPP)	°C	20
Pour point	°C	23

further fractionation. The distillation curve is also shift towards a lower temperature range, especially for the last heaviest fractions, as the new diesel fuel does not contain any heavy molecules, unlike the hydrotreated WCO.

Table 9.6 summarizes the fuel properties of white diesel, showing that this biofuel is a light diesel fuel with no impurities (*i.e.* S, N, O and H_2O). Furthermore, it has an impressively high cetane index that exceeds 77 (compared to the value of *ca.* 50 found for petroleum-derived diesel) and a net heating value of 49 MJ/kg (*vs.* 42–43 MJ/kg of diesel). Furthermore, white diesel shows high oxidation stability (longer than 22 h, well exceeding the 6 h maximum limit of the EN 14214 standard for FAME) as well as negligible acidity, as inferred from the total acid number (TAN). Because of its reduced acidity and increased oxidation stability, "white diesel" does not need antioxidant additives to meet a satisfactory level of oxidation stability. In addition, it is particularly resilient to extended transportation and storage times.[46]

White diesel presents two drawbacks: inferior cold-flow properties and lower densities compared to conventional diesel. Indeed, the cold-flow properties (CFPP) and pour point are particularly high due to the paraffinic nature of white diesel.[43] As this fuel is produced *via* a single hydroprocessing step, triglycerides undergo saturation and deoxygenation, rendering mostly *n*-paraffins, which increase the CFPP and pour point. This problem is possible to overcome by a second hydroisomerization step (which will significantly increase the production costs) or by direct blending with conventional diesel, depending on the national specifications in place. Blending white diesel with conventional diesel is a useful solution for fixing the fuel density. Figure 9.5 shows the effect of mixing "white diesel" with diesel upon the pour point and density. White diesel can be mixed with up to 30 vol% diesel, while still providing a fuel that meets EN590 specifications for pour point and density.[59]

The suitability of the total liquid product of WCO or its fractions as a suitable alternative diesel fuel was also explored in detail.[47] The mid-distillate fractions of the total liquid product were separated by fractionating distillation and were blended in order to produce fuels with properties comparable to those of fuels diesel fuel. These fuels were analyzed according to EN 590. It was found that all fractions within the diesel range (*i.e.* 423–648 K) are suitable to blend with conventional diesel.

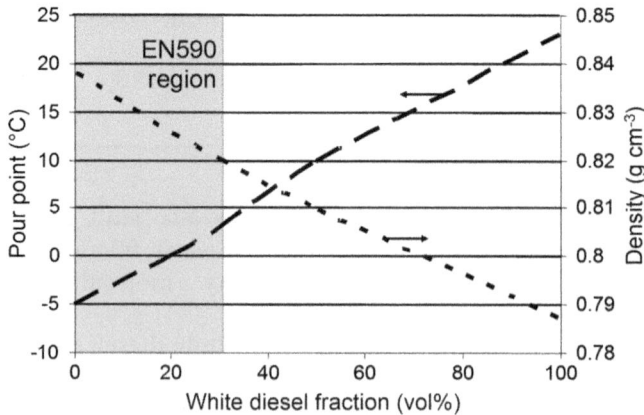

Figure 9.5 Pour point (dashed) and density (dotted) curves found for white diesel and conventional diesel blends.[59]

9.3.2   Technology Demonstration

The single-step process for hydrotreating of WCO to diesel was demonstrated on a pilot plant in order to evaluate both scale-up and applicability of the white diesel technology to the industrial scale in addition to producing enough white diesel for evaluation tests in diesel engines.[48] On that basis, 2 m^3 of "white diesel" were produced from 100% WCO in the Center for Research and Technology Hellas (CERTH), as part of the BIOFUELS-2G project cofounded by the EU program LIFE $+$.[48] In demonstration, 2.5–3 ton of WCO were collected from local restaurants and households after being extensively used for frying. This feedstock was utilized without any pretreatment with the exception of mechanical filtering to remove any food particles remaining in the oil after frying (see Table 9.5 for properties of WCO).

Figure 9.6 provides a simplified schematic representation of the process for production of white diesel. WCO was fed in the continuous large-scale hydroprocessing pilot plant (HDS1) of CERTH with a capacity *ca.* 15 L/day, which consists of a gas–liquid feed system, two reactors connected in series (indicated by "1" in Figure 9.6), and a product separation system for the fractionation of gas and total liquid products (Figure 9.6, operation units 2–4). The process was performed in the presence of an optimal commercial catalytic hydrotreating $NiMo/Al_2O_3$ catalyst,[45] which was presulfided according to the catalyst provider procedures. Hydrogen was introduced to the feed system maintaining 1:500 v/v ratio (WCO/H_2), which was identified as the optimal feed mix ratio.[44]

When the mixture of WCO and H_2 enters the reactor section, it is subjected to triglycerides cracking into FFAs, heteroatom (S, N, O) removal and

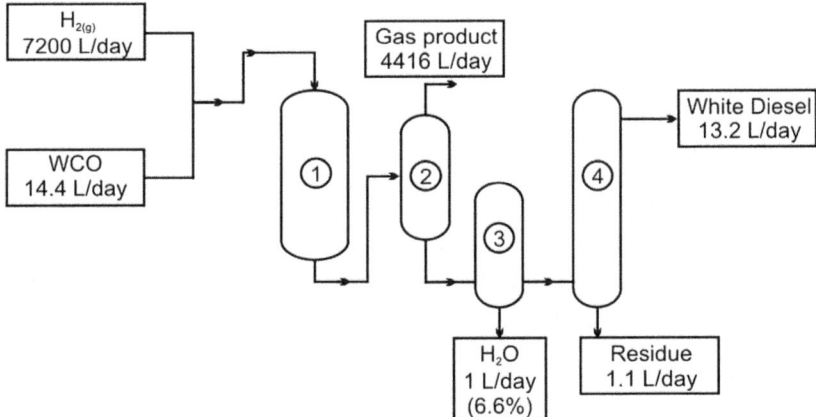

Figure 9.6   Overall pilot-plant process for WCO conversion into "white diesel" developed by Center for Research and Technology Hellas (CERTH), as part of the BIOFUELS-2G project cofounded by the EU program LIFE $+$. Operation units: (1) hydrotreating rector, (2) high-pressure, low-temperature separator, (3) sedimentation, (4) fractionation.
Adapted from Bezergianni and Dimitriadis.[60]

saturation. The product exits the reactor in gaseous state and moves through the separation system in which the byproducts (CO, CO_2, H_2S, H_2O, NH_3, C_1–C_6) and H_2 excess are separated from the desired total liquid product. On an industrial scale, the gas product after cleaning is compressed, and then recycled as a main source of H_2 together with fresh H_2 (as shown in Figure 9.3, operation unit 7). As part of the BIOFUELS-2G project,[41] renewable hydrogen, produced by water electrolysis powered by solar energy,[42,49] is proposed to be used in the process.

The total liquid product contains mainly paraffins and isoparaffins in the diesel range,[43] but also water from decarbonylation and HDO. As the water content is significant (*ca.* 6.6 vol%) and immiscible with the organic phase, an aqueous layer is formed that is easily removed in a precipitation vessel. The resulting water-free organic product (Table 9.5) is finally fractionated in order to collect only the paraffins with boiling points within the diesel range. The overall process has a remarkable 91.67 vol% yield of refined white diesel. The higher boiling point paraffins form a heavy residue (7.64 vol%), which could be used as a blend-in component of industrial fuel or even as a lubricant compound. The fractionation step was demonstrated *via* the batch fractionation unit HYDIS of CPERI/CERTH (40 L per batch capacity).

9.3.3 Coprocessing with Petroleum Fractions

Coprocessing of petroleum fractions with WCO is currently under investigation. This process option enables the incorporation of WCO in an existing refinery without large infrastructure investments. In particular, the integration of residual lipid containing feedstocks in an existing refinery is appealing since it can expedite the large-scale integration of such biobased feedstocks in the fuel market.

The hydrotreatment of heavy atmospheric gas-oil (HAGO) with WCO was studied, aiming to examine the feasibility of integrating WCO into the HAGO hydrotreatment process of an existing refinery.[50] The feedstocks –100% HAGO, 90/10 HAGO/WCO and 70/30 HAGO/WCO – were studied at different temperatures (583, 603 and 623 K), and employing the same commercial hydrotreating catalyst ($NiMo/Al_2O_3$) utilized for HAGO hydrotreatment. The results indicate that heteroatom removal is favored by increasing temperature and is not affected by the presence of WCO in the feedstock, as shown in Figure 9.7. Furthermore, WCO favors the overall conversion, which is not affected significantly by the hydrotreating temperature. Regarding the effect on cold-flow properties, WCO does not exert a negative effect on pour point; however, higher hydrotreating temperatures render hybrid fuels with a lower pour point. As expected, the hydrogen consumption proportionally increases with hydrotreating temperatures when WCO is processed with the HAGO feedstock.

The hybrid diesel production by incorporating WCO in HAGO catalytic hydrotreatment is expected to significantly increase the sustainability of the fuel product. In the case of 70/30 HAGO/WCO mixtures, the resulting hybrid

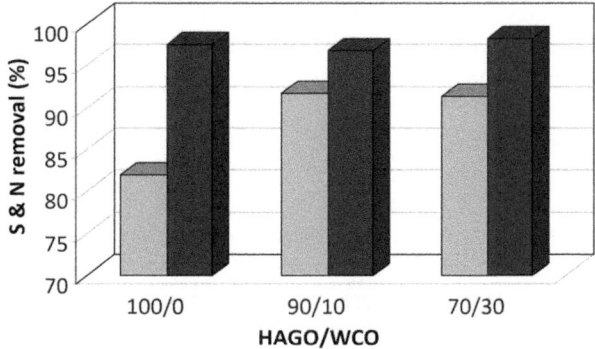

Figure 9.7 Sulfur (black) and nitrogen (gray) removal (in %) of catalytic hydroprocessing of HAGO/WCO mixtures. All experiments were conducted at 623 K, 8.3 MPa of H_2, LHSV: 1.0 h^{-1}, and H_2/oil: 505.9 NL/L. Adapted from Bezergianni and Dimitriadis.[50]

diesel generates 21.4% less equivalent CO_2 emissions than the conventional diesel produced from catalytic hydrotreating of HAGO.[51]

9.4 Conclusions

Hydroconversion processes of vegetable oils and animal fats are undergoing a rapid development, since the hydroprocessing of hydrocarbon streams (*e.g.* petroleum, mineral coal or shale oil), is already present in existing refineries with different purposes, such as sulfur removal or upgrading of heavier streams. As far as the coprocessing technologies are concerned, some have also been tested and used in industrial units. The coprocessing poses no problem to the integrity of the catalytic fixed bed of the HDT unit, simultaneously increasing the yield of hydrotreated liquid products by selective hydroconversion of triglycerides into *n*-paraffins. The levels of hydrogenation of unsaturated hydrocarbons and removal of heteroatom-containing contaminants from the refinery stream of petroleum hydrocarbons are maintained or even improved by the coprocessing with triglycerides.

Different types of vegetable oils and animal fats can be used in hydroconversion. Waste cooking oil (WCO) has also been explored as an alternative residual-based feedstock. Results clearly demonstrate that WCO is a convenient feedstock either for standalone or coprocessing units. Independent of the feedstock, hydroconversion processes lead to biofuels with very low nitrogen and sulfur contents, linear hydrocarbons, and a near absence of aromatic compounds. Such characteristics are extremely favorable for the cetane number (CN) in addition to the reduction of particulate emissions. The formation of a large fraction of linear hydrocarbons and the near absence of aromatics are detrimental to the cold-flow properties, lubricity and density. Essentially, the same problem exists with the kerosene fraction. The freezing point specification of jet fuel (lower than −47 °C) is needed to ensure that the fuel remains as a liquid even at low temperatures experienced by high-altitude flights. This requirement thus places a limit on the amount of linear hydrocarbons in the aviation fuel.[57] For this reason, the hydroisomerization characteristics of the catalyst to be used must be enhanced.[58]

References

1. G. Knothe, Historical Perspectives on Vegetable Oil-Based Diesel Fuels, http://www.oakland.edu/energy/inform_Nov_2001.pdf.
2. F. Ma and M. A. Hanna, *Biores. Technol.*, 1999, **70**, 1–15.
3. B. Milne, *Biofuels Int*, 2011, **5**, 32–33.
4. L. Lin, Z. Cunshan, S. Vittayapadung, S. Xiangqian and D. Mingdong, *Appl. Energy*, 2011, **88**, 1020–1031.
5. S. K. Hoekman, A. Broch, C. Robbins, E. Ceniceros and M. Natarajan, *Renew. Sustain. Energy Rev.*, 2012, **16**, 143–169.
6. J. P. Wauquier, *Crude Oil Petroleum Products Process Flowsheets*, Editions Technip, 1995.
7. M. Tatur, H. Nanjundaswamy and D. Tomazic, *SAE Int. J. Fuels Lubr.*, 2009, **2**, 89–103.
8. K. Varatharajan and M. Cheralathan, *Renew. Sustain. Energy Rev.*, 2012, **16**, 3702–3710.
9. A. Monyem and J. H. Van Gerpen, *Biomass Bioenergy*, 2001, **20**, 317–325.
10. B. R. Moser, *Renew. Energy*, 2001, **36**, 1221–1226.
11. G. Knothe, *Fuel Process. Technol.*, 2007, **88**, 669–677.
12. R. Altin, S. Cetinkaya and H. S. Yucesu, *Energy Conver. Manag.*, 2001, **42**, 529–538.
13. F. D. Gunstone, Vegetable Oils in Food Technology: Composition, Properties, and Uses, in *Chemistry and Technology of Oils and Fats Series* vol. 6, ed. Taylor & Francis Group, 2002, ISBN 0953194922, 9780953194926.
14. F. D. Gunstone and B. G. Herslöf, Lipid glossary 2, in *Oily Press Lipid Library vol. 12*, ed. The Oily Press, 2000, ISBN 0953194922, 9780953194926.
15. T. V. Choudhary and C. B. Phillips, *Appl. Catal. A: Gen.*, 2011, **397**, 1–12.
16. G. W. Huber, P. O'Connor and A. Corma, *Appl. Catal. A: Gen.*, 2007, **329**, 120–129.
17. J. R. Gomes, USPTO Apl., US2006-0186020 A1, 2006.
18. E. Furimsky, *Catal. Rev. - Sci. Eng.*, 1983, **25**, 421–458.
19. J.-C. Duchet, J.-C. Lavalley, S. Housni, D. Ouafi, J. Bachelier, M. Lakhdar, A. Mennour and D. Cornet, *Catal. Today*, 1988, **4**, 71–96.
20. R. G. Egeberg, N. H. Egebjerg, L. Skyum and P. Zeuthen, *ERTC*, Berlin, Germany, 2009.
21. R. Egeberg, N. H. Egebjerg, S. Nystrom, *NPRA Annual Meeting*, Phoenix, AZ,USA, 21–23 March, 2010.
22. L. A. Sousa, J. L. Zotin and V. Teixeira da Silva, *Appl. Catal. A: Gen*, 2012, **449**, 105–111.

23. C. C. Chang and S.-w. Wan, *Ind. Eng. Chem.*, 1947, **39**, 1543–1548.
24. M. Stumborg, D. W. Soveran, W. K. Craig, W. Robinson and K. Ha, In: *Energy from Biomass and Wastes XVI*, D. L. Klass, ed., 1993, 721, Institute of Gas Technology, Chicago.
25. M. Stumborg, A. Wong and E. Hogan, *Biores. Technol.*, 1996, **56**, 13–18.
26. J. R. Gomes and J. L. Zotin, *Rio Oil and Gas*, Rio de Janeiro, Brazil, September, 2012.
27. J. R. Gomes, USPTO, Apl., US Patent 2010/0270207, 2010.
28. F. Fernandes, J. R. Gomes and L. Mignaco, *American Chemical Society 235th National Meeting*, New Orleans, LA, 2008.
29. J. Walendziewski, M. Stolarski, R. Luzny and B. Klimek, *Fuel Process. Technol.*, 2009, **90**, 686–691.
30. F. Baldiraghi, G. Faraci and S. Guanziroli, *ERTC Biofuel Workshop*, Barcelona, 2007.
31. S. Honkanen, *ERTC 14th Annual Meeting*, Berlin, Germany, 9–11 November, 2009.
32. Y. Scharff, From Renewable Lipids to Transportation Fuels, paper presented at *WRA Biofuels, 6th annual conference*, Amsterdam, 2011, Available from: http://core.theenergyexchange.co.uk/agile_assets/1539/12.10_Yves_Scharff,_AXENS.pdf (last viewed December, 15, 2012).
33. L. R. Abhari, USPTO Pat., US 7,846,323 B2, 2010.
34. V. J. Kröger, Neste Oil renewable fuel: Leading the way forward, paper presented at *Simpósio Nacional de Biocombustíveis de Aviação*, September, 14, 2012, Brasília, Brazil.
35. *TRC Thermodynamic Tables - Non-Hydrocarbons*. ed. M. Frenkel, National Institute of Standards and Technology, Boulder, CO, Standard Reference Data Program, Publication Series NSRDS-NIST-74 (1955-2010), Gaithersburg, MD.
36. F. A. Twaiq, A. R. Mohamed and S. Bhatia, *Micropor. Mesopor. Mater.*, 2003, **64**, 95–107.
37. Z. Helwani, M. R. Othman, N. Aziz, W. J. N. Fernando and J. Kim, *Fuel Process. Technol.*, 2009, **90**, 1502–1514.
38. X. Meng, G. Chen and Y. Wang, *Fuel Process. Technol.*, 2008, **89**, 851–857.
39. K. Jacobson, R. Gopinath, L. C. Meher and A. K. Dalai, *Appl. Catal. B: Environ.*, 2008, **85**, 86–91.
40. S. Bezergianni, S. Voutetakis and A. Kalogianni, *Ind. Eng. Chem. Res.*, 2009, **48**(18), 8402–8406.
41. S. Bezergianni and A. Kalogianni, *Biores. Technol.*, 2009, **100**, 3927–3932.
42. S. Bezergianni, A. Dimitriadis, A. Kalogianni and P. A. Pilavachi, *Biores. Technol.*, 2010, **101**, 6651–6656.
43. S. Bezergianni, A. Dimitriadis, T. Sfetsas and A. Kalogianni, *Biores. Technol.*, 2010, **101**, 7658–7660.
44. S. Bezergianni, A. Dimitriadis, A. Kalogianni and K. G. Knudsen, *Ind. Eng. Chem. Res.*, 2011, **50**, 3874–3879.
45. S. Bezergianni, A. Kalogianni and A. Dimitriadis, *Fuel*, 2012, **93**, 638–641.

46. S. Bezergianni and L. Chrysikou, *Biores. Technol.*, 2012, **126**, 341–344.
47. D. Karonis, S. Bezergianni, D. Chilari and E. Kelesidis, *Proc. Am. Chem. Soc., Div. Fuel Chem.*, 2010, **55**, 1.
48. www.biofuels2g.gr [Last viewed December 15, 2012].
49. C. Ziogou, D. Ipsakis, F. Stergiopoulos, S. Papadopoulou, S. Bezergianni and S. Voutetakis, *Int. J. Hydrogen Energ.*, 2012, **37**(21), 16591–16603.
50. S. Bezergianni and A. Dimitriadis, *Fuel*, 2013, **103**, 579–584.
51. S. Bezergianni, A. Dimitriadis and L. Chrysikou, Diesel Sustainability Improvement Perspectives by Incorporation of Waste Cooking Oil in Existing Refinery of Thessaloniki, In: *Proceedings of 5th International Scientific Conference on Energy and Climate Change*, Athens, Greece, 11–12 October 2012, 2, 124–129.
52. D. Hufschmidt, L. Bobadilla, F. Romero-Sarria, M. A. Centeno, J. A. Odriozola, M. Montes and E. Falabella, *Catal. Today*, 2010, **149**, 394–400.
53. A. P. Ferreira, D. Zanchet, J. C. S. Araújo, J. W. C. Liberator, E. Falabella, F. Noronha and J. M. C. Bueno, *J. Catal.*, 2009, **263**, 335–344.
54. J. Ruiz, F. Passos, J. M. Bueno, E. Falabella, L. Mattos and F. Noronha, *Appl. Catal.. A, Gen.*, 2008, **334**, 259–267.
55. G. Arzamendi, P. Diéguez, M. Montes, J. Odriozola, E. Falabella and L. Gandía, *Chem. Eng. J*, 2009, **154**, 168–173.
56. J. Alvarez, G. Valderrama, E. Pietri, M. J. Pérez-Zurita, C. U. Navarro, E. Falabella and M. Goldwasser, *Top*, 2011, **154**, 170–178.
57. E. Falabella, F. Bellot Noronha and A. Faro Jr., *Catal. Sci. Technol.*, 2011, **1**, 698–713.
58. R. P. Silvy, A. C. B. Santos and E. Falabella, *Oil Gas J.*, 2010, **108**, 5170498171.
59. S. Bezergianni, A. Dimitriadis and L. Chrysikou, *Fuel*, 2014, **118**, 300–307.
60. S. Bezergianni and A. Dimitriadis, Catalytic Hydrotreating of Waste Cooking Oil for Renewable Diesel Production, In: *Proceedings of the 9th International Colloquium Fuels - Conventional and Future Energy for Automobiles*, Esslingen, Germany, 15–17 January 2013.

CHAPTER 10

Catalytic Hydrogenation of Vegetable Oils

AN PHILIPPAERTS,* PIERRE JACOBS AND BERT SELS*

KU Leuven, Center for Surface Chemistry and Catalysis, Kasteelpark
Arenberg 23 bus 2461, 3001 Heverlee, Belgium
*Email: An.philippaerts@biw.kuleuven.be; Bert.sels@biw.kuleuven.be

10.1   Introduction

10.1.1   Vegetable Oils

Vegetable oils comprise a mixture of triacylglycerol molecules (*i.e.* tri-glycerides) that are composed of a glycerol backbone and three fatty acid chains. The fatty acids are dissimilar in chain length. They usually contain an even number of C atoms, predominantly ranging from C_{12} to C_{20}. According to the number of double bonds present in the chain, fatty acids are generally classified into three groups: (1) saturated, (2) monounsaturated, and (3) polyunsaturated fatty acids (*i.e.* containing more than one double bond in the carbon chain). Regarding the notation, stearic acid (S), which contains 18 carbon atoms and 0 double bonds, is denoted as C18:0; in turn, linoleic acid (L), which contains 18 carbon atoms and two double bonds, is thus denoted as C18:2. Often, the position of the double bonds, counting from the acid group, is added to the notation together with their configuration (*cis* or *trans*). For instance, linoleic acid is referred to as C18:2 *c*9,*c*12 as the two double bonds are both in the *cis* configuration and located at the positions 9 and 12.[1] A list of common fatty acids is given in Table 10.1.

RSC Energy and Environment Series No. 13
Catalytic Hydrogenation for Biomass Valorization
Edited by Roberto Rinaldi
© The Royal Society of Chemistry 2015
Published by the Royal Society of Chemistry, www.rsc.org

Table 10.1 Some examples of common C18 fatty acids.

Chemical structure	Name	Notation
	stearic acid stearate	C18:0 S
	oleic acid oleate	C18:1 *c*9 O
	elaidic acid elaidate	C18:1 *t*9 E
	linoleic acid linoleate	C18:2 *c*9*c*12 L
	linolenic acid linolenate	C18:3 *c*9*c*12*c*15 Ln

The fatty acids occur at differing positions on the glycerol backbone, *i.e.* at central (*sn*-2) or terminal (*sn*-1/3) positions. Therefore, vegetable oils may be regarded as a very complex mixture of numerous triacylglycerol molecules. The chain length and number, configuration and position of the C=C bonds of the fatty acids as well as their connectivity on glycerol backbone play major roles in the chemical and physical properties of the vegetable oils and fats, and hence determine their application potential.[1] Although food applications encompass the major uses of vegetable oils,[2] a significant proportion is also used for industrial applications (*e.g.* lubricants, plasticizers, coatings, paints, cosmetics, pharmaceuticals, bioplastics and biodiesel).[3–5]

Nature anticipated these various applications of vegetable oils, providing us with vegetable oils of many different chemical compositions. For instance, both highly saturated oils (*e.g.* palm, palm kernel, and coconut oils) and polyunsaturated oils (*e.g.* soybean, rapeseed and sunflower oils) occur in nature.[2] The overall number of double bonds in the oil is typically defined by the iodine value (IV). IV corresponds to the iodine weight (in grams) that is consumed by 100 g of oil or fat through the addition of I_2 to the carbon–carbon double bonds present in the fatty acid chains. Hence, a high IV implies that a high content of unsaturation is present in the oil.

Although vegetable oils with varying chemical compositions are naturally available, they are often catalytically modified to broaden or specify their use. Hydrogenation is the most important catalytic process performed on vegetable oils by the food industry. Nonetheless, fully and partially hydrogenated fatty acids and oils also find important applications in the chemical industry.[4,6] Full hydrogenation of vegetable oils may be desirable for some applications (*e.g.* cosmetics, pharmaceutical products, surfactants, to mention just a few).[7] Fully hydrogenated fats are also used in interesterification processes with liquid vegetable oils, in order to obtain *trans*-free semisolid fats with specific melting properties.[8] In turn, vegetable oils are partially hydrogenated for two main purposes.[1] The first is to increase the stability of the oil by selective partial hydrogenation of polyunsaturated fatty acids (*e.g.* linolenic acid, C18:3), which are especially prone to autoxidation. Importantly, the full hydrogenation to saturated fatty acids and isomerization to *trans* fatty acids should be minimized because of their high melting points they may not meet the specifications for applications in food or chemical industry. For instance, significant levels of saturated or *trans* fatty acids may lead to cloudy edible oils or lubricants with poor low-temperature characteristics. The second purpose is the conversion of liquid oils into semisolid fat products for use in various food applications, mostly margarines and shortenings. Therefore, hydrogenated fats require a specific melting profile, depending on their application.

10.1.2 Catalytic Hydrogenation of Vegetable Oils

The catalytic hydrogenation process of fatty oils was patented by Normann in 1903.[9] At the beginning of the 20th century the industrial hydrogenation of fish and vegetable oils to various oil and fat products for food applications rapidly grew since this process allowed the production of less-expensive alternatives to animal fats. Industrial processes for hydrogenation of vegetable oils are often conducted in a batch reactor at elevated temperatures (453–473 K) under a low hydrogen pressure (0.05–0.25 MPa) using a solid nickel catalyst. In the last decades, hydrogenated vegetable oils have been associated with negative health aspects due to their potentially high level of *trans* isomers, which have been shown to be a risk factor for coronary heart diseases.[10,11] In 2003, Denmark was the first country to regulate the amount of *trans* fatty acids in food products, limiting their level to 2%.[12] In the European Union, there is still no general legislation regulating the presence of *trans* fats in food products. However, an EU report on the presence of *trans* fats in foods, accompanied with a legislative frame, is expected by December 2014.[13] In the meantime, the presence of hydrogenated fats should be mentioned on the food packaging, although this does not give any information about the amount of *trans* isomers present, as some *trans* isomers (*e.g.* vaccenic acid, C18:1 *trans*-11) naturally occur in several food products. In the United States, there is also no general regulation on the maximum amount of *trans* fatty acids allowed in food products, but the amount of *trans* fats should be labeled on the food packaging since 2006.[14]

Accordingly, hydrogenated fats have increasingly been replaced by other modified fats (*e.g.* fractionated and interesterified fats). The modified fats are mainly based on tropical oils, (*e.g.* palm oil), as they do not contain polyunsaturated fatty acids, and hence they do not require a stabilization process. There is currently a debate on the health issues of these tropical *trans*-free fats, as they might increase the bad cholesterol levels due to the presence of the shorter saturated fatty acids.[15,16] The use of vegetable oils with a low content of polyunsaturated fatty acids (*e.g.* high oleic sunflower) is another solution to avoid chemical hydrogenation. Despite this, hydrogenation of vegetable oils remains an important process in the food industry, since it is the only process capable of converting unstable, liquid vegetable oils into stable, (semi-)solid fat products with defined physical properties. To provide hydrogenated fats free of *trans* isomers, new catalysts and catalytic concepts have been developed, as will be addressed in this chapter.

10.2 Activity and Selectivity in the Catalytic Hydrogenation of Vegetable Oils

Hydrogenation and isomerization of vegetable oils are simultaneous reactions explained by the Horiuti–Polanyi mechanism. This mechanism is schematically illustrated in Figure 10.1 for the hydrogenation of oleic acid (C18:1 *c*9). In the presence of a Ni catalyst, the first step (a) involves the dissociative H_2 chemisorption and sorption of the double bond of the unsaturated fatty acid as a diadsorbed species on the Ni surface. After addition of a hydrogen atom, a half-hydrogenated intermediate is formed (b), which can either accept a second hydrogen atom, leading to the saturation of the double bond, (c) or lose a hydrogen atom, resulting in the geometric and positional isomers (d) and (e).[17–20]

Regarding full hydrogenation of vegetable oils, only the catalytic activity is important for the process. However, in the partial hydrogenation, selectivity is the key point. Hence, the term "selectivity" may be referred to as (1) *polyene selectivity* (*i.e.* with respect to the hydrogenation of polyunsaturated to monounsaturated fatty acids), (2) *geometric isomer selectivity* (*i.e.* regarding the formation of *trans* isomers), and (3) *triglyceride selectivity* (*i.e.* accounting for the distribution of fatty acids within the triglyceride molecules). The process selectivity can be tuned by the catalyst type, hydrogenation conditions, and oil feedstock, as will be discussed in Section 10.2.2.

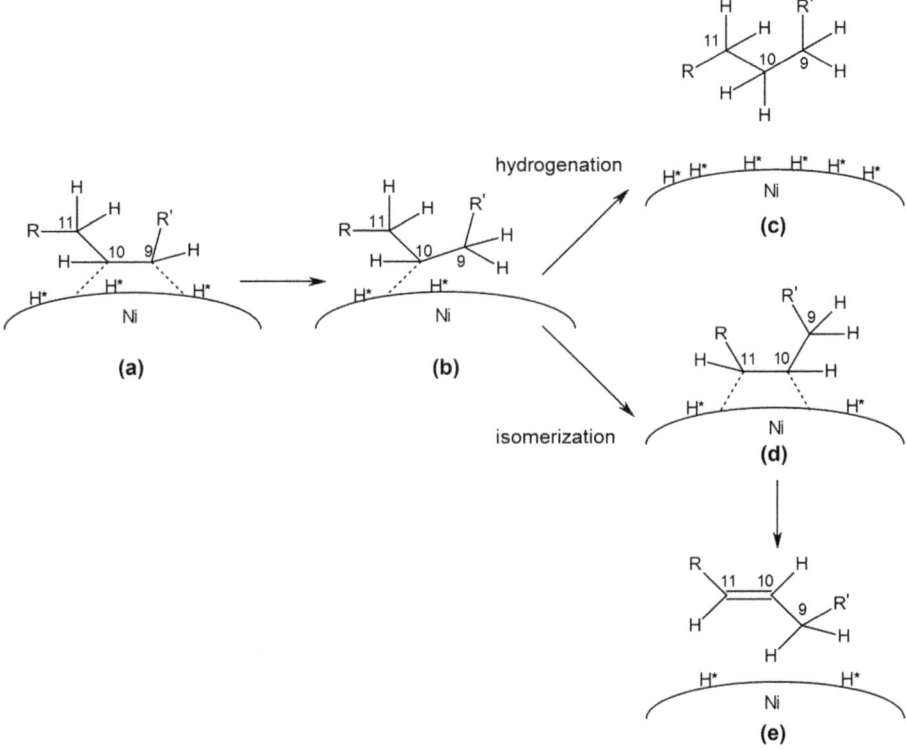

Figure 10.1 Horiuti–Polanyi mechanism explaining the hydrogenation and isomerization of oleic acid over a Ni catalyst: (a) Sorption of the double bond on the catalyst surface forming a diadsorbed complex; (b) Formation of a half-hydrogenated intermediate after reaction with an adsorbed hydrogen atom; (c) Should the concentration of hydrogen near the catalyst surface be high, a second hydrogen atom will react with the half-hydrogenated intermediate, leading to the formation of the saturated product (stearic acid); (d) Should the concentration of hydrogen near the catalyst surface be low, a hydrogen atom will be removed from the half-hydrogenated intermediate, leading again to readsorption of a diadsorbed specimen. Formation of various precursors of geometric and positional isomers is possible at this stage. (e) Desorption of geometric and positional isomers.

The selectivity of the partial hydrogenation of vegetable oils defines the application of the final product. For instance, stabilized cooking oils or lubricants are produced by hydrogenation of polyunsaturated fatty acid chains in order to obtain a liquid product with an IV of 105–110, with minimal formation of saturated and *trans* fatty acids. Some food applications demand semisolid fats with specific melting profiles. For example, shortening or baking fats have typical melting point of 30–40 °C and an IV of 65–85.[21] Fully saturated triglycerides (IV < 1) are generally undesired for food products, as their high melting points lead to a sandy mouth sensation due to the presence of precipitating high melting point solids.

10.2.1 Selectivity Concepts

10.2.1.1 Polyene Selectivity

This selectivity concept was proposed by Bailey in 1949[22] and further developed by Albright in 1965.[23] For determining the polyene selectivity, a simplified scheme of consecutive first-order and irreversible reactions, omitting isomerization,[24] is given as:

$$[C18\!:\!3] \xrightarrow{k_3} [C18\!:\!2] \xrightarrow{k_2} [C18\!:\!1] \xrightarrow{k_1} [C18\!:\!0] \tag{10.1}$$

where, k_1, k_2 and k_3 are the corresponding hydrogenation rate constants.

Hence, the polyene selectivity is defined as:

Linoleic acid selectivity: $S_{Lo} = k_2/k_1$
Linolenic acid selectivity: $S_{Ln} = k_3/k_2$

The polyene selectivity may be calculated from the kinetic equations for the consecutive first-order reactions by iteration procedures using eqns (10.2)–(10.4).[25] This procedure implies that both ratios are conversion independent.

$$[C18\!:\!3]_t = [C18\!:\!3]_0 e^{-k_3 t} \tag{10.2}$$

$$[C18\!:\!2]_t = [C18\!:\!3]_0 \left(\frac{k_3}{k_2 - k_3}\right)\left(e^{-k_3 t} - e^{-k_2 t}\right) + [C18\!:\!2]_0 e^{-k_2 t} \tag{10.3}$$

$$[C18\!:\!1]_t = [C18\!:\!3]_0 \left(\frac{k_3}{k_2 - k_3}\right)\left(\frac{k_2}{k_1 - k_3}\right)\left(e^{-k_3 t} - e^{k_1 t}\right)$$

$$- [C18\!:\!3]_0 \left(\frac{k_3}{k_2 - k_3}\right)\left(\frac{k_2}{k_1 - k_2}\right)\left(e^{-k_2 t} - e^{-k_1 t}\right) \tag{10.4}$$

$$+ [C18\!:\!2]_0 \left(\frac{k_2}{k_1 - k_2}\right)\left(e^{-k_2 t} - e^{-k_1 t}\right) + [C18\!:\!1]_0 e^{-k_1 t}$$

High linoleic acid selectivity (S_{Lo}) indicates a much faster hydrogenation of diunsaturated to monounsaturated fatty acids, compared to hydrogenation of monounsaturated to saturated fatty acids. Therefore, high levels of monounsaturated fatty acids in the triglyceride accumulate before significant levels of saturates with stearic acid chains are formed. Similarly, high linolenic acid selectivity (S_{Ln}) indicates the selective hydrogenation of linolenic to linoleic acid. Consequently, before significant levels of saturated tails in the triacylglycerol are formed, oxidative-labile linolenic and

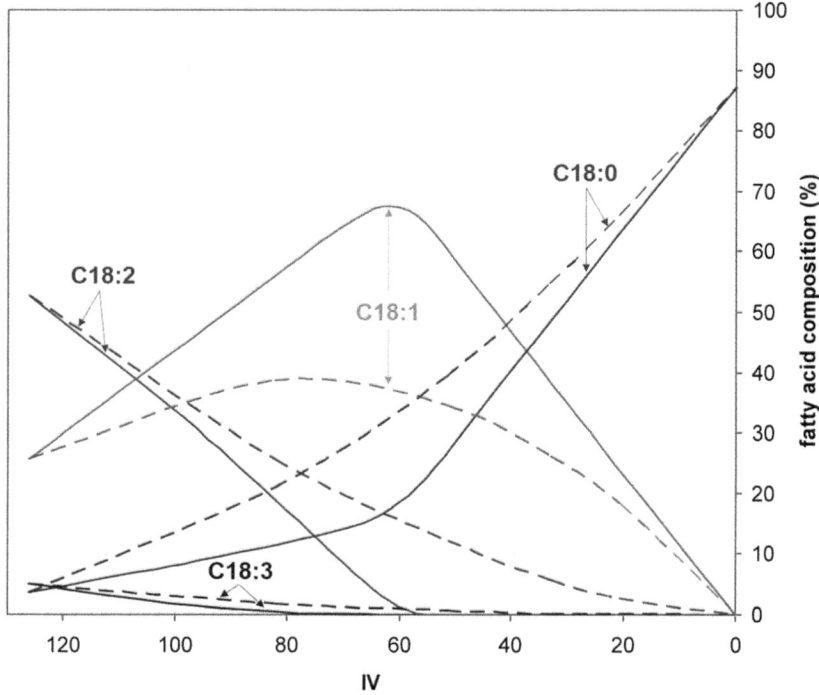

Figure 10.2 Theoretical fatty acid composition curves for the hydrogenation of soybean oil with a starting composition of 5.0% C18:3, 52.7% C18:2, 25.7% C18:1, 3.6% C18:0; 12.6% C16:0; 0.4% C22:0 and an IV of 126. The full lines represent the changes in C_{18} fatty acids during a catalytic hydrogenation process with high polyene selectivity ($S_{Ln} = 2$ and $S_{Lo} = 50$). The dotted lines represent the changes in C18 fatty acids during a catalytic hydrogenation process with low polyene selectivity ($S_{Ln} = 1.2$ and $S_{Lo} = 1.6$).

linoleic acid will be selectively hydrogenated to monounsaturated fatty acids, leading to high S_{Lo} and S_{Ln} values. These observations are also illustrated in Figure 10.2 in which the theoretical fatty acid compositions are plotted for two different catalytic hydrogenations of soybean oil as a function of the IV. The full lines represent the changes in C_{18} fatty acids during a catalytic hydrogenation process with high polyene selectivity values ($S_{Ln} = 2$ and $S_{Lo} = 50$). The dotted lines represent the changes in C_{18} fatty acids during a catalytic hydrogenation process with low polyene selectivity values ($S_{Ln} = 1.2$ and $S_{Lo} = 1.6$). It is clear that in the case of the process with high polyene selectivity values, linolenic acid (C18:2) is more selectively converted and much higher levels of C18:1 can be accumulated prior to further hydrogenation rendering stearic acid (C18:0). Nevertheless, to interpret the results it is important to be aware of the simplifications and assumptions made in this model (*e.g.* *cis–trans* isomerization is neglected).[24]

10.2.1.2   Geometric Isomer Selectivity

The selectivity toward the formation of *trans* fatty acids is expressed by the specific isomerization index (SII), which is the number of *trans* double bonds formed per hydrogenated double bond.[26] The SII is calculated using:

$$\text{SII} = \frac{[trans]_t - [trans]_0}{IV_0 - IV_t} \times 100 \qquad (10.5)$$

where $[trans]_0$ and $[trans]_t$ stand for the concentration of *trans* fatty acids in the substrate and in the hydrogenated product at a given time t, respectively; IV_0 and IV_t correspond to the iodine values found for the substrate and hydrogenated product at a given time t, respectively.

It is noteworthy that *trans* fatty acids are not exclusively formed by a chemical hydrogenation process. Considerable amounts of *trans* isomers also naturally occur in meat. There is current discussion in the literature on the health effects of *trans* fats, formed during partial hydrogenation (*e.g.* elaidic acid, C18:1 *trans*-9) and those of natural occurrence (*e.g.* vaccenic acid, C18:1 *trans*-11).[27,28]

10.2.1.3   Triglyceride Selectivity

Fats with nearly the same fatty acid distribution show distinguished physical properties as a result of the different triglyceride distribution. In spite of this observation, there is no unambiguous definition of triglyceride selectivity in the literature. In 1976, Coenen defined the triglyceride selectivity (S_T) referring to the degree of randomness of double bond hydrogenation within a single triglyceride molecule.[26] Compared to the other selectivity concepts, the triglyceride selectivity is much more difficult to express in quantitative terms. Should the amount of tristearate (SSS) be of importance, eqn (10.6) can be used:[26]

$$S_T = \frac{[C18:0] - [SSS]}{[C18:0] - [C18:0]^3} \qquad (10.6)$$

where [C18:0] and [SSS] represent the concentrations of stearic acid and tristearate, respectively.

Also, the term "positional selectivity" is occasionally used in order to express the dependence of the rate of hydrogenation of an unsaturated fatty acid on its position (*sn*-2 or *sn*-1/3) within the triglyceride molecule.[29] To evaluate such selectivity, it is important to take the natural positional distribution within vegetable oils into account. Interestingly, polyunsaturated fatty acids are mainly located at the central (*sn*-2) position, while saturated fatty acids are almost exclusively encountered at the terminal (*sn*-1/3) positions on the glycerol backbone. In contrast, animal fats demonstrate generally a more random distribution of fatty acids within the triglyceride molecules. Accordingly, Dijkstra proposed the triglyceride selectivity as "the phenomenon that the rate of hydrogenation of an unsaturated fatty acid in a triglyceride molecule depends upon the chemical identity of the other fatty acids in the molecule."[29]

10.2.2 Influence of the Hydrogenation Conditions

The selectivity types defined for the hydrogenation of vegetable oils are significantly affected by reaction conditions.[30-34] An overview of the effects of several variables on the selectivity of the hydrogenation of vegetable oils is given in Table 10.2. Typically, under conditions that the concentration of hydrogen near the catalyst surface is low, high polyene (S_{Ln}, S_{Lo}), geometric isomer (SII) and triglyceride (S_T) selectivity are achieved. The higher polyene selectivity is explained by the stronger adsorption ability of polyunsaturated fatty acids on the catalyst surface, compared to the monounsaturate counterparts. Under low H_2 pressures, the preferential adsorption of polyunsaturated fatty acid chains hinders the hydrogenation of the monounsaturated counterparts. Consequently, S_{Ln} and S_{Lo} assume high values. Under such conditions, the removal of hydrogen atoms from the half-hydrogenated intermediate is also favored, as explained by the Horiuti–Polanyi mechanism (Figure 10.1). Hence, isomerization will be preferred over hydrogenation resulting in higher SII levels.

In the presence of commercial Ni catalysts, a significant effect of pressure on the triglyceride selectivity is also noted.[35] Under high H_2 pressures (3.5 MPa), considerable quantities of di- and trisaturated triglycerides are formed (*i.e.* low S_T). Conversely, under low H_2 pressure (0.35 MPa), no

Table 10.2 Effect of reaction variables on the performance during hydrogenation of fatty oils (↑ and ↓ stand for high and low values of the properties, respectively).

Increasing parameter	H_2 concentration near catalyst surface	Activity	S_{Ln}/S_{Lo}	SII	S_T
Hydrogenation conditions					
Hydrogen pressure	↑	↑	↓	↓	↓
Agitation	↑	↑	↓	↓	↓
Temperature	↓	↑	↑	↑	↑
Catalyst concentration	↓	↑	↑	↑	↑
Oil feedstock					
Oil unsaturation	↓	↑	↑	↑	↑
Oil refining	↑	↑	↓	↓	↓
Catalyst properties					
Catalyst activity	↓	↑	↑	↑	↑
Pore size	↓	↑	↑	↑	↑
Pore length	↑	↓	↓	↓	↓
Surface area	↓	↑	↑	↑	↑
Metal dispersion	↓	↑	↑	↑	↑

trisaturates are produced (*i.e.* high S_T). Obviously, the likelihood of full hydrogenation within the same triglyceride molecule increases with the H_2 pressure, and thus leads to low triglyceride selectivity.

Evidently, a decrease in hydrogen pressure and agitation combined with an increase in catalyst loading leads to decreased hydrogen concentrations near the catalyst surface, and therefore to high polyene and isomerization selectivity. It is clear from Table 10.2 that an increase in temperature leads to lower hydrogen levels near the catalyst surface. Obviously, the higher demand for hydrogen caused by the exponential increase in activity by applying a higher temperature dominates the higher solubility of hydrogen in the oil at elevated temperatures. Consequently, the catalyst surface becomes poor in hydrogen, and the values for the different types of selectivity tend to increase.

10.2.3   Influence of the Oil Feedstock

Polyunsaturated oils with a high unsaturation degree (high IV) are generally hydrogenated much faster compared to monounsaturated oils.[21,36] The high hydrogenation rates lead to depletion of hydrogen at the catalyst surface, and hence to high polyene, geometric and triglyceride selectivity (see Table 10.2). Also, the refining grade of the vegetable oil has an influence on the catalytic performance. Impurities present in the oil (*e.g.* water, free fatty acids, phospholipids, soaps, sulfur compounds and pigments) can act as catalyst poisons.[36] Soaps, chlorophyll and sulfur compounds are known to promote the formation of *trans* isomers.[34] This result is explained through the partial poisoning of the catalytic Ni sites by these impurities, resulting in a low H_2 coverage on the catalyst surface, and therefore increasing the isomerization selectivity.

10.2.4   Influence of the Catalyst Properties

Traditionally, the hydrogenation process is conducted with a supported Ni catalyst. The specific characteristics of the support (*e.g.* pore architecture) as well as the location and dispersion of Ni clusters exert major effects on the catalyst performance. Generally, all properties of the catalyst that increase the catalytic activity (*e.g.* high surface area, high metal dispersion, large pore sizes, *etc.*) increase the polyene, geometric and triglyceride selectivity because of the depletion of adsorbed hydrogen atoms on the catalyst surface.

With an average size of 1.5–2 nm, triglycerides are bulky molecules. Therefore, mass-transfer limitations are of importance when assessing the catalytic performance of different materials in the hydrogenation of vegetable oils.[32] Obviously, catalysts with large pore sizes are preferred to allow for fast diffusion of the bulky triglyceride molecules toward the catalytic sites. In the absence of mass-transfer limitations (*i.e.* under chemical regime conditions), polyunsaturated fatty acids will be preferably hydrogenated, resulting in high polyene selectivity. The low probability of full hydrogenation of the triglyceride molecule is conducive to high triglyceride selectivity. In contrast, the likelihood of full hydrogenation of unsaturated fatty acids increases under the diffusional regime, as found for catalysts with small pore sizes.[26] Moreover, since the catalytic activity of these materials is low due to the mass-transfer limitations, the concentration of hydrogen adsorbed on the catalyst surface builds up, suppressing the formation of *trans* isomers. Also, the average pore length should also be considered in accordance with the theory of Knudsen diffusion. Indeed, lengthy pores hinder the mass transfer of triglyceride molecules, and hence have a negative effect on the catalytic activity as well as on the polyene, geometric and triglyceride selectivity.[32]

10.3 Trends in the Catalytic Hydrogenation of Vegetable Oils

The future of partially hydrogenated oils and fats in food industry relies upon the minimization of the levels of *trans* fatty acids to zero, without impairing other hydrogenation selectivity aspects (*i.e.* polyene and triglyceride selectivity). For the production of biolubricants, high polyene selectivity is needed, while a limited level of *trans* fatty acids is acceptable. It is apparent from Table 10.2 that low SII values are always related to low S_{Ln}, S_{Lo} and S_T values and *vice versa*. It is not obvious how to break the connection between geometric (SII) and the various types of hydrogenation selectivity (S_{Ln}, S_{Lo} and S_T).

Extensive research aiming at the suppression of the *cis–trans* isomerization occurring in the hydrogenation of fatty oils has been published. Low *trans* levels can be obtained by either using alternative catalysts (*e.g.* copper-based and precious-metal catalysts)[37,38] or alternative hydrogenation techniques (*e.g.* catalytic transfer hydrogenation,[39] electrocatalytic hydrogenation,[40–43] and hydrogenation under supercritical conditions).[44,45] However, only a few examples of hydrogenation processes, which combine low *trans* levels with an acceptable hydrogenation selectivity, are reported in the literature. In fact, to produce low *trans* fatty acid levels, the hydrogenation processes are conducted under conditions that often result in low polyene, and likely low triglyceride selectivity, although the analyses of triglycerides are, unfortunately, seldom reported. This observation implies that considerable levels of linolenic acid remain unconverted and/or (tri)saturates are formed, minimizing the application potential of the hydrogenated products.

Table 10.3 compares the performance of the most promising catalysts in the hydrogenation of soybean oil at an IV of 70–80. In these examples, soybean oil was hydrogenated to semisolid fats, rendering low levels of polyunsaturates, saturates, *trans* isomers, through the use of new types of catalyst, and applying appropriate hydrogenation conditions.

Table 10.3 Performance of different catalysts in the hydrogenation of soybean oil.

| Catalyst | T (K) | P(H$_2$) (MPa) | IV | Composition (wt%) | | | | | |
				C18:0	C18:1	C18:2	C18:3	*Trans*	Ref
Pricat 9910	453	0.3	81	5	72	10	1	34	49
Pricat 9910[a]	345	2	85	12	54	21	0	10	49
Pricat 9910[a]	323	2	69	23	50	15	0	7	49
Nano-Pt/silica	323	0.4	70	26	40	17	0.7	4.5	75
Pt/alumina	338	6	65	36	29	21	1.6	1.3	78
Pt/ZSM-5	338	6	63	25	50	11	0.2	3.2	78

[a]Pretreated with soybean oil under N$_2$ for 2 h.

10.3.1 Modified Ni Catalysts

Ni catalysts are active in hydrogenation and they allow highly unsaturated fatty oils to be partially hydrogenated, rendering products with excellent linolenic and linoleic selectivity. In general, soybean oil hydrogenation at temperatures of 438–473 K under H$_2$ pressures of 0.05-0.2 MPa results in products with S_{Ln} and S_{Lo} about 2–3 and 60, respectively. However, Ni catalysts are also characterized by a high specific isomerization index (SII) possibly because of its moderate activity in activating molecular hydrogen. When starting from soybean oil, typically around 50 wt% *trans* fatty acids are formed at an IV of 80.[46–48] Under higher hydrogen pressures, a lower content of *trans* fats are formed but also compromise S_{Ln} and S_{Lo}.

An interesting option to decrease the *cis–trans* isomerization is the use of a low-temperature hydrogenation, typically at 393 K. Under these conditions, the polyunsaturated fatty acid content of the oil can be lowered with only a minimal increase of saturates and *trans* fatty acids.[21] However, a major drawback is the poor activity of Ni catalysts at such low temperatures.

Interestingly, a recent patent describing the use of a modified Ni catalyst, which shows exceptionally high activities at very low temperatures (323–348 K), was published.[49] The catalyst was obtained by simply pretreating a commercial Ni catalyst in a vegetable oil under N$_2$ at 383 K for 2 h. The pretreated catalyst is active in the hydrogenation of soybean oil at 345 K under a H$_2$ pressure of 2 MPa. The process is claimed to render only 10 wt% of *trans* isomers at an IV of 85. In addition, linolenic acid is fully converted, and only 12 wt% stearate is formed. At the reaction temperatures as low as 323 K, 7 wt% *trans* fatty acids are formed at an IV of 69. However, the polyene selectivity is negatively affected, as indicated by the high stearate content (23 wt%, Table 10.3). The chemical concepts on the basis of these results is unclear. Understanding the interaction among impurities and the catalyst may be worth investigating.

10.3.2 Copper-Based Catalysts

In the 1960s and 1970s, the hydrogenation of vegetable oils with Cu and Cu-chromite catalysts were intensively studied because of their exceptionally high polyene selectivity.[50–53] In fact, extremely high selectivity to linoleic acid is obtained, since no stearic acid is formed. Consequently, these catalysts are ideal for increasing the vegetable oil stability without an increase in the content of saturated fatty acids. However, no attention was given to the *trans* content of the partially hydrogenated samples, since the adverse health aspects of these isomers were unknown at that time. Nowadays, it is known that the hydrogenation yields similar levels of *trans* fatty acids in the presence of copper catalysts, compared to the commercial Ni process under similar hydrogenation conditions. Despite their high polyene selectivity, copper-based catalysts have not been used on an industrial scale due to their very low activity and short lifetime.[54,55] Furthermore, residual Cu species, leached from the catalyst, lead to peroxidation of the oil product.[56,57]

More recently, hydrogenation of vegetable oils in the presence of copper catalysts has gained renewed attention. More active Cu/SiO_2 catalysts were prepared by variations in the preparation and activation method, producing highly dispersed Cu species.[58] In addition to metal dispersion, the morphology of Cu species seems to influence activity as H_2 dissociation occurs preferentially on stepped surfaces.[59,60] The highly dispersed Cu/SiO_2 catalysts show good performance in the hydrogenation of various vegetable oils.[58] In the hydrogenation of rapeseed oil at 453 K and 2.0 MPa H_2, the content of C18:1 increased from 53 to 74% (decrease in IV from 110 to 80), while the content of C18:0 rose from 2 to 4%. Such partially hydrogenated oils are excellent for industrial applications, such as biolubricants. The relatively high content of *trans* fatty acids formed (*e.g.* 15% at IV 80) makes these hydrogenated oils unsuitable for the food industry. Also, fatty acid methyl esters (FAME) with various compositions were selectively hydrogenated to monounsaturated methyl esters in the presence of Cu/SiO_2 catalyst.[61] FAME enriched in C18:1 esters are suitable for diverse oleochemical applications (*e.g.* lubricating oils, casting and coating compositions).

A recent patent claimed the use of copper catalysts, thermally treated in the presence of H_2O_2, for the hydrogenation of soybean oil.[62] The linolenic acid content of soybean oil was selectively reduced, generating low quantities of saturated fatty acids and with relatively low levels of *trans* isomers. The improvement in activity of Cu-based catalysts by a heat treatment at 623 K was already reported in the late 1960s.[54] The increase in activity was ascribed to structural changes of the catalyst. X-ray diffraction data of commercial Cu-chromite catalysts showed that the heat-treated catalysts contained higher levels of divalent copper oxide and copper chromate than the untreated catalyst.

10.3.3 Precious-Metal Catalysts

Unlike nickel catalysts, precious-metal catalysts activate hydrogen already at very low temperatures. Hence, they are conducive to low *trans* levels. Table 10.4 compares the activity, linoleic selectivity (S_{LO}) and specific

Table 10.4   Performance   sequence   of   precious-metal
                 catalysts in the hydrogenation of vegetable oils.

Performance aspect	Sequence
Activity	Ru < Ir < Pt < Rh < Pd
S_{Lo}	Ir < Ru < Pt < Rh < Pd
SII	Pt < Ir < Ru < Rh < Pd

isomerization index (SII) achieved in the hydrogenation of vegetable oils in the presence of precious-metal catalysts.[63–66]

Currently, most of the research in the field focuses on Pd catalysts because of their high activities and superior linoleate selectivity. Under typical hydrogenation conditions (high temperature and low hydrogen pressure), Pd is even more selective than commercial Ni catalysts, although also higher *trans* levels are reported.[67–69] However, Pd is active at temperatures as low as 303–373 K. When hydrogenations are performed at low temperatures, low *trans* levels can be obtained along with acceptable polyene selectivity.[70,71]

The activity and selectivity of Pd catalysts can also be tuned by modifications in catalyst support, location and dispersion of the Pd clusters. Pd nanocomposite carbon catalysts, consisting of a carbon support of large pores with a high specific surface area (300–400 m^2/g), was developed for the efficient partial and full hydrogenation of fatty oils.[72] The large pores (50–80 nm) enable the fast transport of oil and hydrogen molecules to the highly dispersed palladium nanoparticles (2–3 nm). In the partial hydrogenation of sunflower oil at 379 K under H$_2$ pressures of 0.6 MPa, excellent polyene selectivity was achieved (70% C18:1 at IV 75). The content of stearic acid rose from 3 to 15%. The amount of *trans* isomers was not reported.

Sol-gel entrapped Pd catalysts, consisting of small Pd nanoparticles (4–6 nm) encapsulated in an organosilica matrix, show promising results regarding the selective hydrogenation of vegetable oils (diluted in THF/MeOH) under ambient H$_2$ pressure ("balloon conditions") at room temperature.[73] Interestingly, high hydrogenation selectivity and no *cis–trans* isomerization were observed under these conditions. These results were explained by the preferential entrance of the hydrophobic tails of the triglyceride molecule into organosilica sol-gel cages where the Pd nanoclusters are supported.

There is general agreement that Pt catalysts produce the lowest amount of *trans* fatty acids during fatty oil hydrogenation.[63,67,68,74] Nonetheless, Pt shows a very poor hydrogenation selectivity resulting in the formation of high levels of (tri)saturates.[74] A patent by BASF claimed the use of a Pt catalyst that performs well in the partial hydrogenation of vegetable oils, providing hydrogenated fats with low levels of *trans* isomers and saturates. The catalyst consists of an aggregate of three components (*i.e.* a solid support, 1–12-nm sized Pt nanoparticles, and surfactant or polymer). The hydrogenation of soybean oil at 323 K under H$_2$ pressure of 0.4 MPa resulted in 2.9 wt% of *trans* isomers at an IV of 110, and 4.5 wt% at an IV of 70.

The content of stearate was 9.9 and 25.9 wt%, respectively (Table 10.3). No information on the triglyceride composition of the hydrogenated fat samples was provided.[75]

Recently, Pt supported on ZSM-5 was described as a catalyst with an unusual selectivity in the hydrogenation of both model triglycerides and soybean oil.[76–78] ZSM-5 is characterized by a uniform pore system with diameters ranging from 0.3 to 0.8 nm. At first sight, this material does not seem to be a suitable catalyst support for the hydrogenation of bulky triglyceride molecules. Yet, a shape-selective Pt/ZSM-5 catalyst, containing small Pt nanoclusters mainly inside of the narrow zeolite pores, shows adequate activities in the hydrogenation of triglyceride molecules. More importantly, very low *cis–trans* isomerization as well as favorable hydrogenation selectivity were obtained from the hydrogenation of triglycerides at low temperature and high hydrogen pressure (Table 10.3). In the hydrogenation of soybean oil at 338 K and under a H_2 pressure of 6 MPa, a 3.2 wt% yield of *trans* isomers and 5% SSS at an IV of 65 were achieved. For comparison, a 16% SSS was achieved in the presence of a Pt/alumina catalyst at the same conversion and under the same reaction conditions.

Obviously, the low *trans* levels, obtained by Pt supported on ZSM-5, can be explained by the use of Pt as a highly active metal in addition to the hydrogenation conditions (*i.e.* low temperature and high hydrogen pressure) under which hydrogenation is favored over isomerization according to the Horiuti–Polanyi mechanism. However, the low trisaturate levels are unusual under such conditions. Furthermore, an unprecedented "positional" selectivity was observed in which the fatty acid chains located at the central (*sn*-2) position are preferably hydrogenated over those located at the terminal positions on the glycerol backbone (*sn*-1/3). This distinct type of selectivity suggests a preferred adsorption mode of triglyceride molecules on the zeolite surface. In a *tuning fork* conformation, the central fatty acid is faced toward the opposite side of the outer fatty acids. Such a conformation allows the central fatty acid chain to enter into the small pores of the ZSM-5, whereas the other two fatty acid chains remain interacting with the external surface of the zeolite. Consequently, only the double bonds of the central fatty acid are hydrogenated by the internal Pt clusters. As a result, a new type of "pore mouth" catalysis, leading to a very low content of trisaturates, was proposed. Such a "positional" selectivity is considered to be negligible in conventional hydrogenations using Ni catalysts.[29]

10.4 Conclusions and Perspectives

Catalytic hydrogenation of vegetable oils is an established technology in the food industry and shows ever-increasing importance in the chemical industry. Both full and partial hydrogenation of double bonds in the fatty acid chains are industrially performed. Currently, the partial hydrogenation of vegetable oils faces new challenges, especially for food applications, but also for specific chemical applications, the *cis–trans* isomerization, occurring as a competitive side reaction, should be avoided. As *trans* fatty acids have a significantly higher melting point when compared to that of corresponding *cis* isomers, they are undesired in some industrial applications (*e.g.* liquid biolubricants). Current goals indicate *trans* levels below 8 wt% for such industrial applications. In contrast, for food applications, such *trans* levels should essentially be zero, which in practice means lower than 1 wt% in the consumed product. A final challenge in the partial hydrogenation of vegetable oils is a control over the triglyceride distribution. For semisolid applications (*e.g.* margarines and shortenings), the triglyceride composition exerts a marked effect on the physical properties of the hydrogenated fats. Importantly, the formation of high melting point trisaturates is undesired. The various types of selectivity (*i.e.* polyene selectivity, specific isomerization index and triglyceride selectivity) are controlled by hydrogenation conditions, feedstock composition, and by the specific properties of the catalyst. Nevertheless, all parameters that increase the hydrogenation selectivity (polyene and triglyceride selectivity) also increase the specific isomerization index and this leads to high *trans* levels in the hydrogenated product.

The commercially available catalysts are unsatisfactory for obtaining hydrogenated fats with excellent hydrogenation selectivity and low *trans* levels. Therefore, hydrogenated fats are becoming ever more replaced by zero-*trans* interesterified and fractionized fat blends in food products. Nevertheless, catalytic hydrogenation still compromises major benefits over alternative processes.[79] First, selective catalytic hydrogenation is the only process available that is able to stabilize polyunsaturated oils (*e.g.* soybean, canola, and sunflower oils). Of note is that soybean oil is widely available in the United States as a byproduct from the feed industry, while canola and sunflower oil are more common in Europe, although there is a tendency to gradually replace them by more expensive, oxidation-stable variants.

Current progress in catalyst design illustrates innovative potential for hydrogenated vegetable oils. Most importantly, the advances in the field are mandatory to secure the continuity of the hydrogenation technologies for food and industrial applications. In effect, catalysts recently described in the literature are conducive to excellent hydrogenation selectivity combined with very low *cis–trans* isomerization. Moreover, hydrogenation of triglyceride molecules was demonstrated to be possible with a specific positional selectivity, paving the way towards the synthesis of hydrogenated fats with defined triglyceride compositions.

References

1. R. C. Hastert, in *Introduction to Fats and Oils Technology*, (ed.) P. J. Wan, AOCS Press, Illinois, 1991, p. 330.
2. R. D. O'Brien, *Fats and Oils: Formulating and Processing for Applications*, CRC Press, Boca Raton, 2009.
3. U. Biermann, U. Bornscheuer, M. A. R. Meier, J. O. Metzger and H. J. Schäfer, *Angew. Chem. Int. Ed.*, 2011, **50**, 3854–3871.

4. H. Wagner, R. Luther and T. Mang, *Appl. Catal. A: Gen.*, 2001, **221**, 429–442.
5. B. Smith, H. C. Greenwell and A. Whiting, *Energy Environ. Sci.*, 2009, **2**, 262–271.
6. G. Buehler, in *Fatty Acids* in Industry, eds. R. W. Johnson and F. E., Marcel Dekker, Inc., New York, Editon edn., 1988.
7. T. Schaaf and H. Greven, *Lipid Technol*, 2010, **22**, 31–35.
8. J. Farmani, M. Safari and M. Hamedi, *Eur. J. Lip. Sci. Technol.*, 2009, **111**, 1212–1220.
9. W. Normann, *GB 190301515*, 1903.
10. P. M. Clifton, J. B. Keogh and M. Noakes, *J. Nutr*, 2004, **134**, 1848–1848.
11. D. Mozaffarian, M. B. Katan, A. Ascherio, M. J. Stampfer and W. C. Willett, *New Engl. J. Med.*, 2006, **354**, 1601–1613.
12. S. Stender, J. Dyerberg, A. Bysted, T. Leth and A. Astrup, *Atherosclerosis Supp.*, 2006, 7, 47–52.
13. Regulation (EU) No 1169/2011, *Off. J. Eur. Union*, **304**, 18–63.
14. FDA, US Department of Health and Human Services, 21 CFR Part 101 [Docket No. 94P-0036], Washington, Editon edn., 2003, p. 254.
15. J. E. Hunter, J. Zhang and P. M. Kris-Etherton, *Am. J. Clin. Nutr.*, 2010, **91**, 46–63.
16. R. Micha and D. Mozaffarian, *Lipids*, 2010, **45**, 893–905.
17. J. Horiuti and M. Polanyi, *Trans. Faraday Soc.*, 1934, **30**, 1164–1172.
18. A. E. Bailey, *J. Am. Oil Chem. Soc.*, 1949, 644–648.
19. L. F. Albright, *J. Am. Oil Chem. Soc.*, 1963, **40**, 16.
20. L. F. Albright, *J. Am. Oil Chem. Soc.*, 1970, **47**, A91.
21. H. B. W. Patterson, *Hydrogenation of Fats and Oils: Theory and Practice*, AOCS Press, Illinois, 1994.
22. A. E. Bailey, *J. Am. Oil Chem. Soc.*, 1949, **26**, 596–601.
23. L. F. Albright, *J. Am. Oil Chem. Soc.*, 1965, **42**, 250–253.
24. A. J. Dijkstra, *J. Am. Oil Chem. Soc.*, 2010, **87**, 115–117.
25. R. O. Butterfield, *J. Am. Oil Chem. Soc.*, 1969, **46**, 429–431.
26. J. W. E. Coenen, *J. Am. Oil Chem. Soc.*, 1976, **53**, 382–389.
27. F. A. Kummerow, *Prev. Control*, 2005, **1**, 157–164.
28. A. L. Lock, P. W. Parodi and D. E. Bauman, *Aust. J. Dairy Tech.*, 2005, **60**, 134–142.
29. A. J. Dijkstra, *Inform*, 1997, **8**, 1150–1158.
30. A. E. Bailey, R. O. Feuge and B. A. Smith, *Oil Soap*, 1942, 169–176.
31. L. F. Albright and J. Wisniak, *J. Am. Oil Chem. Soc.*, 1962, **39**, 14.
32. M. W. Balakos and E. E. Hernandez, *Catal. Today*, 1997, **35**, 415–425.
33. N. Hsu, L. L. Diosady, W. F. Graydon and L. J. Rubin, *J. Am. Oil Chem. Soc.*, 1986, **63**, 1036–1042.
34. B. Drozdowski and E. Szukalska, *Eur. J. Lipid Sci. Technol.*, 2000, **102**, 642–645.
35. G. R. List, W. E. Neff, R. L. Holliday, J. W. King and R. Holser, *J. Am. Oil Chem. Soc.*, 2000, 77, 311–314.
36. S. Schmidt, *Eur. J. Lipid Sci. Technol.*, 2000, **102**, 646–648.

37. A. J. Dijkstra, *Eur. J. Lipid Sci. Technol.*, 2006, **108**, 249–264.
38. A. Beers, *Lipid Technol.*, 2007, **19**, 56–58.
39. A. Smidovnik, A. Stimac and J. Kobe, *J. Am. Oil Chem. Soc.*, 1992, **69**, 405–409.
40. G. J. Yusem and P. N. Pintauro, *J. Am. Oil Chem. Soc.*, 1992, **69**, 399–404.
41. W. D. An, J. K. Hong, P. N. Pintauro, K. Warner and W. Neff, *J. Am. Oil Chem. Soc.*, 1998, **75**, 917–925.
42. W. An, J. K. Hong, P. N. Pintauro, K. Warner and W. Neff, *J. Am. Oil Chem. Soc.*, 1999, **76**, 215–222.
43. K. Mondal and S. B. Lalvani, *Chem. Eng. Sci.*, 2003, **58**, 2643–2656.
44. J. W. King, R. L. Holliday, G. R. List and J. M. Snyder, *J. Am. Oil Chem. Soc.*, 2001, **78**, 107–113.
45. C. A. Piqueras, G. Tonetto, S. Bottini and D. E. Damiani, *Catal. Today*, 2008, **133**, 836–841.
46. D. Jovanovic, R. Radovic, L. Mares, M. Stankovic and B. Markovic, *Catal. Today*, 1998, **43**, 21–28.
47. M. Gabrovska, J. Krstic, R. Edreva-Kardjieva, M. Stankovic and D. Jovanovic, *Appl. Catal. A Gen.*, 2006, **299**, 73–83.
48. I. Karabulut, M. Kayahan and S. Yaprak, *Food Chem.*, 2003, **81**, PII S0308-8146(0302)00397-00397.
49. H. van Toor, G. J. van Rossum, M. B. Kruidenberg, *US 7,498,453 B2*, 2009.
50. A. de Jonge, J. W. E. Coenen and C. Okkerse, *Nature*, 1965, **206**, 573–574.
51. S. Koritala and H. J. Dutton, *J. Am. Oil Chem. Soc.*, 1966, **43**, 556–558.
52. S. Koritala and H. J. Dutton, *J. Am. Oil Chem. Soc.*, 1966, **43**, 86–89.
53. O. Popescu, S. Koritala and H. J. Dutton, *J. Am. Oil Chem. Soc.*, 1969, **46**, 97–99.
54. K. J. Moulton, D. J. Moore and R. E. Beal, *J. Am. Oil Chem. Soc.*, 1969, **46**, 662–666.
55. J. A. Heldal and P. C. Moerk, *J. Am. Oil Chem. Soc.*, 1982, **59**, 396–398.
56. L. E. Johansson and S. T. Lundin, *J. Am. Oil Chem. Soc.*, 1979, **56**, 981–986.
57. R. E. Beal, K. J. Moulton, H. A. Moser and L. T. Black, *J. Am. Oil Chem. Soc.*, 1969, **46**, 498–500.
58. N. Ravasio, F. Zaccheria, M. Cargano, S. Recchia, A. Fusi, N. Poli and R. Psaro, *Appl. Catal. A Gen.*, 2002, **233**, 1–6.
59. F. Boccuzzi, A. Chiorino, G. Martra, M. Gargano, N. Ravasio and B. Carrozzini, *J. Catal.*, 1997, **165**, 129–139.
60. F. Boccuzzi, S. Coluccia, G. Martra and N. Ravasio, *J. Catal.*, 1999, **184**, 316–326.
61. F. Zaccheria, R. Psaro, N. Ravasio and P. Bondioli, *Eur. J. Lipid Sci. Technol.*, 2012, **114**, 24–30.
62. R. Sleeter. *US 2006/0241313 A1*, 2006.
63. M. Zajcew, *J. Am. Oil Chem. Soc.*, 1960, **37**, 473–478.
64. G. C. Bond, G. Webb, P. B. Wells and J. M. Winterbottom, *J. Catal.*, 1962, **1**, 74–84.
65. B. Nohair, C. Especel, P. Marécot, C. Montassier, L. C. Hoang and J. Barbier, *C.R. Chimie*, 2004, 7, 113–118.

66. E. S. Jang, M. Y. Jung and D. B. Min, *Compr. Rev. Food Sci. Food Saf.*, 2005, **4**, 22–30.
67. G. Cecchi, J. Castano and E. Ucciani, *Rev. Fr. Corps Gras*, 1979, **26**, 391–397.
68. J. Edvardsson, P. Rautanen, A. Littorin and M. Larsson, *J. Am. Oil Chem. Soc.*, 2001, **78**, 319–327.
69. K. Belkacemi, A. Boulmerka, J. Arul and S. Hamoudi, *Top. Catal.*, 2006, **37**, 113–120.
70. V. I. Savchenko and I. A. Makaryan, *Platinum Met. Rev.*, 1999, **43**, 74–82.
71. N. Hsu, L. L. Diosady and L. J. Rubin, *J. Am. Oil Chem. Soc.*, 1988, **65**, 349–356.
72. I. L. Simakova, O. A. Simakova, A. V. Romanenko and D. Y. Murzin, *Ind. Eng. Chem. Res.*, 2008, **47**, 7219–7225.
73. V. Pandarus, G. Gingras, F. Béland, R. Ciriminna and M. Pagliaro, *Org. Proc. Res. Dev.*, 2012, **16**, 1307–1311.
74. P. H. Berben, F. Borninkhof, B. Reesink and E. Kuijpers, *Inform*, 1994, **5**.
75. A. E. W. Beers and P. H. Berben, *WO 2006/121320 A1*, 2006.
76. A. Philippaerts, S. Paulussen, S. Turner, O. I. Lebedev, G. Van Tendeloo, H. Poelman, M. Bulut, F. De Clippel, P. Smeets, B. Sels and P. Jacobs, *J. Catal.*, 2010, **270**, 172–184.
77. A. Philippaerts, S. Paulussen, A. Beersch, S. Turner, O. I. Lebedev, G. van Tendeloo, B. Sels and P. Jacobs, *Angew. Chem. Int. Ed.*, 2011, **50**, 3947–3949.
78. A. Philippaerts, A. Breesch, G. de Cremer, P. Kayaert, J. Hofkens, G. van den Mooter, P. Jacobs and B. Sels, *J. Am. Oil Chem. Soc.*, 2011, **88**, 2023–2034.
79. A. Philippaerts, P. A. Jacobs and B. F. Sels, *Angew. Chem. Int. Ed.*, 2013, **52**, 5220–5226.

CHAPTER 11

Hydrogenolysis of Lignocellulosic Biomass with Carbon Monoxide or Formate in Pressurized Hot Water

ULF SCHUCHARDT*[a] AND JEAN MARCEL R. GALLO*[b]

[a] Institute of Chemistry, State University of Campinas,
Postal Box 6154, 13083-970, Campinas, SP, Brazil; [b] Department of
Chemistry, Federal University of São Carlos, Postal Box 676, 13565-905,
São Carlos, Brazil
*Email: ulf@iqm.unicamp.br; jean@ufscar.br

11.1 Introduction

The liquefaction of biomass by hydrogenolysis in water using CO was first reported by Fischer and Schrader in 1921.[1] In this method, carbon monoxide was used as a reaction promoter, allowing for liquefaction and deoxygenation of biomass.[2,3] Fischer and Schrader suggested that CO dissolved in water is in equilibrium with formic acid.[4] Hence, formate is formed under neutral or alkaline conditions. In order to prove this concept, they processed several solid substrates (*e.g.* lignite, peat and wood) with formate in a water slurry. High conversion and improved yields of heavy oils were achieved, demonstrating the importance of formate species for the liquefaction process.[4]

In 1971, Appell *et al.*[5] published an extensive report on the conversion of cellulosic wastes, sewage, sludge, wood, lignin and bovine manure

RSC Energy and Environment Series No. 13
Catalytic Hydrogenation for Biomass Valorization
Edited by Roberto Rinaldi
© The Royal Society of Chemistry 2015
Published by the Royal Society of Chemistry, www.rsc.org

into oils. The liquefaction was performed in water under CO pressures of 6.6–12.7 MPa at temperatures of 523–623 K. The addition of alkali or alkali-earth carbonates (*e.g.* 5 wt% Na_2CO_3) improved the conversion of cellulose (from 63 to 88%), and the oil yield (from 15 to 40%), compared to reference experiments performed in the absence of bases. Surprisingly, the conversion of low-rank coals at higher temperatures under H_2 pressures in the presence of a catalyst was found to be less productive than the conversion using carbon monoxide in combination with 5 wt% Na_2CO_3 in pressurized hot water.[6]

Solvent liquefaction is another interesting method using hot pressurized water.[7] Even though catalysts might enhance the yield of liquid products, high temperatures (573–873 K) and elevated pressures (2–20 MPa) are invariably required.[7] In the early 1980s, a process of biomass thermo-chemical liquefaction based on the treatment of biomass in pressurized hot water was developed by Shell. The method known as HTU (hydrothermal upgrading) subjects plant biomass to temperatures from 573 to 633 K under working pressures of 10–18 MPa at a residence time of 5 to 20 min.[8] Under the high-severity conditions, biomass is deoxygenated mainly by decarboxylation and dehydration.[8] The product generated is known as "bio-crude" that usually contains between 10-18 wt% of oxygen (versus 40–45 wt% found for raw biomass).[8]

The reaction mechanism for the biomass deoxygenation in pressurized hot water is not well understood.[8] However, it is known that this reaction medium exhibits very particular physical and chemical properties.[9,10,18] Notably, depending on temperature and pressure, pressurized hot water can facilitate either free-radical or ionic reactions, and the latter reaction class is apparently needed for biomass liquefaction.[9] As shown in Table 11.1, subcritical water (523 K and 5 MPa) or supercritical water (673 K and 50 MPa) will present a pK_w value as low as 11.2 and 11.9, respectively. These pK_w values are much lower than the one for ordinary water ($pK_w = 14$), indicating that under these conditions water will be more ionized and, therefore, ionic

Table 11.1 Properties of water under different conditions. Adapted with permission from Ref. 9.

	Ordinary water	Subcritical water	Supercritical water	
T/K	298	523	673	673
Pressure/MPa	0.1	5	25	50
Dielectric constant	78.5	27.1	5.9	10.5
pK_w	14	11.2	19.4	11.9
Density/g cm^{-3}	1	0.80	0.17	0.58
Heat capacity/kJ kg^{-1} K^{-1}	4.22	4.86	13.0	6.8
Dynamic viscosity/mPa	0.89	0.11	0.03	0.07
Heat conductivity/mW m^{-1} K^{-1}	608	620	160	438
Reaction mechanism[a]	-	Ionic	Radical	Ionic

[a]Reaction mechanism favored in biomass liquefaction reactions (relative to ordinary water).

reaction mechanisms are more likely to take place.[9] In contrast, free-radical reactions are predominant in supercritical water at low density (523 K and 25 MPa), since the pK_w value is significantly high ($pK_w = 19.4$) under this condition.[9]

11.2 Mechanism of Lignocellulosic Biomass Liquefaction with CO in Water

While the liquefaction and deoxygenation of lignocellulosic biomass has been demonstrated to be efficient in the presence of CO in pressurized hot water, few studies have been performed to understand the reaction mechanism. Unlike the HTU process, which results in the biomass deoxygenation through decarboxylation, the process performed in pressurized hot water in the presence of CO predominantly leads to hydrogenolysis.[5] Therefore, several questions on how the reducing agent would be produced *in situ* by the presence of CO and water (eqn (11.1)) remain unanswered. A first hypothesis was the reaction of CO with water though the water-gas shift reaction (WGS), producing carbon dioxide and molecular hydrogen. Accordingly, H_2 would be the reducing agent. However, even under the high-severity conditions, this reaction is very unlikely to happen in the absence of a catalyst. In fact, WGS takes place at high temperature (*e.g.* 623 K) in the presence of a catalyst (*e.g.* iron oxide).[20] Furthermore, the hydrogenolysis using H_2 requires a selective catalyst (*e.g.* copper-based catalysts) to take place in the reaction system.[11,12]

$$CO + H_2O \leftrightarrows CO_2 + H_2 \tag{11.1}$$

Appell *et al.*[6] suggested that water, CO and alkaline salts react, possibly forming alkali formate intermediates, which transfer hydrogen as a hydride to the substrate. They also evaluated the extent of the WGS reaction during cellulose conversion at different temperatures. The shift reaction consumed 9.8% CO at 523 K, 37.8% at 573 K, and becomes the predominant reaction at 623 K, consuming 79.8% CO.[6] Note that the WGS is not conducive to the liquefaction, since the H_2 produced is not used for the hydrogenolysis in the absence of a catalyst.

Schuchardt and Matos[13] reported that the hydrogenolysis of sugarcane bagasse with carbon monoxide/base or formate is possible at temperatures as low as 453 K, when certainly no WGS reaction is occurring. Furthermore, they showed that at temperatures below 473 K, formate was more efficient than the carbon monoxide/base system. This observation was explained by the slow formation of formate in the second system.

For the carbohydrate fraction of the biomass, Appell *et al.*[6] proposed a mechanism for the hydrogenolysis in water using CO, as described by:

$$Na_2CO_3 + H_2O + 2\,CO \leftrightarrows CO_2 + 2\,HCOONa \tag{11.2}$$

$$\underset{\underset{HO}{}\;\underset{OH}{}}{\overset{R_1\;\;\;R_2}{\diagup\!\diagdown}} \;\rightleftharpoons\; \underset{\underset{OH}{}}{\overset{R_1\;\;\;R_2}{\diagup\!\diagdown}} \;\rightleftharpoons\; \underset{\underset{O}{}}{\overset{R_1\;\;\;R_2}{\diagup\!\diagdown}} \tag{11.3}$$

$$\underset{\underset{O}{}}{\overset{R_1\;\;\;R_2}{\diagup\!\diagdown}} + HCOO^- \longrightarrow \underset{\underset{O^-}{}}{\overset{R_1\;\;\;R_2}{\diagup\!\diagdown}} + CO_2 \tag{11.4}$$

$$\underset{\underset{O^-}{}}{\overset{R_1\;\;\;R_2}{\diagup\!\diagdown}} + H_2O \;\rightleftharpoons\; \underset{\underset{OH}{}}{\overset{R_1\;\;\;R_2}{\diagup\!\diagdown}} + OH^- \tag{11.5}$$

In the presence of water and a weak base (*e.g.* sodium carbonate), CO will react rendering sodium formate (11.2), which thus works as a hydrogen donor. In the first step, the monosaccharide suffers a dehydration of vicinal hydroxyl groups, forming enol species, which tautomerize to ketones (11.3). In the second step, hydrogen is transferred from the formate to the carbonyl group, thus forming CO_2 and reducing the organic functionality (11.4 and 11.5).[3] Notably, formate was often used by Sasson's group[14] as a reducing agent in organic synthesis. They provided a technoeconomic analysis showing the potential viability of formate salts as replacements for H_2 in selected industrial processes.

Decarboxylation of biomass molecules is another role of the formate ion that leads to the cleavage of biomass into smaller fragments through the hydrothermal processing. The experimental observations suggest that the reaction of formate with the hydroxyl groups of hemicellulose, cellulose and lignin renders formic esters. These formic esters can be easily obtained from cellulose at room temperature under acidic conditions.[15,16] Under neutral or slightly basic conditions, higher temperatures are required to favor the esterification. Also lignin is expected to form formic esters.[17] The formic esters are then decomposed to the hydrogenated products and carbon dioxide, as displayed by:

$$\underset{\underset{H}{}}{\overset{R_1}{\underset{R_2}{\diagup\!\diagdown}}}\!\!-\!O\!-\!\overset{}{\underset{O}{\diagdown}} \;\rightleftharpoons\; \overset{R_1}{\underset{R_2}{\diagup\!\diagdown}} + CO_2 \tag{11.6}$$

At temperatures lower than 573 K, carbon dioxide is detected as the main gaseous product formed by decarboxylation in addition to formate decomposition.[17] Only a small amount of molecular hydrogen is detected in the reaction gases, confirming that the water-gas shift reaction is occurring to a very limited extent.[17] The hydrogenated products obtained by decarboxylation of the formic esters show a strong reduction of their oxygen content (typically 20 to 25%) and a much smaller average molar weight (typically 250 to 400 Da, as determined by vapor-pressure osmometry).[13]

Very little research has been directed toward understanding the effect of the base on the performance of biomass liquefaction in the presence of the water/CO system. Karagoz *et al.*[19] showed the conversion and yield of liquid products from the hydrothermal liquefaction of pinewood to be affected by the nature of the base and follows the trend: $NaOH < Na_2CO_3 < KOH < K_2CO_3$. When no base is used, a 42% yield of solid residues is obtained. Minowa *et al.*[20] showed that alkali species inhibits the formation of char from biocrude. However, the direct use of formate salts instead of CO/base gives better results irrespective of their alkali cation.[16]

11.3 Hydrogenolysis of Cellulose

Appell *et al.*[5] reported the reaction of cellulose and carbon monoxide in water. In the presence of 5 wt% of sodium carbonate, similar conversion (88%) at 523, 573 and 623 K was achieved. The oil yields were low (16 to 30 wt%). The reaction temperature had a significant influence on the viscosity and oxygen content of the product. At 523 K, a bitumen-like solid with 20% oxygen was obtained. However, at 573 and 623 K, oils with 13 and 10% oxygen, respectively, were formed. The drawback of this system was the relatively high extent of the WGS reaction, which is a disadvantage, as the *in situ* formed H_2 and the substrate do not react under the experimental conditions.

In order to increase the oil yield, the authors also investigated the effect of CO pressure (Table 11.2), water/cellulose ratio, and the presence of additives. The oil yield is strongly dependent on the CO pressure. At 573 K, the increase in the initial CO pressure (from 0.69 to 4.14 MPa at 298 K or from 7.93 MPa to 12.69 MPa at the reaction temperature) enhanced the oil yield from 24% (at 78% conversion) to 40% (at 90% conversion).[5] The increase in the water/ cellulose ratio (from 1:1 to 4:1) improved the oil yield (from 14.5 to 23%) at a reaction temperature of 623 K.

The use of organic liquids as additives considerably improved the oil yields (Table 11.3). When anthracene oil or isoquinoline were added to the

Table 11.2 Effect of the CO pressure on the process performance. Reaction conditions: $NaHCO_3$, 523 K, 1 h. Adapted with permission from Ref. 5.

Initial Pressure/MPa	Operating Pressure/MPa	Conversion/%	Oil yield/%
0	6.61	78	24
0.69	7.93	76	24
1.38	9.52	83	32
2.07	10.20	82	32
2.76	10.34	84	34
3.45	11.31	87	35
4.14	12.69	90	40

Table 11.3 Effect of organic liquids on cellulose liquefaction. Conditions: Initial CO pressure of 8 MPa, 653 K, 15 min. Adapted with permission from ref. 5.

Solvent ratio			Operating		
Water	Anthracene	Isoquinoline	Pressure/MPa	Conversion/%	Oil yield/%
1	-	-	40	73	22
1	2	-	29	95	51
1	2 (2 h)	-	29	96.5	53
4	-	1	29	98	57

reactions at 653 K, the conversion increased from 73 to 96.5% and the oil yield from 22 to 53%.[5] Interestingly, the operating pressure lowered (from 40 to 29.6 MPa) in the experiments performed in the presence of anthracene oil or isoquinoline.

Another important factor for the success of biomass liquefaction in the water/CO system is the nature of the base used.[5] Sodium carbonate or hydroxide as well as the potassium counterparts are the most effective, while ferrous sulfate and tin chloride only marginally increase the oil yields. As discussed before, the reaction of bases with carbon monoxide forms formate species, *i.e.* the real reducing agent. As carbon dioxide and small organic acids are also formed during the reaction, the pH value of the medium decreases; however, the pH value always stays at 5 or a higher value in the presence of 5 wt% base. In the liquefaction of newsprint paper at 523 K using 20 wt% sodium bicarbonate, the aqueous phase could be recycled for at least 8 times with only a marginal decrease in conversion and oil yield.[5]

Under the optimal conditions, the oil obtained from cellulose at 573 K showed an elemental composition of 78.5% C, 8.0% H and 13.4% O, respectively. For comparison, purified cellulose presents 45.6% C, 6.9% H, and 47.5% O, respectively.[5] Overall, the results clearly show that the hydrogenolysis with CO/base in water strongly reduces the oxygen content of the oil, and increases the hydrogen content, demonstrating that the hydrogen from formate species is incorporated in the resulting oil composition. The infrared spectra of the oil showed the characteristic bands found for aliphatics comprising ether linkages, carbonyl and hydroxyl groups.[5,21] ^{1}H-NMR spectra of the oil mostly showed signals of methylene and methyl groups, with a large portion in alpha or beta position to oxygen or carbonyl groups.[14,22] Only about 4% of the hydrogen content was associated with unsaturated carbon atoms.[14]

11.4 Hydrogenolysis of Lignin

El-Saied[22] studied the hydrogenolysis of lignin from rice straw, sugarcane bagasse and cotton stalks at a temperature of 643 K under an initial CO pressure of 7 MPa. For both rice straw and bagasse, a 63% conversion was achieved, rendering 48% oil yield. The carbon and hydrogen content of the oils was 70 and 6.8%, respectively.

Table 11.4   Summary of the findings of the liquefaction of eucalyptus lignin in a continuous-flow reactor. Reaction conditions: 10% $NaHCO_2$, initial pH 9 Adapted with permission from ref. 17.

Temperature/K	Pressure/MPa	Conversion/%	Oil Yield/%	Oil elemental composition/%		
				C	H	O
523	10	44.6	38.3	62.5	6	31.3
543	10	50.5	44.3	63.7	6.1	30
573	10	56.5	50.1	64.7	6.1	29
573	8	54.5	47.9	64.1	6.2	29.5
573	12	65.4	61	66.2	6.1	27.5
573[a]	10	49.9	42.9	63.4	6.1	30.3
573[b]	10	59.8	54.3	65.4	6.2	28.2

[a]5 wt% $NaHCO_2$.
[b]15 wt% $NaHCO_2$.

Schuchardt et al.[17] studied the hydrogenolysis of hydrolytic eucalyptus lignin with formate in water, using batch and continuous-flow reactors (Table 11.4). In the batch reactor operating at a temperature of 543 K under an Ar pressure of 13 MPa, 49% conversion of lignin with a 45% oil yield was achieved. In a continuous-flow reactor operating at 573 K and 12 MPa, 65% conversion of lignin with an oil yield of 61% was reached. In both experiments, the elemental composition of the oil was 67% C and 6.3% H. The oil obtained in the batch reactor was distilled under a reduced pressure of 0.175 kPa and analyzed by GC-MS. Thirty seven compounds could be identified, which correspond to 58% of the distillate. The major components were guaiacol and 2,6-dimethoxyphenol comprising 11 and 22% of the distillate, respectively. Relative to the starting lignin, these values correspond to approximately 2 and 4% yield of guaiacol and 2,6-dimethoxyphenol, respectively. These yields are not as high as those observed in the alkaline hydrolysis of hydrolytic eucalyptus lignin in a dioxane–water mixture, where guaiacol (8%) and 2,6-dimethoxyphenol (5%) were obtained.[23] Microwave pretreatment of the lignin–formate–water mixture was shown to increase the conversion of hydrolytic eucalyptus lignin and the oil yield by 11 to 14%.[24] However, ultrasound treatment reduced the conversion due to promotion of lignin reticulation.[24]

11.5   Hydrogenolysis of Biomass

In 1976, Oelert and Siekmann[25] published their results on the hydrogenolysis of coals and biomass with carbon monoxide in water at 653 K. They used dry leaves and pine needles, obtaining 85 to 90% conversion with oil yields of 64 to 73%; however, the yield calculations were based on the amount of residue and not on the initial weight of substrate. The elemental composition of the oil was 78% C, 8.6% H and 12% O. They did not identify the components of the oils. The authors tried to replace CO with H_2 under the same reaction conditions, but in this case the oil yields were much smaller.

Table 11.5 Summary of the results of the liquefaction of lignocellulose. Adapted with permission from ref. 22.

Pressure/ MPa	Time/ min	T/K	Conversion/ %	Oil Yield/ %	Oil elemental composition /%		
					C	H	O
7	0	643	72.7	49.1	70.5	7.4	20.4
7	0	643	62.4	51.2	71.0	6.9	20.3
7	5	643	68.4	46.8	72.5	7.5	18.4
7	10	643	86.1	70.1	73.8	7.9	16.7
7	10	643	74.9	52.1	77.1	7	14.3
7	15	643	87.0	71.0	75.5	7.5	15.3
7	30	643	69.0	54.2	75.2	8.7	14.3
4	0	643	58.6	33.8	71.4	6.9	20.3
4	10	643	71.8	53.5	76.2	7.7	14.2
7	0	523	41.6	20.6	67.5	7	24.1
7	0	573	61.2	44.5	70.4	7.3	20.6
7	0	713	91.6	80.2	71.0	6	21.2

The hydrogenolysis of lignocellulose from waste black liquor was studied by El-Saied in the temperature range of 523–713 K, using an initial CO pressure of 7 MPa (Table 11.5).[22] The author reported conversions of 69–87% and oil yields of 49–71%, using the amounts of oil and residue for the calculation of the yields. The addition of a small amount of calcium hydroxide increased the conversion to 90.7% and the oil yield to 79.4%.[22] In all cases, the reaction was found to be rather quick, reaching the maximum oil yield within 10 to 15 min at 643 K. Increasing the initial CO pressure from 4 to 7 MPa improves the conversion and oil yield by approximately 15%. Irrespective of the severity of the process, the oils had a similar elemental composition (70–77% C, 7–8% H and 15–20% O), compared to the oils obtained from other raw materials.

In 1982, Schuchardt and Matos published a paper on the hydrogenolysis of sugarcane bagasse using sodium formate, and argon to pressurize the system (Table 11.6).[13] The authors found that the reaction takes place at considerably lower temperature. The highest oil yield obtained with the water/formate/argon system was 64% (at 96% conversion) at 513 K and under an argon pressure of 5 MPa (measured at 298 K). The pressure was shown to have an important effect on the oil yield. Under an initial argon pressure to 12 MPa, the oil yield increased to 75% (at full conversion) at 453 K. This result clearly shows that formate species facilitates the use of a considerably lower reaction temperature when a high enough pressure is applied to the system.

The best compromise between oil yield and its quality was reached using the system Ca(OH)$_2$/CO/water at 623 K under an initial CO pressure of 5 MPa, which led to the *in situ* generation of formate species (Table 11.6). The oil yield was 72% (at full conversion) with an elemental composition of 70.3% C and 7.6% H. The average molecular weight was 307 Da.[13] However,

Table 11.6   Effect of different reaction systems on the conversion of sugarcane
bagasse and the yield of heavy oil. Reaction systems: (A) HCO_2Na
using Ar as a pressurizing gas; (B) $Ca(OH)_2$ under CO pressure; (C)
$Ca(OH)_2$ under Ar pressure, 15 min. Adapted with permission from
ref. 13.

T/K	Initial Pressure/ MPa	Final Pressure/ MPa	Conv./%	Yield based on carbon/%	Oil elemental composition /%			Average molar weight/ Da
					C	H	H/C	
623	5.0	32.0	100	43	74.3	8.1	1.31	225
573	5.0	30.0	100	51	70.1	7.9	1.35	252
513	5.0	22.0	96	64	62.4	6.9	1.33	321
453	5.0	16.0	85	50	64	7.2	1.35	385
413	5.0	12.0	72	40	68.7	7.8	1.36	397
623	5.0	38.0	100	72	70.3	7.6	1.3	307
573	5.0	32.0	100	68	67.5	7.3	1.3	282
513	5.0	22.0	93	63	61.4	6.5	1.27	316
453	5.5	14.5	80	43	63.6	6.8	1.28	406
413	5.5	12.5	34	24	65.3	7.5	1.38	392
633	0.5	3.2	87	48	64.5	6.1	1.13	297
573	0.5	3.0	83	40	60.9	6	1.18	307
513	0.5	2.2	51	28	62.3	6.1	1.18	287
453	0.8	2.1	27	17	63.3	6.8	1.29	331
413	0.8	1.9	22	15	65.4	7.1	1.3	376

these conditions are very drastic. In the presence of formate/Ar at 513 K, the
obtained results seem to be more interesting.

In contrast to the pressure effect upon the process, the concentration of
formate in the feed did not exert a significant effect on the product yield. For
example, using 1 wt% in the feed led to an oil yield of 65%. An increase in the
formate concentration to 30 wt% improved the oil yield by just 5%. Interestingly, higher concentrations of formate reduced the oil yield, as more water-
soluble (unextractable) products were formed.[13] Carbon dioxide was the only
product detected in the gas phase. In the heavy oils (extracted with chloroform),
hydrocarbons, aliphatic alcohols, phenols and carboxylic acids were detected.
In the water residue, carboxylic acids (*e.g.* acetic acid, oxalic acid and proto-
catechuic acid), in addition to unreacted sodium formate, were identified.

Schuchardt and Matos[13] tested several metal salts as additives for the
hydrogenolysis of bagasse using formate at 453 K under an initial Ar pressure of 9 MPa. The effect of the additives was generally modest. In absence of
additives, a 59% conversion was obtained, rendering a 40% oil yield. In the
presence of 30 wt% $NiSO_4 \cdot 7H_2O$, a 77% conversion was obtained, leading to
a 44% oil yield. Similar findings were obtained for experiments using
$SnCl_2 \cdot 2H_2O$, $FeSO_4 \cdot H_2O$ or $Pd(CH_3CO_2)_2$ as an additive.

Kuznetsov *et al.*[26] demonstrated molten alkali formates to be active in
wood liquefaction under atmospheric pressures. Furthermore, the composition of the formate–alkali mixtures significantly affected the oil yield

and composition. Using this system at 753 K, the highest oil yield reported was 15.2% using sodium formate–KOH in a ratio of 0.91. The product with lowest oxygen content (8.4%) was obtained using a sodium formate/KOH ratio of 0.6 at 723 K. Upon analyzing the composition of the products, the authors found that 6.5 and 14.0 wt% yields of light distillate (b.p. < 473 K) were obtained, depending on the parameters of the wood liquefaction process. The heavier distillate fraction (b.p. 473–623 K) corresponded to 72.5–81.0 wt% of the obtained oil. In turn, the coke residue was between 5.2–13.3 wt%. Kuznetsov *et al.*[26] also reported that potassium formate is more stable than sodium formate, and therefore, allows for hydrogenation of wood at higher temperatures. The yield of liquid product reportedly increased using potassium formate instead of sodium formate. However, no numerical values were given.

11.6 Conclusions

Liquefaction of biomass with CO/base or formate in water achieve moderate to full conversions and oil yields exceeding 65%. Liquid pressurized hot water is a solvent with properties that can be tuned by temperature and pressure. Hence, either ionic or radical reactions can be promoted. Under conditions optimized for ionic reactions, both the reaction intermediates and products are stabilized. The formate species are proposed to form formic esters with the biomass. These intermediates undergo decarboxylation yielding aliphatics comprising ether linkages, carbonyl and hydroxyl groups. Moreover, the ketones are hydrogenated by the formate species to the corresponding alcohols.

The results obtained for the liquefaction of biomass with CO/base or formate are comparable with other methods, such as the Catchlight Energy Process[7] and the HTU process.[8] The advantage of the CO/base system and the HTU process over the Catchlight Energy Process is the use of water at moderate temperatures instead of a toxic solvent (*e.g.* dimethoxyphenols) and high temperatures. Compared to the HTU process, the CO/base system offers the possibility to work at even lower temperatures. The results obtained by Schuchardt and Matos[13] show that formate in water is active in biomass liquefaction at temperatures as low as 453 K.

Even though indicative results had been obtained from the liquefaction of biomass with CO/base or formate, there are very few reports on this system after the 1990s. In this field, systematic studies are still needed to show the viability and technical feasibility of this method as an entry process for the production of transportation fuels.

References

1. F. Fischer and H. Schrader, *Brennst.-Chem.*, 1921, **2**, 257–261.
2. M. Balat, *Energy Sources Part A*, 2008, **30**, 649–659.

3. A. Demirbas, *Energy Convers. Manag.*, 2000, **41**, 633–646.
4. H. J. Hodsman, *J. Soc. Chem. Ind.*, 1931, **50**, 249–249.
5. H. R. Appell, Y. C. Fu, S. Friedman, P. M. Yavorsky, I. Wender and C. Pittsburgh Energy Research, *Converting organic wastes to oil: a replenishable energy source*, U.S. Bureau of Mines, [Washington, D.C.], 1971.
6. H. R. Appell, I. Wender and R. D. Miller, *Chem. Ind. (London)*, 1969, 1703.
7. CA2804581A1, CN103124780A, EP2591068A1, WO2012005784A1, WO2012005784A8.
8. F. Goudnaan, B. van de Beld, F. R. Boerefijn, G. M. Bos, J. E. Naber, S. van der Wal and J. A. Zeevalkink, in *Progress in Thermochemical* Biomass Conversion, ed. A. V. Bridgwater, Blackwell Science Ltd, Birmingham, Editon edn., 2008, pp. 1312–1325.
9. Y. Yu, X. Lou and H. W. Wu, *Energy Fuel*, 2008, **22**, 46–60.
10. A. Kruse and E. Dinjus, *J. Supercrit. Fluids*, 2007, **39**, 362–380.
11. J. H. Sinfelt, D. J. C. Yates and J. L. Carter, *J. Catal.*, 1972, **24**, 283.
12. M. A. Dasari, P. P. Kiatsimkul, W. R. Sutterlin and G. J. Suppes, *Appl. Catal. A*, 2005, **281**, 225–231.
13. U. Schuchardt and F. D. P. Matos, *Fuel*, 1982, **61**, 106–110.
14. H. Wiener, B. Zaidman and Y. Sasson, *Int. J. Hydrogen Energy*, 1989, **14**, 365–370.
15. T. Fujimoto, S. I. Takahashi, M. Tsuji, T. Miyamoto and H. Inagaki, *J. Polym. Sci. Pol. Lett.*, 1986, **24**, 495–501.
16. M. Schnabelrauch, S. Vogt, D. Klemm, I. Nehls and B. Philipp, *Angew. Makromol. Chem.*, 1992, **198**, 155–164.
17. U. Schuchardt, J. A. R. Rodrigues, A. R. Cotrim and J. L. M. Costa, *Bioresour. Technol.*, 1993, **44**, 123–129.
18. A. Demirbas, *Appl. Energy*, 2011, **88**, 17–28.
19. S. Karagoz, T. Bhaskar, A. Muto, Y. Sakata, T. Oshiki and T. Kishimoto, *Chem. Eng. J.*, 2005, **108**, 127–137.
20. T. Minowa, F. Zhen and T. Ogi, *J. Supercrit. Fluids*, 1998, **13**, 253–259.
21. F. T. d. Silva and U. Schuchardt, *Ciência e Cultura Supl.*, 1980, **32**, 385.
22. H. El-Saied, *J. Appl. Chem. Biotechn.*, 1977, **27**, 443–452.
23. U. Schuchardt and A. R. Gonçalves, *Biomass for Energy and Industry*, Elsevier Applied Science in London, 1990.
24. A. R. Goncalves and U. Schuchardt, *Appl. Biochem. Biotechnol.*, 2002, **98**, 1213–1219.
25. H. Oelert and R. Siekmann, *Fuel*, 1976, **55**, 39–42.
26. B. N. Kuznetsov, V. I. Sharypov, S. A. Kuznetsova, V. E. Taraban'ko and N. M. Ivanchenko, *Int. J. Hydrogen Energy*, 2009, **34**, 7051–7056.

CHAPTER 12

Reactor Technology and Modeling Aspects for the Hydrogenation of Components from Biomass

TEUVO KILPIÖ, VICTOR SIFONTES, KARI ERÄNEN, DMITRY YU. MURZIN* AND TAPIO SALMI

Åbo Akademi, Department of Chemical Engineering, Process Chemistry Center, Laboratory of Industrial Chemistry and Reaction Engineering, FI-20500 Turku/Åbo, Finland
*Email: dmurzin@abo.fi

12.1 Introduction

Carbohydrates constitute three quarters of the available renewable biomass,[1] and are thus a key target in the field of green chemistry and process technology. However, not only should the carbohydrates become an important industrial feedstock but also in order to guarantee a smooth transition to a biomass-based industry, the products should be developed having the structural features of the alternative raw materials in mind, as opposed to seeking routes to reproduce exactly the same chemicals obtained from other technologies, *e.g.* the petrochemical industry.[2,3]

The key points for the successful performance of catalytic hydrogenation of components from biomass are the catalyst development, catalyst screening and characterization, kinetic and mass transfer studies, as well as

RSC Energy and Environment Series No. 13
Catalytic Hydrogenation for Biomass Valorization
Edited by Roberto Rinaldi
© The Royal Society of Chemistry 2015
Published by the Royal Society of Chemistry, www.rsc.org

evaluation of the catalyst durability and transport phenomena. Based on a deep understanding of these phenomena, the decision of the production technology can be taken: continuous or discontinuous; slurry, fixed-bed or structured reactors. In this chapter, the reactor technology and modeling aspects for the hydrogenation of sugars are discussed in detail.

12.2 Reactor Technology

12.2.1 Slurry Reactor

Conventional technology for the production of sugar alcohols utilizes (semi)batch slurry reactors. In effect, a finely dispersed, supported or sponge metal catalyst (catalyst particles <0.1 mm) is suspended in a batch of an aqueous sugar solution, to which hydrogen is continuously added so that the pressure is kept constant. Among the reported operating conditions, the typical H_2 pressures are 3–18 MPa and the temperature ranges from 353–423 K.[4,5] Hydrogen pressure is one of the limiting factors of the process. Accordingly, alternative solvents with better hydrogen solubility, such as ethanol, have been proposed.[6] The alternative solvents are, however, in most cases not economically competitive.

12.2.1.1 Experimental Equipment and Procedures for Hydrogenation

Hydrogenations can be successfully performed in laboratory-scale batch reactors (150–500 mL) equipped with baffles, a gas entrainment impeller, a sampling line with a sintered filter, a cooling coil, a heating jacket, a temperature and stirring rate controller, a pressure controller and microprocessor, and a bubbling chamber for pretreatment of sugar solutions. A typical experimental setup is shown in Figure 12.1 (see Section 13.4.5 for practical aspects on the stirrer selection).

A typical experimental procedure can be sketched as follows. A given catalyst weight is loaded in a reactor vessel and flushed with nitrogen to purge oxygen from the system. In a bubbling chamber, the sugar solution is flushed with nitrogen and, immediately afterwards, with hydrogen. The preheated sugar solution is injected into the reactor vessel, and the system is adjusted to the desired operating conditions. Samples are withdrawn from the system and analyzed with an appropriate method (*e.g.* by HPLC).

A preliminary set of experiments is carried out by varying the stirring speed, the catalyst weight, and the catalyst particle size in order to screen the conditions under which the intrinsic kinetics prevail, and therefore the mass-transfer limitations are suppressed. Tests with varying stirring speeds indicate at which stirring rate the kinetic curves start to coincide, indicating the suppression of gas–liquid and liquid–solid mass-transfer limitations.

A general test for mass-transfer phenomena and catalyst durability comprises experiments carried out with different ratios of catalyst weight to

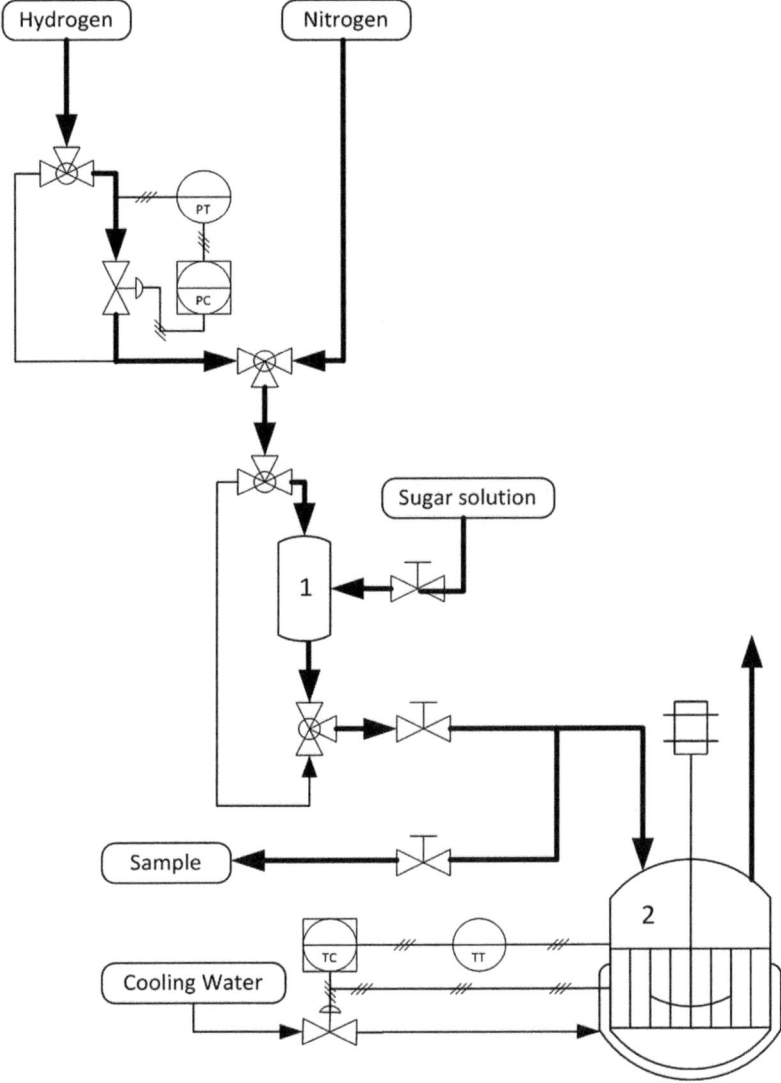

Figure 12.1 Flowsheet of a semibatch reaction system for hydrogenation: (1) bubbling chamber, (2) reactor vessel.

liquid volume. The concentrations are plotted against a transformed unit (*e.g.* time×catalyst–weight/liquid volume). When the kinetic curves coincide, the system most likely is operating under the kinetic regime with a stable catalyst. Finally, experiments with various catalyst particle sizes reveal the impact of internal diffusion in catalyst pores: the increase in the catalyst particle slows down the apparent reaction rate, since the internal diffusion resistance becomes prominent. Altogether, small catalyst particles, high stirring speeds, and small amounts of catalyst guarantee operation in the regime of intrinsic kinetics.

12.2.1.2   *Advantages and Disadvantages of Slurry Reactors*

The main advantages that slurry reactors can offer are the more efficient mixing and the opportunity of using much smaller catalyst particle sizes (<100 micrometers) than those used in practice in fixed beds (1 mm and more). The use of small catalyst particles implies that the mass transfer resistance in catalyst pores is efficiently suppressed in slurry reactors. The catalyst may be present even without any kind of support, such as the sponge catalyst. In stirred tanks, mixing is freely adjustable. In bubble columns, the gas feed generates the mixing. Therefore, the mixing can be freely adjustable by the gas recycling. With efficient mixing applied and small particles used, slurry reactors can be forced to operate under practically no gas–liquid, liquid–solid or internal mass-transfer limitation. Suitable mixing and the selection of catalysts with small particles are the two essential means to avoid heat-transfer resistances inside particles and localized hot spots in the reactor. Regarding heat and mass transfer, stirred tanks are attractive alternatives, compared to other reactors, especially on a small scale.[7–11]

Slurry reactors can be operated continuously, but a broad residence time distribution (RTD) is a drawback. The operation of continuously stirred tanks and bubble columns can be far from the plug flow. The backmixing of a liquid is extensive even when two or three stirred tanks are operated in series. This implies that RTD inevitably becomes broad and, consequently, it is hard to achieve high product yields. It is straightforward to model a continuous lab-scale unit; however, due to the broad RTD, intrinsic kinetics can be better and more accurately obtained from experiments and modeling of a semibatch unit.

With respect to modeling and scale-up, stirred tanks, cascades of stirred tanks and bubble columns are straightforward to model when ideal mixing can be assumed and gas–liquid mass transfer does not limit the productivity. However, the scale-up of stirred tanks is a very demanding task, since in the larger units the stirrer diameter becomes large. Consequently, the fluids have to face very uneven fields (the strongest forces being found close to the tips of stirrers). Furthermore, the force fields on a large scale are very different compared to small-scale operation. For industrial applications, one simply cannot afford to provide as much power per unit volume as on a lab and pilot scale. Hence, the mass and heat transfers may start to play a more dominant role in the process.

In large units, gas–liquid mass transfer easily becomes rate limiting. Since liquid is the continuous phase and bubbles are the dispersed phase, the interfacial surface area for mass transfer depends on the bubble-size distribution. The residence time of bubbles depend on their size, which dictates their rising velocity. Large bubbles have a short residence time, while the smallest ones stay in the liquid phase for a longer time. Accordingly, gas recirculation becomes an attractive alternative that is quite often used. Ideal mixing of liquid and particles in large stirred tanks becomes increasingly difficult to achieve, but modeling can be successfully performed by dividing the reactor into different compartments. Computational fluid dynamics (also known as CFD) could be demanded in order to determine the flow rates of liquid arriving or leaving each compartment. Clearly, the difficulties of scaling up stirred-tank reactors are an obvious drawback of the slurry reactor technology.

12.2.2   Fixed-Bed Reactors

Continuous fixed-bed reactors are practical for catalyst screening, since they do not demand frequent filling and emptying, compared to batch reactors. They are especially well suited for studying the catalyst stability and de-activation processes. Fixed-bed reactors are also used in order to eliminate the possible losses of catalyst. A fixed-bed reactor construction particularly designed for screening of catalysts and reaction conditions is displayed in Figure 12.2. The construction consists of parallel tubular units where dif-ferent catalysts can be examined under varying conditions. Since the system is operated continuously, it gives valuable information about the catalyst stability, easily revealing any deactivation effects (*i.e.* the reactant conversion declines with time-on-stream if the catalyst deactivates). This kind of in-formation cannot be obtained easily from batchwise operation.

One of the main advantages of fixed-bed reactors is that no catalyst sep-aration is required. Furthermore, catalyst loading can be much higher in fixed-bed reactors than in slurry reactors, making them more compact and smaller in size. An ideal plug-flow pattern is often targeted, since it is well known to be the most productive way of operation, especially for systems that require high yields. A long and cocurrently operating fixed-bed reactor can approach a plug-flow behavior, which is favorable for most kinds of reaction kinetics.

A fixed-bed reactor operating in trickling flow regime is a popular choice because of simple construction (random catalyst packing), low investment, and low operational costs. Furthermore, the fluid dynamics in fixed-bed reactors can approach that of the plug flow. Notably, these reactors should be designed carefully to achieve their maximum performance.[12–14] The liquid flow in trickle beds is gravity driven down on the surfaces of catalyst particles. Due to the gravity-driven flow, local turbulence is not so intense that, on the one hand, leads to low operation costs; but on the other hand, a less-turbulent flow also implies that nonideal flow fields may emerge.

12.2.2.1   Experimental Design in Fixed-Bed Technology

Desirable targets. The desirable products fall either into the category of value-added materials or platform chemicals (key intermediates for production of a wide range of compounds). Developing a new commercially viable

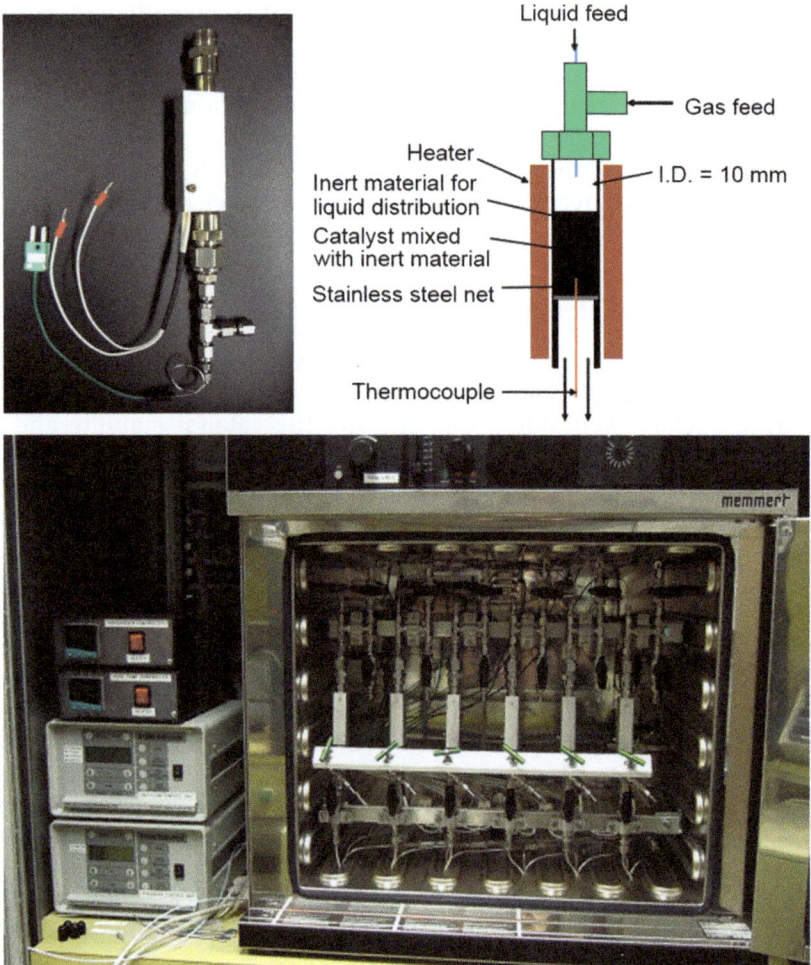

Figure 12.2   A parallel fixed-bed unit for studies of catalyst and reaction conditions.

production process or an efficient way to produce a new chemical compound is usually an extensive task.

Catalyst selection. The catalyst selection and preparation are the most demanding tasks. In most cases, only a few catalysts are used and the aim is to reveal how the productivity and the selectivity change under varying operating conditions. The experts rely on the most up-to-date literature and experience on systems of a similar nature when making the crucial decisions.

Concentration range. The initial reactant concentration ranges should be selected to avoid complete conversion, since complete conversion or equilibrium conversion hide the kinetic information. Diluted streams are often used in kinetic studies to rule out undesirable heat- and mass-transfer effects.

Operation conditions. The operation conditions (temperature, pressure, liquid, and gas feed flow rates) are selected according to the following reasons. The first is that an increase in temperature typically accelerates not only the desired reaction but also the side reactions. Since the activation energies may differ, product selectivity becomes temperature dependent, presenting an opportunity for controlling the process selectivity. Therefore, careful attention should be paid in the selection of the proper temperature ranges for the operation of a continuous reactor. For example, in glucose hydrogenation the previous knowledge of selectivity loss restricts the upper limit of temperature to 403 K. Generation of multiple side products has to be avoided in sugar hydrogenation, since the side products make the subsequent separation steps very difficult. The partial pressure of H_2 is targeted to be high in order to improve its solubility in the liquid phase. In addition, the liquid flow should be adjusted to provide reasonable residence times for reactions to progress in the short reactors to a significant extent.

Reactor size. A small reactor is ideal for catalyst screening. Small catalyst particles are thus preferred on a lab-scale, since they are known to create better flow profiles. However, the use of too small particles should be avoided as they may result in a severe pressure drop.

Particle geometry. The use of small catalyst particles could be afforded if it does not generate an unacceptable pressure drop in a short reactor. Typically, sieved fractions of catalysts are used on lab-scale experiments. Even though the catalyst particles originate from a sieved fraction, it does not mean that the particles are all of exactly the same size. The external surface area of the catalyst particle is proportional to the second power, and volume to the third power of the characteristic dimension. Therefore, if the largest particles are twice the size of the smallest ones, the external surface area of the largest ones will be roughly four times and the volume in order of magnitude eight-times the volume of the smallest particle. Thus, the smallest and the largest particles show particular properties. Mass-transfer limitations typically arise first with the largest particles. Therefore, for area-based phenomena, such as mass and heat transfer, it would be logical to use an area-based average size. For volume-based phenomena, such as heat production, it would be more logical to use a volume-based average.

Randomly packed fixed-bed reactors have a very complicated geometry, especially when the catalyst has a distribution of particle sizes, and the particles have irregular shapes. Modeling of heat conduction in a system of highly complex detailed geometry, including *e.g.* the catalyst pore system, is very challenging. The detailed geometry can only be taken into account in cases of simpler regular geometries (*e.g.* by using computational fluid dynamics).

Feed distribution. The liquid feed should be evenly distributed at the reactor entrance. In lab-scale reactors, a fine sand bed serves as a proper feed distributor. If the liquid is initially well distributed, the assumption of complete external wetting of the catalyst particles is then justified.

Start-up procedure. The start-up procedure of a reactor for the operation under a trickling flow regime is essential to establish proper liquid flow fields. Flooding the bed beforehand is recommended, as it improves the liquid flow field and permanently reduces channeling. Either full flooding of the bed or operation in the pulse flow regime, for a while during start-up, have been noted to be the ways of establishing permanent fine flow fields.[13] In laboratory reactors, used by the authors, the initial flooding and fine sand feed distributors were used to establish fine flow paths. Thus, complete wetting was assumed. Fixed-bed reactors are most often operated in a trickling flow regime. The known flow regimes range from bubble-flow, trickling-flow, pulse-flow, spray-flow up to mist-flow regimes.

The pulse-flow regime is even more favorable than the trickling flow regime, since it results in a more even distribution of flow and prevents the formation of localized hot spots. However, the pulse flow regime would demand high superficial velocities for liquid and gas, and thus leading to short residence times, when a short reactor is used. Some studies have been focused on generating a pulse flow artificially at low flow rates simply by making fluctuations in the feed flow.[14,15] The reactor has been designed to operate in a way that periods of high and low liquid hold-up would take their turn repeatedly with adjustable period lengths. If this method could be done successfully, a major leap forward in flow dynamics would be made, overcoming one of the most important flow problems that the operation in the trickling-flow regime faces. Nonetheless, the propagation of pulsed flow inside the fixed bed continues to be rather problematic. The fluctuations in liquid flow tend to be dampened quickly.[18] Periodic liquid-feed modulation has been used and reported to lead to a productivity improvement in styrene hydrogenation on a pilot-scale.[16] For catalytic biomass conversion, periodic modulation could provide major benefits (*e.g.* more even liquid hold-up, less channeled flow and better wetting), particularly in hydrogenation processes.

Catalyst dilution. Temperature gradients can be suppressed by diluting the catalyst bed with conductive inert particles and by making sure that both the particles and the support material conduct heat well. Bed dilution directly decreases the produced heat per unit volume. However, the catalyst bed should not be very dilute in order to avoid catalyst bypassing. Small particles are preferred for dilution since they improve catalyst-wetting characteristics, although they can lead to a large pressure drop.[14] Catalyst dilution has been studied for sugar hydrogenation by making preliminary experiments in our laboratory with short and long reactors. Similar productivity per catalyst weight was obtained. This result indicated that the flow fields are not impaired and the catalyst sites are not bypassed.

Residence-time distribution (RTD). A residence-time distribution typically occurs in fixed beds. RTD can be revealed by using inert, nonreactive tracer components. The standard way of modeling the liquid RTD is to use the single-parameter, axial-dispersion model. The extent of axial dispersion is characterized by Peclet number, which is zero for complete backmixing and infinite for plug flow. Typically, the order of magnitude value of Peclet number is less than 10 in short fixed beds used on a lab-scale, which indicates moderate backmixing. It is essential to include the axial-dispersion coefficient in the models, since any degree of backmixing impairs the productivity for nonzero-order reactions. The model leads to misinterpretation of the results, when axial dispersion occurs and is not included in the reactor model. In fixed beds, stagnant zones sometimes cause a tail to the distribution, which the axial-dispersion model simply cannot predict. The piston-exchange model is then another modeling option. It is a two-parameter model that produces a tailed distribution.[15] In the piston-exchange model, the liquid is divided into two streams, one progressing rapidly and the other one slowly. Mass transfer is assumed to take place in between these two streams.

12.3   Reaction Kinetics

12.3.1   Reaction Mechanism and Rate Equations

Intrinsic reaction kinetics for a heterogeneously catalyzed reaction is often described by several ways (*e.g.* power laws, Langmuir–Hinshelwood and Langmuir–Hinshelwood–Hougen–Watson equations). A standard expression of reaction rate is given by eqn (12.1), which includes the main terms and serves as an example. At low concentrations, the denominator approaches unity. In effect, the rate expression becomes a power law of reactant concentrations. The rate law contains the activity factor, reaction rate constant, Arrhenius-type temperature dependency as well as reaction orders of reactants. Adsorption and desorption terms in the denominator are also concentration and temperature dependent and include both reactants and products. The coefficients i, j and n_i depend on whether a molecular or an atomic adsorption takes place. For dissociative adsorption, the values of n_i are 0.5, whereas for molecular adsorption they are equal to 1.

$$R = \frac{ak \exp\left(-E_A/R_gT\right)C_{A,L}{}^{i}C_{B,L}{}^{i}}{\left(1 + K_A C_{A,L}{}^{n_A} \exp\left(-\Delta H_{ads,A}/R_gT\right) + K_B C_{B,L}{}^{n_B} \exp\left(-\Delta H_{ads,B}/R_gT\right) + K_C C_{C,L}{}^{n_C} \exp\left(-\Delta H_{ads,C}/R_gT\right)\right)^2}$$

$$(12.1)$$

where: R stands for reaction rate; a, activity correction [ ]; k, reaction rate constant, units to express the rate as [mol g^{-1} s^{-1})]; E_A, activation energy of the main reaction [J mol^{-1}]; K_i, adsorption parameter for component, $i = A, B, C$ [L mol^{-1}]; C_i, local concentration of i [mol L^{-1}]; n_i, adsorption exponent for i [ ]; $\Delta H_{ads,i}$, adsorption energy for i [J mol^{-1}]; R_g, general gas constant [J mol^{-1} K^{-1}]; and T, temperature [K].

The reaction-rate expressions often include both rate and adsorption constants as parameters. The adsorption model (dissociative or non-dissociative) is generally assumed based on the literature or former experience. Therefore, as a denominator, the rate expression includes adsorption/desorption terms, which are based on either molecular or atomic adsorption and desorption steps of each reactant and product. Adsorption or desorption of each component is concentration dependent. For concentrated solutions, adsorption can reduce the reaction rates and suppress the effective reaction orders. Power-law rate equations can be used in a preliminary description of the reaction rates, if no detailed information about the reaction mechanisms is available.

The catalytic hydrogenation of sugar molecules is a complex process, for which several kinetic approaches have been proposed, such as competitive adsorption of sugar and hydrogen as well as noncompetitive adsorption of them. The latter model can be motivated by the fact that the sugar and hydrogen molecules are very different in size. The concept has been further developed, by considering a semicompetitive adsorption model that assumes some interstitial sites always exist between the adsorbed sugar molecules. Therefore, the interstitial sites are accessible to H_2.[16]

The reaction activation energies are obtained from experiments conducted at various temperatures. The safest way to obtain the values is to carry out the experiments in a vigorously stirred slurry reactor, using small catalyst particles, in order to perform the reaction under a truly kinetic regime. If the activation energy is obtained from a system limited by mass transfer, the value will not correspond to an intrinsic activation energy. Generally, the temperature dependency of reactions is much stronger than that of mass transfer. Therefore, surprisingly low values of the activation energies indicate that mass-transfer limitations may be present.

12.3.2   Comparison of Reaction Rates in Continuous Fixed-Bed and Batch Reactors

From the modeling perspective, an ideal plug-flow reactor resembles much an ideally mixed batch reactor. By replacing time with space-time, a plug-flow reactor can be modeled by using similar mass-balance equations to those used in the batch unit. Then, the space-time represents the location how far the observation point has been progressed along the flow. However, when having a multiphase flow in a fixed bed, the situation demands corrections. The first regards the fact that the liquid only partly occupies a tube. This can be handled by considering the liquid hold-up in the mass balances. If gas is abundantly available so that the liquid can be treated as a saturated solution, the analogy is still noticeable.

The flow nonidealities, caused by either a low superficial liquid velocity or too large particles, are the main reasons that lead to differences in the values found for the reaction rates obtained from fixed-bed and batch reactor experiments. Flow nonidealities could be minimized by using a catalyst with very small particles and by increasing the superficial liquid velocity. The latter option is seldom the best choice, since it increases the pressure drop, and decreases the yields due to the reduction of the residence time.

More complications appear in experiments performed on a lab-scale, when a catalyst shows a distribution of particle sizes instead of a single particle size. Low flow velocities can cause mass-transfer limitations and become the reason behind more severe, location-dependent deactivation. In theory, if sites are not deactivated differently, the intrinsic kinetics on the active metal sites of the particles should be the same for both batch and continuous operation, when using the same catalyst. The question is whether the active sites are reached by the reactants and how active the sites are. Catalyst deactivation in a fixed-bed reactor was found to be location dependent on the hydrogenation of sugars,[17] since it was originating from the organic reactant. At the feed entrance, the concentration of the organic reactant is the highest and it is there where the most severe deactivation takes place. In a continuous ideally stirred tank reactor, the deactivation is uniform.

12.3.3 Catalyst Stability and Deactivation

Unfortunately, deactivation of a heterogeneous catalyst is expected under the process conditions. There are factors accounting for the decrease in activity: catalyst poisoning, coke deposition, sintering caused by heat, formation of volatile compounds, to mention just a few. Coking often takes place in the presence of hydrocarbons at elevated temperatures. In this process, polymeric compounds are first formed, which undergo several successive chemical transformations, leading to carbon deposits on the catalyst surface. Coke simply blocks the pores making it impossible for reactants to penetrate further into the particle and to reach the active sites. The processes accounting for coking can be modeled. For cases following Langmuir–Hinshelwood or Langmuir–Hinshelwood–Hougen–Watson kinetics, coking is modeled by ordinary differential equations having only the numerical solution as an option. When coking follows first-order kinetics, an analytical expression can be deduced for the catalyst activity.

12.3.4 Activation Energy

The interpretation and comparison of the activation energies from reaction experiments is challenging. Many studies report values that were obtained from experiments carried out outside the intrinsic kinetic regime, which implies that they have been conducted under various kinds of mass-transfer control. Accordingly, gas–liquid, liquid–solid or internal mass-transfer limited the productivity. Compared to the intrinsic kinetic rate constants, mass-transfer coefficients and pore diffusivity typically depend far less on the temperature. Then, if the given activation energies show a very weak temperature dependency, some of the mass-transfer steps may be rate limiting.[9]

Considering that in many studies regarding reaction rates, the information, on whether the experiments had been conducted in an intrinsic kinetic (chemical) regime, is not provided, only the forms of the kinetic expressions can be taken from the literature, but not necessarily the absolute numerical values of rate constants. To carry out the very first kinetic experiments, the utilization of both efficient mixing and very small catalyst particles is the best way to confirm whether neither external mass transfer nor pore diffusion limits the reaction rate. An isothermal semibatch slurry reactor is well suited for this kind of study, since the mixing conditions in this type of reactor can easily be adjusted, and small particles in the slurry have their total surface free for pore diffusion. Another kind of device for these studies is a reactor where the catalyst is fixed in the stirrer. This setup is advantageous because it resembles a fixed bed in some respects, although it would be operating far away from the trickling regime. Importantly, the activation energy can be obtained from these results. When the experiments are successfully carried out in the kinetic regime, the obtained activation energies can be valid even for extrapolation.[12]

12.4   Mass-Transfer Effects

The mass-transfer effects appearing in a three-phase system are illustrated in Figure 12.3, which shows qualitatively, how the reactor concentration decreases due to mass-transfer resistances according to the film theory. The gas-phase reactant has to diffuse all the way from the gas phase into the porous solid to reach the reaction sites. Gas–liquid and liquid–solid mass transfer is expressed by power laws of dimensionless numbers, which depend on geometry and physical properties of both phases. The steady-state internal diffusion is described by an ordinary differential equation, and it depends strongly on the kinetic reaction rate, the effective diffusivity and the particle shapes and sizes.

12.4.1   Gas–Liquid and Liquid–Solid Mass Transfer

For reactions with fast intrinsic rates, gas–liquid mass transfer is rate limiting when intense mixing is not applied to the reaction system or when the reaction is performed under a low H_2 pressure. Accordingly, an increase in

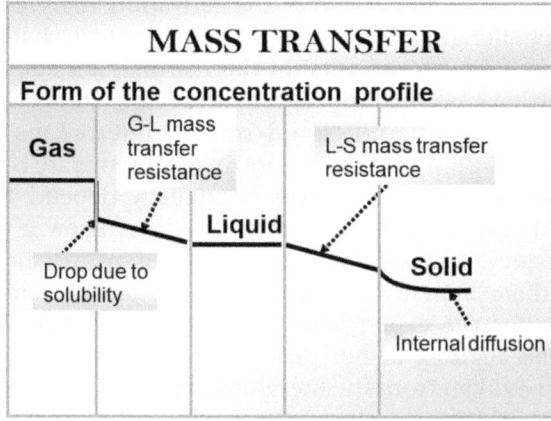

Figure 12.3   Mass-transfer steps in a three-phase reactor.

pressure is conducive to the kinetic regime, since it improves the mass flux at the gas/liquid interface, and thus elevates the levels of H_2 next to the catalyst surface. Gas–liquid mass-transfer limitations are often overcome by providing a large interfacial area for mass transfer or by using high superficial velocities to enhance the mass-transfer coefficient. High flow rates may demand a long reactor to provide a reasonable residence time. High flow rates can also be obtained by liquid circulation, but then the advantage of the plug-flow nature may partly be lost. For a low yield system, liquid circulation may still be acceptable. Mass-transfer area depends strongly on particle sizes and shapes, since liquid can be thought to flow as a film on the surfaces of the particles. Accordingly, we adopt and recommend, as a good practice, the use of a low flow interaction regime in continuous experiments on sugar hydrogenation on a lab-scale.

A collection of correlation equations of mass-transfer coefficients is reported in the literature.[14] It is apparent from these expressions that very different values for the gas–liquid mass-transfer coefficients are found for the range of very low superficial velocities.[19,20] The united gas-liquid mass-transfer coefficient is a function of diffusivity, characteristic particle dimension and various dimensionless numbers.

Gas-liquid mass transfer for nonreactive and reactive systems has been compared and observed to deviate from each other.[21] The mass-transfer coefficients obtained from reactive systems were much higher compared to those from simple absorption experiments. The reason behind the low mass-transfer coefficients obtained from absorption experiments was attributed to the presence of stagnant liquid zones, which after a while become saturated with gas, and hence did not participate in the mass transfer at all. In reactive cases, the saturation did not take place so easily, since the reaction consumes the dissolved gas.

Liquid–solid mass transfer depends strongly on the prewetting procedure. This observation corroborates the finding that the initial wetting is conducive to the hydrodynamic properties of the bed. Hence, the extent of channeling is generally shortened, and the generated flow fields are usually retained.

12.4.2 Effectiveness Factor and Internal Diffusion in Catalyst Pores

The evaluation of the effectiveness factor is the ultimate goal of every study on internal diffusion and heat conduction.[22] The effectiveness factor is a direct measure of how effectively the catalyst particle and active sites are used. With negligible internal concentration and temperature gradients, the effectiveness factor becomes 1, *i.e.* the diffusion is rapid enough to provide reactants to active sites, and to guarantee that the surface concentrations prevail within the whole particle. Typically, in cases where the arriving liquid feed stream is saturated with the reactant gas components, the highest reaction rates will be reached close to the feed entrance where the concentrations of the reactants are the highest. In the locations where reactions proceed at the highest rates, the reactants may not be able to diffuse as fast as the reactions could consume them, and thus the internal mass transfer becomes rate limiting (*i.e.* resulting in a low value for the effectiveness factor).

In order to mitigate the internal mass-transfer limitations, catalysts with small particles are often used on lab-scale reactors. This choice usually does not lead to a marked pressure drop when short fixed-bed reactors are operated. However, the pressure drop on large-scale reactors is an issue associated with high operating costs. Therefore, the use of small particles simply cannot be afforded on an industrial scale. Accordingly, the particle shape becomes the primary factor to increase the surface to volume ratio.[17] Catalyst suppliers have also special geometries available, which compared to spheres or cylinders, can provide a variety of external surface to volume ratios. For a multitude of catalyst particles originating from a simple cylindrical basic structure, a comparison of the ratios was reported.[7] The ratio of surface area to volume increases as follows: discs, cylindrical extrudates, rings, hollow extrudates, wagon wheels and miniliths. Remarkably, the ratio for the miniliths reaches a value as high as 20.

Another factor that directly determines the degree of limitation posed by the internal mass transfer within a catalyst particle is the length of the diffusion path. Accordingly, short diffusion paths are highly desirable in order to minimize the losses of the effectiveness factor caused by internal diffusion. Diffusion paths can be made short by two means. The first is by selecting proper geometries for a particle. The second is by keeping the active sites close to the external surface of the particles (*e.g.* as found for an eggshell type of catalyst).

Effective diffusivity is the property that is needed for the assessment of the impact of the pore diffusion on the process performance. It is typically calculated by making a porosity and tortuosity correction to the molecular diffusivities of components in liquid. It is specified as being directly dependent on the porosity, which makes high porosity a highly desirable attribute for the catalyst particle. Since trickle-bed reactors operate continuously, the catalyst particles have to be very durable. This requisite sets an upper limit to the porosity. Importantly, tortuosity is the other correction factor that can be used in order to compensate for the shortcomings stemmed from complicated pore structures and twisted diffusion paths within the catalyst. Generally, the uniform wetting of the external surface of a catalyst particle leads to pores uniformly wetted through capillary forces. In effect, the particle diffusion is then assumed to take place in this stagnant liquid.

Internal concentration gradients are not directly comparable with each other if the particle geometry is not similar. For example, the comparison between a spherical and a flake-shaped particle clearly shows important features. Considering that the geometry of a flake-shaped particle is divided using constant steps and the same is done for a spherical particle in the radial direction, the volume of elements remain constant throughout the flake-shaped particle, whereas the volume of the sphere elements decreases rapidly upon progressing further inside the particle. Therefore, a much more severe concentration drop can be afforded within a sphere, still yielding the same effectiveness factor found for flake-shaped particles.

12.5 Physical Properties of Gas Mixtures and Solutions

Physical properties of the solution or gas mixture are dependent on the composition of the solution and the operating conditions (*i.e.* temperature, pressure). When studying new systems, the challenge is to find the properties for a multicomponent mixture under specified process conditions. The approaches consist in either the use of an extrapolation for the estimation of the property value, or the use of values from a system that from the composition resembles that of the current case. Linear extrapolation works well for liquid densities. For highly nonlinear liquid viscosities, one can use nonlinear functions, presented in reference books, for extrapolation.[23] Should the values be unavailable in the literature, a resembling mixture with a known viscosity can be taken as a starting value for an extrapolation in order to estimate the needed values. A resembling mixture in cases of organic compounds means a mixture of compounds that have almost the same chain length and the same functional groups.

12.5.1 Diffusion Coefficients

The theory of liquid-phase diffusion is summarized in the literature.[23] A review of various methods for estimating the diffusion coefficients in liquid phase is provided by Poling *et al.*[23] No general agreement exists on the estimation method, but the selection of the method depends on the diffusing molecule and the solvent. For diffusivity in liquid phase, the Wilke–Chang equation has been compared favorably with other expressions and has been the most popular choice.[24] The diffusivity of a component is dependent on the size of the molecule. Binary diffusion coefficients were systematically used in all the studies even for multicomponent liquid mixtures. This approach was considered acceptable especially when working with dilute solutions or when the pure reactant has reacted to produce a product of almost the same chain length.

12.5.2 Gas Solubility

Correlation equations based on experimental data for gas solubility are provided by Fogg and Gerrard.[25] Notably, the measurements were performed mostly on neat solvents. The dissolution of organic molecules in water can considerably change the gas solubility. Therefore, the determination of the gas solubility is recommended when working with a new system.

12.5.3   Heat-Transfer Parameters and Reaction Enthalpies

Simultaneous solutions for heat and mass balances are required, whenever the reaction rates in the system are temperature dependent and the heat generation is significant. Heat can be transferred inside the reactor by means of both conduction and convection. When having a three-phase reactor, the thermal conductivity is material dependent since the conductivities of each phase differ from one another. The catalyst support can be made from a highly thermal-conductive material. Typically, the liquids conduct heat well, whereas gases most often act as insulators. In order to simplify the conduction modeling, effective conductivities are often used. They represent a kind of mean value for the thermal conductivity of the three phases.

For particle heat transfer, a mean thermal conductivity may be used only if the conductivity of the catalyst is close to the conductivity of the liquid. If the conductivity of the support material and the catalyst are superior to the conductivity of the liquid, most of the heat will be transferred through the better conducting material. Nonetheless, should the support be a material with a low thermal conductivity (*e.g.* active carbon), the heat conduction by the liquid dominates. The catalyst particles have a porous structure, which makes conduction geometry very complicated. Effective conductivity for such a situation was described as a function of the solid and liquid conductivities, and particle porosity.[26] Considering that the porous particles are externally well wetted, the pores can be regarded as fully filled with liquid, and the presence of gas can be totally omitted in the single particle modeling studies. Moreover, since the liquid and solid conductivities are much higher than the gas conductivity, a direct use of this equation to three-phase applications would overestimate the bed conductivity. For example, the estimate for the conductivity of quartz sand (which is typically used in the catalyst bed for sugar hydrogenation) is 0.6 W m^{-1} K^{-1}, water 0.65 W m^{-1} K^{-1} and hydrogen around 0.2 W m^{-1} K^{-1}. The effective thermal conductivity of a fixed bed depends also on the superficial liquid and gas velocities. However, practically no promotion of the effective conductivity is expected for slow flows.

Experimentally, a direct measurement of reaction enthalpies can be carried out in reaction calorimeters. Nonetheless, reaction enthalpies can also be determined either from the heats of formation of individual components (available in the literature) or estimated by using the group contribution methods or even computationally predicted from a compound model chemistry, such as CBS-QB3 and G2 models available in Gaussian (a software suite for computational chemistry). An alternative method is still the use of formation heats of analogous types of compounds (*i.e.* organic compounds having similar chain lengths and functional groups).

12.6   Liquid Flow Effects

Flow direction. The most popular flow arrangement for fixed-bed reactors that operate in the trickling flow regime is cocurrent downward flow.

For such operation mode, flooding does not set limits for the flow rates. Compared to the downward flow mode, the operation mode based on an upward liquid flow leads to higher liquid hold-up (*i.e.* a longer average residence time), larger extent of backmixing, and a more considerable pressure drop. A high liquid hold-up may be beneficial in some cases. However, a high liquid hold-up causes productivity loss due to severe backmixing and marked pressure drop.

Channeling and stagnant zones. In the downward mode, the liquid flows down driven by gravity. Since the liquid flow is not forced, the liquid takes multiple paths of least resistance. This selection creates channeling and stagnant zones. The stagnant zones become easily saturated with the gas reactant, and the reaction progresses until liquid or dissolved reactant is fully consumed. However, if the stagnant zones are not replenished with reactants, such locations eventually become nonproductive. With a well-constructed feed distributor and a short reactor, the problems associated with channeling and stagnant zones are not as critical as that found for a long reactor with a large diameter.

External wetting and wall flow. Complete wetting is often assumed for shallow laboratory reactors, if the bed is initially flooded and a fine sand-bed feed distributor is used. A collection of empirical correlations for wetting efficiency has been given by Lappalainen *et al.*[27] In principle, the parameter values are valid for the data originally used for the fitting. When modeling a new system, a re-estimation of the parameters is then the safest option. The equations point out that the liquid flow is the variable that most influences the wetting efficiency. The external wetting efficiency of particles exerts a direct effect on the catalyst effectiveness factor. Should the particles be only partially and externally wetted, the active sites of the nonwetted parts become unreachable and do not participate in reactions at all. Incomplete wetting (*e.g.* for a sphere,[28] and for various kinds of cylindrical particles)[29] is discussed. Furthermore, the ways that the effectiveness factor of the particles should be corrected, if only partial external wetting takes place, are also presented.

The concentration gradients are observed mainly in the axial direction and often the radial direction is not modeled, since the net flow is directed axially downward. In the literature, one can find experimental studies of radial mass distribution in which the bottom of the reactor is divided into multiple annuli, and the liquid is collected from each of these sections.[30] Through this approach, the wall flow and its contribution to the entire flow can be determined. In laboratory-scale operation, the wall effects can be reduced simply by selecting a small enough particle to reactor-diameter ratio (<0.1). The wall flow is dependent not only on this ratio but also on the particle shape and physical properties of the fluids. Liquids with low surface tension and/or high density suppress the wall effects.

Pressure drop. In the cases of sugar hydrogenation on lab-scale reactors, operating with the low superficial flow velocities and short reactors, the pressure drop is usually only a small fraction of the operating pressure, and can thus be regarded as negligible. However, in an industrial operation, it is of crucial importance. A large pressure drop inside an industrial reactor implies that the energy consumption increases. Furthermore, a decrease in the productivity can also occur by lowering the partial pressure of the gas reactant.

More precise pressure-drop correlations can be obtained by solving momentum and mass balances or even as a result of computational fluid dynamics for a well-specified geometry. A collection of pressure-drop correlations, which can be used for various flow regimes, is also given.[14] An extensive database of pressure drop for fixed beds has been used for a neural network study to obtain a semiempirical model of pressure drop in two-phase flow.[17]

Liquid hold-up. Many types of liquid hold-ups have been described for fixed beds with porous particles. Therefore, one has to be very careful when specifying and when interpreting the reaction rates on the liquid volume basis. The dynamic hold-up is the volume fraction of that part of liquid that flows dynamically. The static hold-up is the volume fraction occupied by stagnant liquid especially in the space between the particles. Internal hold-up is related to the liquid inside pores that is held by capillary forces, whereas external hold-up is the volume fraction of liquid outside the pores.

The widely used liquid hold-up correlations are power-law expressions based on dimensionless numbers. These correlations are tailor-made for each system. Therefore, no guarantee can be given for their level of accuracy when applied to a completely different reaction system under different operating conditions. All these equations confirm the fact that the liquid hold-up depends on the physical properties of the fluids, the properties of the catalyst and the superficial velocities of gas and liquid.

The general idea behind the liquid hold-up correlations is that each dimensionless number represents either an aspect ratio or ratios of some forces known to affect the system and that these ratios determine how much liquid is held in a fixed bed. For example, the Reynolds number represents the ratio of inertia forces to viscous forces and the Galileo number represents gravity forces divided by viscous forces. Power laws represent very simple ways to express the fluid dynamics, but they are still favored for good reasons. They sound scientifically acceptable (after all various kinds of forces generate the flows, even computational fluid dynamics is based on force or momentum balances). They are also easy to incorporate in reactor models and they have only a limited number of parameters. By observing the values of the powers, one can realize the variables for which the hold-up is sensitive and for which it is not. A collection of liquid hold-up correlations can be found in an extensive review article.[14]

12.7 Reactor Modeling

12.7.1 Modeling Steps

The model development should progress gradually from simpler approaches to more advanced ones. The procedures in the model development are discussed as follows.

Modeling of individual phenomenon. Most of the studies of fixed-bed reactors operating at trickling flow regimes have been focused on single or a few separated phenomena. These studies demonstrate that each phenomenon, when looked at closely, becomes a research field of its own. In the modeling approach for hydrogenation of sugars, the models were developed not penetrating deeply into two fields, *i.e.* fluid dynamics and particle technology. If all the phenomena were included, it would inevitably have led to a model task having too many parameters to be identified.

Modeling of experimentally studied simplified system. Intrinsic reaction kinetics of a simplified system can be studied sometimes separately (*e.g.* by using batch reactors with small catalyst particles and vigorous stirring). This study enables the determination of rate parameters independently and the extensive parameter estimation task is then divided into several smaller ones.

Various process conditions and reactor setup aspects are of importance for alleviating problems associated with heat and mass transfer:

1. Dilute substrate concentrations and a short reactor length are conducive to the mitigation of the effects related to the adsorption terms in the reaction equation and to heat and mass transfer. Indeed, working with dilute streams is a recommended practice in initial laboratory tests in order to avoid temperature gradients in the catalytic bed.
2. The effect of impurities is not negligible for the processing of real feeds of biomass. Accordingly, the use of reactants at a high purity grade eliminates the effects of impurities on the reaction performance.
3. Small catalyst particles minimize internal diffusion. Moreover, by the selection of small particles instead of large ones, the internal heat-transfer effects can be reduced, however, at the cost of the increased pressure drop.
4. The use of a slurry reactor with intense mixing is recommended for fundamental studies on the intrinsic kinetics of a reaction. Catalyst loading is one key factor that dictates directly the severity of temperature effects. The dilution of the catalyst bed with inert particles reduces the temperature gradients.
5. In addition, the choice of the material for catalyst support is a key issue, since the thermal conductivity of various support materials can considerably differ. Therefore, the choice of a material having a high thermal conductivity may be useful to overcome problems associated with temperature effects.

Several other criteria for avoiding the generation of serious temperature gradients and means for minimizing them have been presented in the literature.[13]

Sensitivity studies prior to estimation. Sensitivity studies consisting of multiple single-parameter experiments can directly be carried out using suitable softwares (*e.g.* Matlab, Modest and gPROMS). A sensitivity study is very valuable to provide an educated guess on how the system behaves and reacts to individual parameter changes.

12.7.2 Mass Balances for Continuous Fixed-Bed Reactor: Isothermal System

On a lab-scale operation, it is common and easier than on an industrial-scale operation to end up with an isothermal system. This is particularly true if the feed streams are dilute, the reactor is short, the catalyst particles are small, and the superficial velocities of the fluids are low. For the axial-dispersion model, the mass balance for a liquid-phase component can be written as eqn. (12.2)

$$\frac{\partial C_{i,L}}{\partial t} = \frac{1}{\varepsilon_L}\left(-w_L \frac{\partial C_{i,L}}{\partial z} + \varepsilon_L D_{a,L}\frac{\partial^2 C_{i,L}}{\partial z^2} + k_{gl}a_i\left(C_{i,L}^* - C_{i,L}\right) + \varepsilon_L \eta_{e,i} r_i \rho_{cat}\right)$$

(12.2)

where: C_i corresponds to local concentration of i [mol/L]; w_L, superficial liquid velocity, [m s^{-1}]; t, time, [s]; ε_L, liquid hold-up, [ ]; $D_{a,L}$, axial dispersion coefficient in liquid, [m^2 s^{-1}]; z, axial location [m]; $k_{gl}a_i$, united gas–liquid mass-transfer coefficient for component i, [s^{-1}]; $\eta_{e,i}$, effectiveness factor of component i, [ ]; r_i, generation rate for component i, $i = A, B, C$ [mol g$_{cat}^{-1}$ s^{-1}]; ρ_{cat}, weight concentration of the catalyst in bed, [g$_{cat}$ L^{-1}].

The equation is written in a general form for component i in the liquid. For liquid-phase components, the gas–liquid mass-transfer term is not required. The smaller the axial dispersion coefficient, the more the reactor approaches a plug-flow behavior.

12.7.3 Thermal Effects in Fixed Beds

The hydrogenation and hydrodeoxygenation of biomass-derived molecules are exothermic reactions.[16] Since the reaction enthalpies are significant, temperature gradients are generated. Gradients may arise inside the whole reactor or locally within each single catalyst particle. Both axial and radial temperature gradients may emerge inside the fixed bed. By equipping the fixed-bed reactor with a cooling jacket, the axial gradients can be reduced, but the radial temperature gradients may then become more severe. This observation is particularly noted when the reactor is forced to be kept at an axially isothermal state, which is quite often targeted on a small-scale operation as well as when the product selectivity is highly dependent on temperature. It is always necessary to check whether the experiments are carried out under truly isothermal conditions, since reaction rates increase exponentially as a function of temperature. Otherwise, the data can be misinterpreted, resulting in too promising reaction rates.

12.8   Application Examples

Polysaccharides (*e.g.* cellulose, starch, and hemicelluloses) provide a source of simpler mono- and disaccharides that, in turn, serve as a platform for the production of value-added molecules. Notably, among these are the sugar alcohols, which are obtained through the reduction of the carbonyl group in the sugar molecule either by means of chemical agents (*e.g.* sodium borohydride) or by molecular hydrogen in the presence of a homogeneous or heterogeneous catalyst. Solid metal-based catalysts can be applied in an aqueous or alcoholic environment, circumventing the use of stoichiometric reducing agents and the subsequent generation of inorganic waste. Accordingly, heterogeneous catalysts based on Ni, Pd, Pt or Ru are used in the industrial hydrogenation of sugars to sugar alcohols.[31–58]

Sugar alcohols are versatile molecules that have found a wide variety of applications. Higher molecular weight polyols (chains longer than four carbon atoms) are building blocks in the manufacture of polyesters, alkyd resins and polyurethanes.[3] They are also intermediates in the production of pharmaceuticals and starting molecules for the synthesis of ligands.[44] However, their most common use is associated with the fact that polyols possess a sweet taste, but produce a lower glycemic response. Accordingly, their use as "natural" sweeteners is quite desirable and straightforward.[41–44]

Traditionally, sugar hydrogenation is carried out batchwise, but in the recent years, sugar hydrogenation in continuous operation has been demonstrated. For instance, glucose hydrogenation has been examined in trickle beds, monoliths and on activated carbon cloths.[3,44] Furthermore, it is possible to carry out sugar hydrogenation in loop reactors in which the liquid, containing the catalyst, is recirculated through an external loop system.[16] Since the sugar molecules are bulky and the H_2 solubility in water is limited, the diffusion limitation inside the porous catalyst layer becomes severe in porous catalyst layers, as demonstrated experimentally and by numerical simulations.[17]

Two application examples relevant for hydrogenation of components in biomass are considered in this section: hydrogenation of glucose to sorbitol[17] and hydrogenation of arabinose to arabitol.[17] Experiments were carried out both in batch and continuous reactors. In both examples, the goal is to obtain the corresponding sugar alcohols, mannitol, and arabitol with high selectivity and high efficiency. In these studies, carbon-supported ruthenium catalysts were used in both cases. The examples illustrate the research and development strategy: from semibatch slurry reactors towards continuous operation in fixed beds.

12.8.1   The Core of Successful Hydrogenation of Biomass: Solid Catalyst

Several kinds of solid catalysts have been examined in sugar hydrogenation, but two types were found to be superior: skeletal or sponge nickel (Raney nickel) and supported ruthenium catalysts. Sponge nickel still has a strong position in industrial applications, particularly because of its low price and proven technology. However, the use of sponge-nickel catalysts is very much limited to batch processes in slurry reactors. In addition, sponge-nickel catalysts suffer from other serious disadvantages (*e.g.,* they are poisonous and pyrophoric). Therefore, an increasing effort has been put in the search for catalyst alternatives. In this context, ruthenium supported on various materials has emerged as the most promising class of catalysts for sugar hydrogenation. Ruthenium can be supported on solids with high surface area (*e.g.* active carbon and alumina), and therefore used as finely dispersed slurry (particle sizes 0.1 mm or less), conventional pellets and eggshell catalysts (particle sizes 1–10 mm), and as thin layers in structured reactors (*e.g.* catalyst monoliths, solid foams or fibers with a layer thickness <50 micrometers).

Recent studies have demonstrated the hydrogenation of sugars to sugar alcohols to be possible with a very high selectivity (>95 %), provided that the Ru catalyst is properly designed and pretreated, the presence of hydrogen is significant in the liquid phase, and moderate temperatures (typically below 403 K in order to avoid formation of side products *via* caramelization) are used.

The structure sensitivity of glucose hydrogenation, in the presence of Ru catalysts with varying nanoparticle sizes, was demonstrated.[3] A reactivity maximum was found for a certain particle size (3 nm), after which the activity declined slowly.[3] This implies that Ru is active within a broad interval of particle sizes. The oxidation state of ruthenium, under hydrogenation conditions, has been a subject of debate.[3] Practical experience has indicated that a pre-reduction of Ru is not necessary for successful sugar hydrogenation. It has been demonstrated by XPS that Ru coexists in the catalyst both in a metallic and oxidized form, and that the oxidized form can contribute to the hydrogenation.[3]

12.8.2   Glucose Hydrogenation

We have studied the hydrogenation of glucose to sorbitol experimentally as well as by modeling in a slurry reactor and in fixed beds.[17] The temperature and liquid feed flow ranges in fixed-bed experiments were 363–403 K and 1.5–4 mL min^{-1}, respectively. Most of the experiments were conducted using particles within the size range of 125–250 µm, but one of the experiments was performed with larger particle sizes, 330–500 µm.

Three kinds of models were developed. The first is an axial-dispersion model for the reactor with temperature-dependent kinetics. It was used exclusively in the numerical solution of the mass balances. The second model is based on simultaneous solution of mass and energy balances for a radial section of the reactor. This model was chosen to reveal how severe temperature and concentration profiles may arise at the feed entrance section of the reactor. The third model is a particle model also based on the simultaneous solution of mass and energy balances. The aim was to study how large concentration and temperature gradients may emerge within the catalyst particles under the reaction conditions used, and how much the catalyst effectiveness factor can change. Accordingly, the focus was on the effect of the catalyst particle size. According to the modeling results, the observed radial temperature gradients were small (within 1 K) both in the feed entrance section and within the catalyst particles, even with the maximum concentrations, due to the moderate reaction enthalpy. The simulations verified that the isothermal assumption was not far away from reality in each individual experiment.

The semibatch experiments carried out in a slurry reactor were modeled using Langmuir–Hinshelwood–Hougen–Watson kinetics with an Arrhenius-type temperature dependence. Since the concentrations were low, the effective reaction orders were known and the kinetic expressions could be extended by incorporating a deactivation term into it. This term also included an Arrhenius-type temperature dependence. The parameter estimation results along with the original data are shown in Figures 12.4.

Based on the experience gained from the semibatch experiments, the parameter estimation was also conducted for the continuous fixed-bed reactor. Since the liquid stream was very dilute (0.1 mol L^{-1} glucose), and mass transfer was not rate limiting, kinetic regime and catalyst deactivation dominated. Catalyst deactivation was immediately detected in the

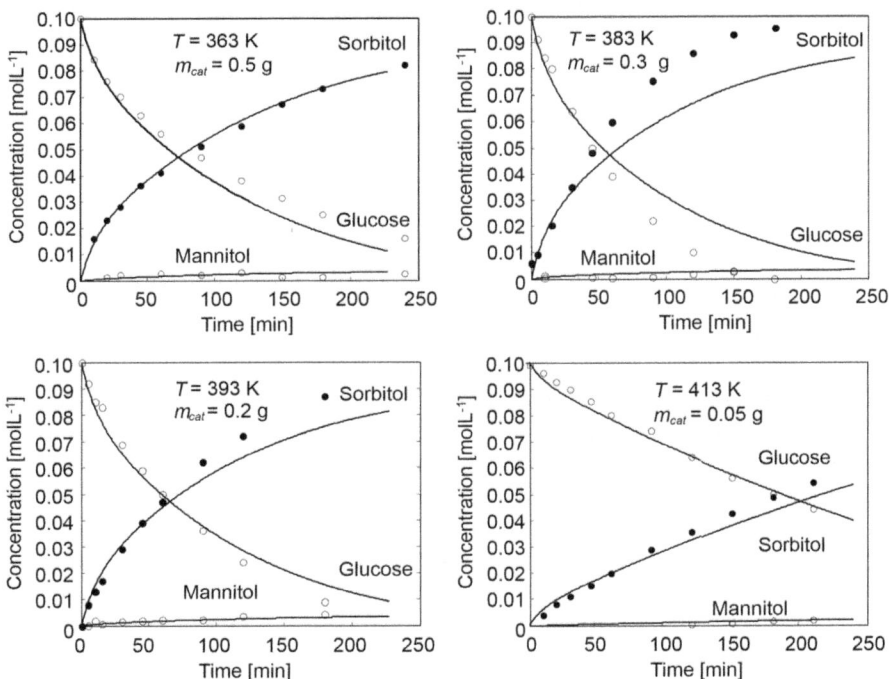

Figure 12.4 Parameter estimation for glucose hydrogenation in a slurry reactor.

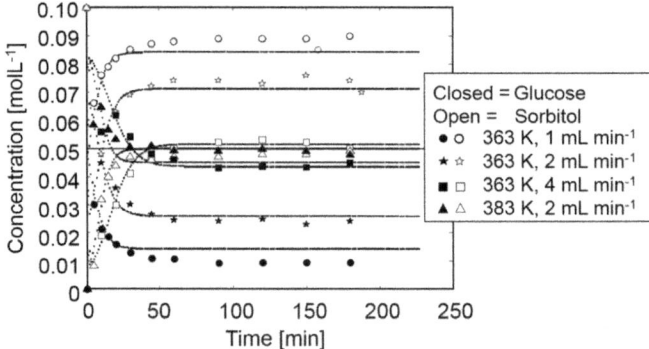

Figure 12.5 Experimental and predicted behavior of a continuous fixed-bed reactor in glucose hydrogenation.

experimental results, and it was more severe than in the semibatch experiments. In continuous operation, the highest concentrations are all the time observed close to the feed entrance. The experimental and calculated concentration trends are shown in Figure 12.5.

To reveal how severe temperature gradients might arise for the first radial section of the continuous reactor, mass and energy balances were simultaneously solved. Since the effective radial conductivity was not exactly known, the case was solved for different values of effective conductivity and example solutions are illustrated in Figure 12.6. This analysis shows, on the one hand, how high temperatures may arise with reasonable values of the conductivity, and on the other hand, how sensitive these temperatures depend on conductivity values.

12.8.3 Arabinose Hydrogenation

Arabinose hydrogenation on a lab-scale fixed-bed reactor was modeled. The catalyst was ruthenium on an active carbon support. The system was experimentally studied within the semibatch slurry reactor at temperatures of 363–403 K, H_2 pressure of 4–6 MPa, and at a rotation speed of 1800 rpm. Very small particles were used to guarantee that the semibatch experiments were conducted in the intrinsic kinetic regime. The main reaction followed closely the Langmuir–Hinshelwood kinetics. The arabitol selectivity was very high and practically no byproducts were formed.

The calculations showed that internal diffusion becomes important for catalyst particles with diameters higher than 100 μm; the effectiveness factor was 0.8 at 373 K, which suggests that either slurry technology or structured reactors with thin catalyst layers would provide the most efficient way to carry out the hydrogenation. This feature is characteristic of the molecules originating from biomass. Since the molecules are relatively large, the internal diffusion is always an important factor, when conventional fixed beds are used with catalyst particles larger than 1 mm. Simulations of the

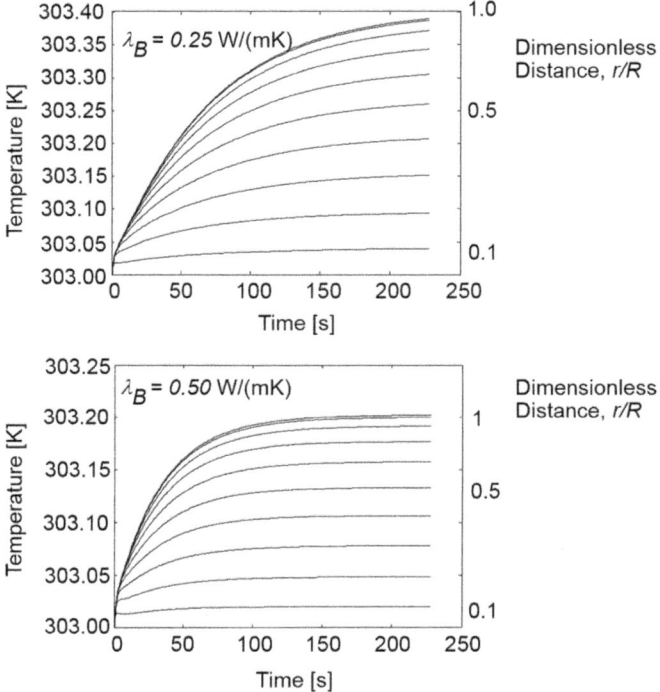

Figure 12.6   Severity of the temperature rise as a function of effective radial conductivity. Each step corresponds to one tenth of the reactor length.

concentration profiles in the catalyst particle are therefore very helpful in the
design of new processes and intensification of existing ones.

The continuous fixed-bed reactor models were written in Matlab. The
governing equations were transformed into dimensionless forms before
solving them and higher-order spatial derivative approximations were used.
An adaptive grid was tested. The reactor model was used for two purposes.
First, it was used for a sensitivity study conducted for a system having larger
(commercial) catalyst particles in a concentrated solution, in order to
demonstrate how severe the mass-transfer restrictions could become. Finally, to obtain a parameter estimation, the experimental data obtained from
the fixed-bed reactor, loaded with crushed catalyst particles, were used. The
parameter estimation and simulation results are presented in Figure 12.7.

The simulation results agree well with the experimental observations. In
the sensitivity study, the gas–liquid mass transfer was modeled using the
Ellman correlation, but in the final parameter estimation, the constant was
optimized to improve the model accuracy. Liquid–solid mass transfer was
also modeled. The major difference, that the inclusion of the liquid-solid
mass transfer effect produced, was that the number of variables and differential equations increased, since the balances had to also be solved for
the liquid on the surface of the particles. As a conclusion, the liquid–solid

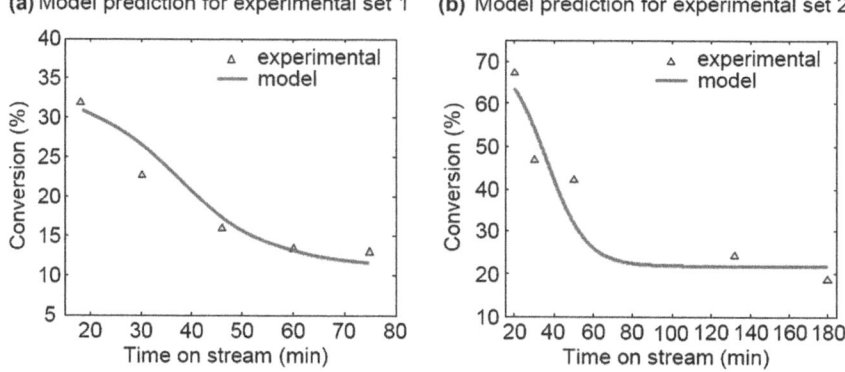

(a) Model prediction for experimental set 1   **(b)** Model prediction for experimental set 2

Figure 12.7   Example results from parameter estimation for the continuous fixed-bed reactor: the effect of catalyst deactivation.

mass-transfer resistance was observed to be less severe than the gas–liquid mass-transfer resistance.

12.9   Conclusions

Hydrogenation of biomass-derived compounds, in particular carbohydrates, has been a topic of extensive research into catalyst development and understanding the mechanistic aspects of the reaction. Since catalysis is a kinetic phenomenon, elucidation of kinetics is a vital part of any catalytic research, including hydrogenation of sugars. In the batchwise reactions, in order to separate the reaction kinetics from transfer phenomena, experiments are carried out at lower catalyst loadings with small catalyst particle sizes and efficient mixing. For fixed-bed lab-scale reactors, which could operate under a significant influence of mass-transfer phenomena (gas-liquid- or liquid–solid-transfer rate or intraparticle transfer rates), a special care should be taken not only in catalyst selection, but also in selecting the reactor size, particle geometry, catalyst dilution and feed distribution.

Mass- and heat-transfer effects are present inside the porous catalyst particles as well as at the surrounding fluid films. In addition, heat transfer from and to the catalytic reactor gives an essential contribution to the energy balance. The core of modeling a two-phase catalytic reactor is the catalyst particle, namely simultaneous reaction and diffusion in the pores of the particle should be considered. Mechanistic interpretations of catalytic results obtained in fixed-bed reactors should be done with a clear understanding about the impact of mass and heat transfer.

Comparison between catalytic performance in slurry and fixed-bed reactors is even more complicated considering flow patterns in fixed-bed reactors, especially for catalysts with a distribution of particle sizes. Not only mass-transfer limitations but also deactivation would then be location specific depending on flow velocities and the concentration of the organic reactant. Adequate modeling of sugar hydrogenation requires knowledge of physical properties of the reaction mixture, such as, for example, liquid viscosity, diffusion coefficients, gas solubility, heat-transfer properties. Information about the liquid flow (directions, wetting, channeling, pressure drop, liquid hold) is needed for reactor modeling, which should progress gradually from simpler approaches to more advanced ones. The modeling approach was considered for two application examples relevant for hydrogenation biomass-derived components, namely hydrogenation of glucose to sorbitol and arabinose to arabitol. The examples illustrated the research and development strategy from semibatch slurry reactors towards continuous operation in fixed beds.

Acknowledgements

The work is a part of the activities of Process Chemistry Center (PCC), a center of excellence financed by Åbo Akademi University. Financial support from the Academy of Finland is gratefully acknowledged (T. Kilpiö, T. Salmi).

References

1. F. Lichtenthaler and S. Peters, *C. R. Chim.*, 2004, 7, 65.
2. F. Lichtenthaler, *Carbohydr. Res.*, 1998, **313**, 69.
3. A. Aho, S. Roggan, O. A. Simakova, T. Salmi and D. Yu. Murzin, *Catal. Today*, 2014, http://dx.doi.org/10.1016/j.cattod.2013.12.031.
4. C. Eisenbeis, R. Guettel, U. Kunz and T. Turek, *Catal. Today*, 2009, **147**, 342.
5. B. Chen, U. Dingerdissen, J. Krauter, H. Rotgerink, K. Mobus, D. Ostgard, P. Panster, T. Riermeier, S. Seebald, T. Tacke and H. Trauthwein, *Appl Catal. A*, 2005, **280**, 17.
6. J.-P. Mikkola, T. Salmi and R. Sjoholm, *J. Chem. Technol. Biotechnol.*, 2001, **76**, 90.
7. R. Krishna and S. Sie, *Chem. Eng. Sci.*, 1994, **49**, 4029.
8. G. Froment and K. Bishoff, in *Chemical Reactor Analysis and Design*, John Wiley & Sons, New York, 1979.
9. O. Levenspiel, *Chemical Reaction Engineering*, John Wiley & Sons, New York, 1999.
10. T. Salmi, J.-P. Mikkola and J. Wärnå, in *Chemical Reaction Engineering and Reactor Technology*. CRC Press, New York, 2009.
11. J. Hagen, in *Industrial Catalysis, A Practical Approach*, Wiley-VCH Verlag GmbH, Weinheim, Germany, 1999, 2nd edn.
12. J. Butt, in *Reaction Kinetics and Reactor Design*, Marcel Dekker Inc., New York, 2000.
13. F. Mederos and J. Ancheyta, *J. Chen, Appl. Catal. A: Gen.*, 2009, **355**, 1.
14. M. Al-Dahhan, F. Larachi, M. Dudukovic and A. Laurent, *Ind. Eng. Chem. Res.*, 1997, **36**, 3292.

15. B. Wilhite, X. Huang, M. McCready and A. Varma, *Ind. Eng. Chem. Res.*, 2003, **42**, 2139.
16. H. V. Sifontes, in *Hydrogenation of L-Arabinose, D-Lactose, D-Maltose and L-Rhamnose*, Thesis, Åbo Akademi Department of Chemical Engineering, Åbo 2012.
17. T. Kilpiö, in *Mathematical Modelling of Laboratory Scale Three-Phase Fixed-bed Reactors*, Thesis, Åbo Akademi, Department of Chemical Engineering, Åbo 2012.
18. P. Silveston and R. Ross, in *Periodic Operation of Chemical Reactors*, Elsevier 2013.
19. K. Metaxas and N. Papayannakos, *Chem. Eng. Technol.*, 2008, **31**, 410.
20. K. Metaxas and N. Papayannakos, *Ind. Eng. Chem. Res.*, 2006, **45**, 7110.
21. B. Koning, in *Heat and Mass Transport in Tubular Fixed Bed Reactors at Reacting and Non-Reacting Conditions*, Thesis, University of Twente, 2002.
22. W. Woodside and J. Messmer, *J. Appl. Phys.*, 1961, **32**, 1688.
23. B. Poling, J. Prausnitz and J. O'Connell, in *Properties of Gases and Liquids*, McGraw-Hill, 2001, 5th edn.
24. C. Wilke and P. Chang, *AIChE J.*, 1955, **1**, 264.
25. P. Fogg and W. Gerrard, in *Solubilities of Gases in Liquids*, Wiley, Chichester, 1991.
26. E. F. Jaguaribe and D. E. Beasley, *Int. J. Heat Mass Trans.*, 1994, **27**, 399.
27. K. Lappalainen, V. Alopeus, M. Manninen and J. Aittamaa, *Ind. Eng. Chem. Res.*, 2008, **47**, 8436.
28. W. Sakornwimon and N. Sylvester, *Ind. Eng. Chem. Process Des. Dev.*, 1982, **21**, 16.
29. D. Tsamatsoulis and N. Papayannakos, *AIChE J.*, 1996, **42**, 1853.
30. M. Kohler and W. Richarz, *Ger. Chem. Eng.*, 1985, **8**, 295.
31. A. Corma, S. Iborra and A. Velty, *Chem. Rev.*, 2007, **107**, 2411.
32. N. Déchamp, A Gamez, A. Perrard and P. Gallezot, *Catal. Today*, 1995, **24**, 29.
33. M. Cunningham and C. Walker, *Maltitol-rich Solutions and their Manufacture*, US Patent 20040224058, 2004.
34. M. Niimi, Y. Hario, Y. Ishii, K. Kataura and K. Kato, *Preparation of Maltose and Maltitol from Starch*, Japanese Patent 02042997, 1990.
35. G. Darsow, *Preparation of Epimer-free Sugar Alcohols*, European Patent 423525, 1991.
36. M. Magara, K. Shimazu, M. Fuse, K. Kataura, J. Osada, K. Kato and K. Yoritomi, *Manufacture of Maltitol with High Purity*, Japanese Patent 01093597, 1989.
37. J. Kondo, T. Miyamoto and M. Asano, *Maltitol from Maltose*, Japanese Patent 53119812, 1978.
38. A. Maisin, A. Lefevre, M. Wauters and A. Germain, *Reactivation of a Catalyst Containing Platinum Group Metals for Hydrogenation of Sugars*, Belgium Patent 8822791980.

39. D. Vanoppen, M. Maas-Brunner, K. Ulrich and J.-D. Arndt, *Preparation and Use of Silicon Dioxide-supported Ruthenium Catalysts for Saccharide Hydrogenation*, German Patent 10128205, 2002.
40. V. Benessere, R. Del Litto, A. De Roma and F. Ruffo, *Coord. Chem Rev.*, 2010, **254**, 390.
41. G. Helmchen and C. Murmann, *Preparation of Optically Active Diphosphine Ligands as Asymmetric Hydrogenation Cocatalyst*, European Patent 885897, 1998.
42. Position of the American Dietetic Association: Use of nutritive and nonnutritive sweeteners, *J. Am. Dietetic Assoc.*, 2004, **104**, 255.
43. H. Mitchell, in *Sweeteners and Sugar Alternatives in Food Technology*, Blackwell Pub., Oxford, Ames, Iowa, 2006.
44. C. Eisenbeis, R. Guettel, U. Kunz and T. Turek, *Catal. Today*, 2009, **147**, 342.
45. J. Wisniak, M. Hershkowitz and S. Stein, *Ind. Eng. Chem. Prod. Res. Dev.*, 1974, **13**, 232.
46. J. Wisniak, M. Hershkow, R. Leibowit and S. Stein, *Ind. Eng. Chem. Prod. Res. Dev.*, 1974, **13**, 75.
47. A. Heinen, J. Peters and H. van Bekkum, *Carbohydr. Res.*, 2000, **328**, 449.
48. M. Makkee, A. Kieboom and H. van Bekkum, *Carbohydr. Res.*, 1985, **138**, 225.
49. B. Chen, U. Dingerdissen, J. Krauter, H. Rotgerink, K. Mobus, D. Ostgard, P. Panster, T. Riermeier, S. Seebald, T. Tacke and H. Trauthwein, *Appl. Catal. A*, 2005, **280**, 17.
50. J. Kuusisto, J.-P. Mikkola, M. Sparv, J. Wärnå, H. Heikkilä, R. Perälä, J. Väyrynen and T. Salmi, *Ind. Eng. Chem. Res.*, 2006, **45**, 5900.
51. J. P. Mikkola, T. Salmi and R. Sjöholm, *J. Chem. Technol. Biotechnol.*, 1999, **74**, 655.
52. E. Crezee, B. Hoffer, R. Berger, M. Makkee, F. Kapteijn and J. Moulijn, *Appl. Catal. A*, 2003, **251**, 1.
53. J. Wisniak and R. Simon, *Ind. Eng. Chem. Prod. Res. Dev.*, 1979, **18**, 50.
54. T. Salmi, J. Kuusisto, J. Wärnå and J.-P. Mikkola, *Chim. Ind.*, 2006, 88.
55. T. Salmi, D. Yu. Murzin J. P. Wärnå J.-P. Mikkola J. E. B. Aumo and J. I. Kuusisto, in *Catalysis of Organic Reactions*, S. R. Schmidt (ed.), CRC Press, 2007, 187.
56. J.-P. Mikkola, T. Salmi and R. Sjoholm, *J. Chem. Technol. Biotechnol.*, 2001, **76**, 90.
57. P. Simms, K. Hicks, R. Haines, A. Hotchkiss and S. Osman, *J. Chromatogr.*, 1994, **667**, 67.
58. P. H. Brahme and L. K. Doraiswamy, *Ind. Eng. Chem. Process Des. Dev.*, 1976, **15**, 130.

CHAPTER 13

Safety and Practical Aspects of Liquid-Phase Hydrogenation

MARCO KENNEMA AND NILS THEYSSEN*

Max-Planck-Institut für Kohlenforschung, Kaiser-Wilhelm-Platz 1,
D-45470 Mülheim an der Ruhr, Germany
*Email: theyssen@kofo.mpg.de

13.1   Introduction

The conversion of biomass-derived molecules, through hydrogenation and hydrogenolysis, is typically carried out at temperatures of 325–573 K under a H_2 pressures of 0.1–20 MPa. Knowing beforehand—the reaction (*i.e.* products, thermodynamics, and kinetics), reactor design, and reactor material—is mandatory not only to achieve the best results possible, but also to guarantee the utmost safety of both the operator and the environment. This chapter is designed to provide an introduction of the practical aspects concerning biomass hydrogenation, with a focus on the safety requirements for reactions conducted at typical temperature and pressure ranges of catalytic hydrogenation. The safety recommendations in this chapter are to be taken as a starting point for the setup or operation of a high-pressure reactor design. However, it is important to point out that local safety regulations may apply, and they cannot be overlooked. This chapter also provides important aspects of mixing and stirring technology for batch reactor vessels.

RSC Energy and Environment Series No. 13
Catalytic Hydrogenation for Biomass Valorization
Edited by Roberto Rinaldi
© The Royal Society of Chemistry 2015
Published by the Royal Society of Chemistry, www.rsc.org

13.2 Infrastructure for the Operation of High-Pressure Reactors

13.2.1 High-Pressure Reactor Boxes

Due to the high rupture potential of pressurized reactors, operators must take the utmost caution to prevent injury in the case of pressure vessel failure. Table 13.1 summarizes the effects of overpressure on the facilities and human tissue.[1]

To alleviate the consequences of an event of reactor failure, the hydrogenation vessel should be installed in an isolated facility called a high-pressure reactor box. The minimum specifications for a pressure box are listed below:[2]

- A 30-cm wall thickness of ferroconcrete.
- Alternatively, a steel plate (10–20 mm) or sand-filled steel plate (6–7 mm) on each side of the box.
- One side of the box should be attached to a collapsible wall in order to reduce the shockwave and allow for fast pressure discharge in case of explosion.

An emergency shut-off switch or button must always be accessible in an observation area, to close the hydrogen supply to the reactor and disconnect the power supply. Depending on the type of reaction conducted in the high-pressure box, proper ventilation should be installed both on the top of the box for the exhaustion of light gases (< 29 g mol^{-1}), and on the bottom of the box for heavy gases (> 29 g mol^{-1}). The area around the reactor should be

Table 13.1   Effect of overpressure on the facilities and human tissue. Adapted from Ref. 1.

Peak Overpressure (MPa)	Maximum wind speed (km/h)	Damages to facilities	Injuries to human tissue
0.07	60	Glass shatters	Light injuries mainly due to shrapnel
0.14	110	Mild damage to houses (*e.g.* doors blown in, major roof damage)	Severe injuries mainly due to shrapnel and debris
0.21	165	Residential structures collapse, trees uprooted	Serious, potentially life threatening injuries occur, fatalities
0.35	260	Most buildings collapse	Ear drum rupture
1	605	Buildings built using reinforced concrete collapse or are severely damaged	Lung damage, significant fatality rate
1.4	810	Heavily built concrete buildings are demolished	100% fatalities

properly ventilated; a minimum of twenty-fold air renewal is recommended.[2] A gas warning system should be in place, with detectors specific to the gases utilized in the experiments. These detectors should be set to gas thresholds, which if reached in the atmosphere; the gas feed to the reactor is automatically shut off, and must be manually reset. Likewise, maximum operation pressure and temperature settings should be set for the reactor with an automatic shut-off of the electricity for heating if these levels are reached. The reactor system should have a complete data-acquisition system, accessible in the observation area to enable easy monitoring of temperature and pressure in the reaction vessel.

13.2.2   Hydrogen-Specific Information

The primary risk of handling hydrogen lies in its broad combustible range in air (4–75%) together with its very low minimum ignition energy of only 0.02 mJ. This value is at least one order of magnitude lower than those of frequently used organic solvents.[3] Such a low energy level can easily be provided by an electrostatic discharge between two surfaces. An electrostatic discharge is often underestimated as a potential ignition source because the human recognition threshold for an electrostatic discharge is much higher (0.5 mJ).[3,4] It is noteworthy that the thermal energy of a hydrogen fire is lower than that of hydrocarbons due to its low infrared intensity. Accordingly, in the case of a hydrogen explosion, it is more likely to be injured by the consequences of the blast rather than the thermal energy.

Since hydrogen is odorless and colorless, the notification of critical leaks is difficult. The high potential for a hydrogen leakage is based on its small molecular size and very high diffusivity. For small reactors, a hydrogen leak check can be performed by simply immersing the batch reactor in a water bath at room temperature to verify whether the system is adequate; however, depending on the pressure and the size of the reactor this test may not be possible. Accordingly, a handprobe connected to a microelectronic hydrogen sensor, capable of detecting an audio signal in the proximity of a leak, is the most appropriate choice for a check of hydrogen leakage in a pressurized reactor vessel.

13.2.3   High-Pressure Bench-Top Reactors

Depending on the laboratory regulations, some small-scale high-pressure reactions may be performed using a bench-top apparatus in a fumehood. Prior to any bench-top reaction be performed, a full understanding of both the thermodynamics and kinetics of the reaction must be undertaken to determine the safe operating conditions. The following precautions must be taken:

- The system must not exceed the safe operation temperature and pressure levels supplied by the manufacturer of the reactor vessel.

The webpage of the National Institute of Standards and Technology delivers an extremely helpful tool to calculate the expected pressure for a pressurized system at a certain temperature.[5]
- The reactants should not be corrosive to the system.
- A blast shield should be placed in front of the reactor.
- As calculated by the thermodynamic and kinetic parameters, large increases in temperature and pressure must be excluded.

13.3 Risk Assessment for Thermal Runaway

Some hydrogenations can proceed quite quickly. Accordingly, care must be taken to guarantee that the reaction is not violent, particularly on a large-scale operation. Since hydrogenations are highly exothermic, a fast reaction in a batch reactor might create overpressure, posing risks to the operator and facilities. Time is the most important factor in the risk assessment of a runaway reaction. The time factors required associated with a runaway reaction are displayed in Figure 13.1 and will be analyzed throughout this section.

Four important temperature levels are considered when determining the thermal potential of a runaway reaction:

The process temperature (T_p): The operation temperature of the reactor at the time of a reaction failure. For nonisothermal reactions this temperature is taken as the highest operation temperature.

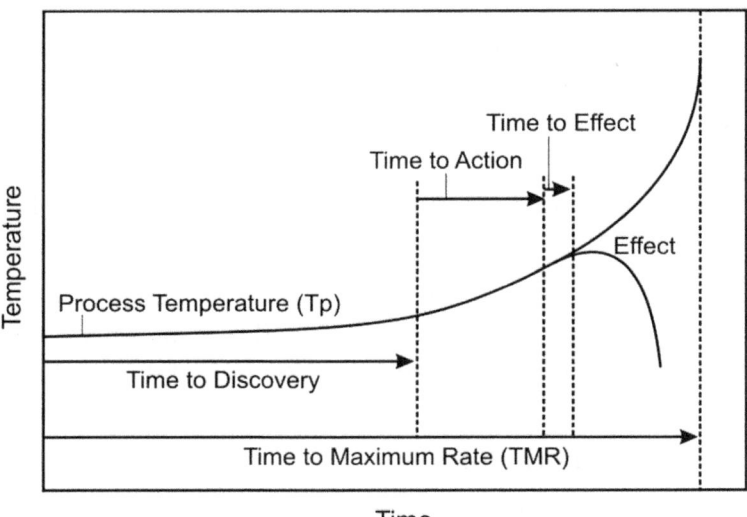

Figure 13.1 Important time parameters to prevent a runaway reaction. Time to discovery: time required before first detection after a malfunction occurs. Time to action: after the detection of a problem, the time required for quenching or cooling the system. Time to effect: the time required before the effect becomes measureable.
Adapted from Refs. 6 and 24.

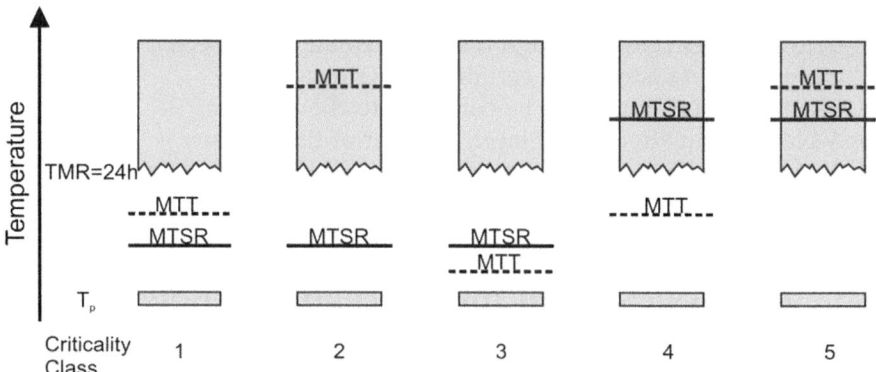

Figure 13.2   Criticality classes for runaway scenarios.
             Adapted from Ref. 9.

Maximum temperature of synthesis reaction (MTSR): this value is dependent on unconverted starting material and is dependent on the process design.

Time to maximum rate under adiabatic conditions (TMR$_{ad}$): Depending on the current reaction temperature, the heat of decomposition, the activation energy and the specific heat capacity of the reactor content.

Maximum temperature for technical reasons (MTT): For open systems, this value is the boiling point of the solvent, for closed systems it is the temperature resulting in the maximum permissible pressure for the system dictated by bursting disks or vents.

The determination of these temperature parameters are thoroughly presented in Refs. 6–12. As displayed in Figure 13.2, the four temperature levels allow for further classification of reaction systems into five classes according to Stoessel.[9]

Classes 1 and 2 correspond to scenarios that secondary reactions and technical limits of the reactor are not reached. In Class 3, the technical limit of the reactor is reached before secondary reactions are initiated. In Class 4, the technical limit of the reactor is the limiting factor, and secondary reactions may be triggered. In Class 5 (*i.e.* the highest-risk scenario), the technical limit of the reactor is reached at the point where secondary reactions have initiated, and the runaway is too fast for a safety barrier.

13.3.1   Reaction Enthalpies

Hydrogenations are exothermic reactions. Therefore, prior knowledge of the reaction enthalpies is mandatory for an initial risk assessment of reactor safety, especially for reactions performed on a large scale. Using computational chemistry, Dixon and coworkers[13] predicted the thermodynamic properties for a variety of biomass-derived molecules using G3MP2 method. This method is one of the most accurate quantum chemistry calculation methods. The predicted properties are within ± 2 kcal mol^{-1} of

Table 13.2 Typical cooling rates and heat loss^a for laboratory and industrial scale equipments. Adapted from Ref. 12.

Vessel	Volume	Cooling time for 1 K	Cooling rate/ K min^{-1}	Heat loss^b/ W kg^{-1}
Test tube	10 mL	11 s	5.5	385
Beaker	100 mL	20 s	3	210
Flask	1000 mL	2 min	0.5	35
Reactor	2.5 m^3	21 min	0.047	3.29
Reactor	5.0 m^3	43 min	0.023	1.61
Reactor	12.7 m^3	59 min	0.0169	1.18
Dewar	1000 mL	62 min	0.0161	1.13
Reactor	25 m^3	233 min	0.0043	0.3

^{a}Temperature difference between inside and outside is 60 K.
^{b}For an 80% fill level with water.

the available experimental values.[13] This work also reported the values of reaction enthalpies computed for several reactions of glucose, glycerol, levulinic acid, succinic acid and γ-valerolactone.

The increased risk for large-scale reactors for hydrogenations is due to the heat-exchange capability of the reactor to be proportional to the surface area of the reactor (r^2), whereas the heat production, to the reactor volume (r^3, where r stands for the reactor radius). Accordingly, the heat-transfer capability is dramatically decreased with the reactor size. Table 13.2 compares the heat-transfer capability of practically relevant laboratory and industrial-scale equipment.

Hydrogenations are often performed under pressure and in a closed system. Therefore, it is important to keep in mind that the amount of heat released from the reactor is equal to the change of the inner energy (ΔU_r) and not only to the reaction enthalpy (ΔH_r). Hence, the released heat amount, q_r, is given by eqn (13.2):

$$q_r = \Delta H_r \text{ (for open systems with constant pressure)} \tag{13.1}$$

$$q_r = \Delta U_r = \Delta H_r - \Delta v_{gas} RT \text{ (for closed systems with constant volume)} \tag{13.2}$$

where, the Δv_{gas} is equal to the change in the number of moles of gas during a reaction ($n_{gas\ products} - n_{gas\ initial}$). Logically, if no gaseous product is formed, Δv_{gas} assumes a negative value, since hydrogenation consumes hydrogen, removing oxygen in the form of $H_2O_{(l)}$. As a result, a heat lower than ΔH_r is released.

13.3.2 Adiabatic Temperature Rise

In an event that the reaction heat cannot be dissipated by heat transport, the maximum temperature reached in a reactor can be calculated using the adiabatic temperature rise (ΔT_{ad}). This is an important value to know in the case of a cooling system failure or a thermal runaway initiated by hot-spot formation. Chemical reactors, especially those exceeding lab scales, can be assumed adiabatic as a first approximation. From the estimated released heat, the adiabatic temperature rise can be calculated from calorimetric data using:

$$\Delta T_{ad} = \frac{q_r}{\bar{c}} = \frac{\Delta H_r}{\bar{c}_p} = \frac{\Delta H_r - \Delta \nu_{gas} RT}{\bar{c}_v} \tag{13.3}$$

where $\bar{c}_p$ is equal to the average specific heat from the reaction mixture and the reactor material. This value can be easily measured by conducting a "reference experiment" (without a catalyst) in a calorimeter. The temperature rise is measured for a known heat input, which is used to calibrate the calorimeter. This can be done most accurately using an electric heat source because the added heat amount can be calculated using:

$$q_r = U_e I t \tag{13.4}$$

In this equation the voltage (U_e) and current (I) are multiplied with time to give (q_r) the heat of reaction $(1 \text{ A V s} = 1 \text{ J})$. This can then be used in combination with the measured change in temperature to calculate $\bar{c}_p$.

The *Maximum temperature attainably by the synthesis reaction* (MTSR) is the sum of the process temperature (T_p) and the adiabatic temperature rise (ΔT_{ad}) as given by:

$$\text{MTSR} = T_p + \Delta T_{ad} \tag{13.5}$$

Equation (13.5) is valid for most of the batch processes, considering that a cooling failure occurs right after the addition of all reactants. Although more realistically a cooling failure will occur throughout the reaction, Equation (13.5) corresponds to the worst-case scenario because the maximum temperature is reduced as the reaction proceeds:

$$\text{MTSR} = T_p + \Delta T_{ad} \cdot (1 - X) \tag{13.6}$$

where, X stands for the fraction of converted starting material. For semi-batch processes in which a key component is added by dosing, the MTSR is calculated in a slightly modified manner. In this situation, a temperature threshold should be set, automatically stopping the addition in the case of a runaway reaction.

The boiling point of the reaction mixture plays an important role in many systems. For adiabatic temperature rises reaching the boiling point of the reactor components, vaporization functions can act as energy-absorbing processes, which decreases the observed ΔT_{ad}. Therefore, the presence of large volumes of solvent increases the safety for a given process by two parameters: the additional heat capacity of the system and the necessary thermal energy amount for its vaporization. Although the solvent can act as a heat sink for the reaction, significant vaporization of the solvent can cause a dramatic pressure increase within the reactor. As discussed in Section 9.2.1, for continuous reactors, the vaporization of the solvent can lead the catalyst bed to dry out, disturbing the reactor fluid dynamics.

13.4 Practical Aspects of Hydrogenation Reactions

13.4.1 Gas Purification and Delivery

The supply of gases is an important part of the design process for a high-pressure reactor. The quantity of gas and the purity are important factors to consider. If the gas purity available from a supplier is not adequate for a reaction, onsite purification can be performed depending on the type of gas required. The purifiers are commercially available in a range of flow rates, (< 0.100 to 1000 L min^{-1}).[14] The maximum pressure operation for gas purifiers is given by flow-rate requirements, but most purifiers operate in the range of 1.0–3.5 MPa, depending on the flow rate of the gas. This pressure is too low for most hydrogenation reactions. As a result, a compressor may be required between the purifier and the reactor.

The following components can be used in combination to increase the practicality and safety of a gas supply to a high-pressure reactor:[2]

- Gas-discharging devices allowing for automatic switching between different cylinders, the cylinders should be easily accessible, as they may need to be switched regularly.
- To maintain the pressure levels, a double-contact manometer can be used.
- Either an electrically or pneumatically driven reciprocating compressor will allow for the maximization of the cylinder contents.
- Using buffer chambers connected to the ring-line system will help maintain a constant pressure.

13.4.2 Materials for the Construction of High-Pressure Reactors

High-pressure reactions require a reaction vessel made of (stainless) steel. The challenge with using stainless steel reactors is that they are susceptible to chemical attack and corrosion. Metal corrosion is already an issue for high H$_2$ pressures at room temperature.[15] This process is called hydrogen embrittlement (HE). The massive adsorption and dissociation of molecular hydrogen on metal surfaces results in localized plastic processes, which accelerate the crack propagation rate of the metal.[4] Thus, the risk of sudden material failure is increased when the reactor operates under high H$_2$ pressures. To alleviate this problem, several special steels have been developed in recent decades.[15,16] Table 13.3 provides a short description of advantages and disadvantages for common reactor alloys. Table 13.4 lists elemental composition for common metal alloys used in reactor vessels. Before selection of a reactor alloy, it is advised that the reaction conditions and system be discussed with the manufacturer to ensure safety and durability of the reactor.

Table 13.3   Advantages,   disadvantages   and   warnings   for   different   alloys (composition listed in Table 13.4).[18]

Material	Advantages	Disadvantages, warnings
Stainless Steel T316	Excellent with most organic compounds, resistant to ammonia and most ammonia compounds	Do not use with halogenated organic compounds
Alloy 20Cb-3	Use with dilute (30% by weight) sulfuric acid, nitric and phosphoric acid, resistant to ammonia and most ammonia compounds	Do not use with halogenated organic compounds
Alloy 400	Stable for chloride salt, fluorine, hydrogen fluoride and hydrofluoric acid	Not suitable for nitric acid or ammonia due to the high copper content
Alloy 600	Excellent resistance to caustic solutions and chlorides at high temperatures and pressures, stable for a wide range of corrosive conditions	Cost is high and is typically only used when required
Alloy B-2	Resistant to reducing acids, hydrochloric, sulfuric and phosphoric	Ferric and other oxidizing ions in >50 ppm cause rapid degradation
Alloy C-276	Broad range of general corrosive resistance for commonly used alloys, corrosive resistance in high pressure and temperature ranges	Sensitive to strong oxidizing conditions
Alloy C-2000	Stable both reducing and mild oxidizing conditions, higher corrosion resistance and broader range of corrosive resistance than C-276 wide operation range for temperature and pressure	Slightly more expensive than C-276, poorer machinability than C-276
Nickel 200	Ultimate corrosion resistance in highly caustic environments	Poor machinability resulting in high cost
Titanium Grade 2	Excellent for use with oxidizing agents	Not suitable for welding, it burns in the presence of oxygen at elevated temperatures
Titanium Grade 4	Higher strength due to higher iron and oxygen content, excellent for use with oxidizing agents, higher operation temperatures and pressures	Not suitable for welding, it burns in the presence of oxygen at elevated temperatures
Zirconium Grade 702	Excellent resistance to hydrochloric and sulfuric acids	Ferric and other oxidizing ions cause rapid degradation
Zirconium Grade 705	Higher working pressure than Zirconium Grade 702	Less resistance to corrosion than Zirconium Grade 702

   In some cases, stainless steel reactors can suffer from "stress corrosion cracking".[2] This type of corrosion is a combination between mechanical and environmental failure in which the tensile stress is well below the yield stress

Table 13.4   Elemental composition of the alloys listed in Table 13.3.[18]

Material	Elemental composition (%)					
	Fe	Ni	Cr	Mo	Mn	Other
Stainless Steel T316	65	12	17	2.5	2.0	
Alloy 20Cb-3	35	34	20	2.5	2.0	Cu 3.5, Nb 1.0 (max)
Alloy 400	1.2	66				Cu 31.5
Alloy 600	8	76	15.5			
Alloy B-2	2	66	1	28	1	Co 1.0
Alloy C-276	6.5	53	15.5	16	1	W 4.0, Co 2.5
Alloy C-2000	3.0	54	23	16	0.5	Co 2.0, Nb 6
Nickel 200		99				
Titanium Grade 2						Ti 99 min
Titanium Grade 4						Ti 99 min
Zirconium Grade 702						Zr 99.2, Hf 4.5 (min.)
Zirconium Grade 705						Zr 95.5 (min.), Hf 4.5 (max.), Nb 2.5

Figure 13.3   Secondary electron micrograph showing a stress-corrosion crack through both plies of a 316L stainless steel bellows.
Reprinted from Ref. 17 with permission from Elsevier.

of the material and chemical attack combined.[17] This corrosion can penetrate deep into the body of the steel and lead to pinholes or cracks with little or no evidence on the external surface, as shown in Figure 13.3. The high speed of this type of reactor corrosion makes it more dangerous than other forms of corrosion, since it can lead to rapid failure of pressurized components. Stress-corrosion cracking starts to appear at temperatures just above 373 K.[2]

Another problem with the use of a metal reactor vessel is posed by the chemical reduction of the surface metal oxide layer under high hydrogen pressures at elevated temperatures. In this situation, the reactor itself may catalyze the reactions. An example of this type of system was given by Schlaf and coworkers[19] in the hydrodeoxygenation of levulinic acid (eqn (13.7)) and glycerol (eqn (13.8)):

$$(13.7)$$

$$(13.8)$$

The conversions described by eqns (13.7) and (13.8) were carried out in a 50 mL Parr® reaction vessel under acidic conditions (40 mmol/L), at 523 K, under a H_2 pressure of 5.5 MPa (measured at 298 K). Under the experiment conditions, the protective chromium oxide layer of the reactor interior was dissolved by the reaction medium. Since the reaction is performed under a H_2 pressure, the oxide layer cannot be reformed. This process occurs only when the reactor vessel is re-exposed to air. Consequently, after removal of the protecting layer of chromium oxide, the exposed stainless steel surface showed activity for hydrodeoxygenation of glycerol and levunilic acid. Remarkably, the results were comparable to those obtained from experiments using a commercial Ru/C catalyst.[19] Altogether, the findings by Schlaf and coworkers[19] alert to the importance of a blank reaction prior to the catalyst test. Unexpected reactions of the metal components with the reaction mixture can potentially decrease the structural integrity of the reactor vessel, resulting in structural failure. For some reactor designs, a glass liner can be inserted into the reaction vessel to protect the metal from the gas or liquid in the reactor. In most cases, a liner is not suitable because the components inside of the reactor cannot be completely protected (*e.g.* thermocouple, stirrer, *etc.*). For this reason, the pressure reactor, seals and accessories, must meet the demands of the reaction with respect to chemical strains (corrosion) and physical stress (temperature and pressure).

13.4.3   Tubing

Many different materials may be used for high-pressure reactors offering a wide variety of different tube sizes and internal diameters. The metal alloys most commonly used as tubing options are summarized in Table 13.3. The use of brass valves or copper lines must be avoided for corrosive gases (*e.g.* hydrogen sulfide, ammonia, vinylchloride, vinyl bromide, dimethylamine, trimethylamine, nitrogen monoxide and dioxide, boron trifluoride, boron trichloride, phosphogene and several others). Cold-drawn tubes are the best options for high-pressure applications and rolled tubes should be avoided. A few important notes are listed below for tubing selection for high-pressure applications:

- Proper training from the manufacturer of the fittings may be recommended before the system is assembled.
- The tubing selected must be softer than or have the same hardness as the fittings used (*e.g.,* a brass fitting should not be used with stainless steel tubing).
- Excessive torque should not be used on nuts since the connections may be permanently damaged.
- The wall thickness should always be double-checked for the pressure requirements.
- The tubing must be checked for scratches, depressions, or any other defects that may cause a seal to fail.

13.4.4 General Design of Batch Reactors

After selecting the proper type of reactor material, a few design considerations may be useful before the reactor is constructed:

- the safety precautions of the reactor (*e.g.* jackscrews, detectors, safety valves);
- the temperature range that the reaction will take place and the heat-control method, including heat exchange;
- the maximum pressure the reactor will operate at and the maximum potential pressure in the event of a malfunction;
- the stirrer type;
- the volume and interior design of the reaction vessel;
- the importance of accessing the interior of the chamber (*e.g.,* for cleaning).

The wall thickness of a reactor vessel will depend on the operating temperature and pressure, inner diameter of the reactor, length, the reactor base (curved or planar), and any additions to both the interior and exterior of the reactor. Any modification of a reactor vessel requires an approval by an engineer. It is important to note the maximum allowable working pressure before selecting a type of reactor. Decreasing the operation temperature of the reactor increases the maximum pressure rating, but the vessel should never be exposed to conditions more than 20% above the maximum allowable working pressure (except for pressure tests at room temperature which are usually done with a liquid medium at $p = 1.5\ p_{max}$). The manufacturer must be contacted to confirm safe operation.

13.4.5   Selection of Stirrer and Stirring Speed

One of the most important parameters affecting the gas-liquid mass transfer in a reactor design is the selection of the proper stirrer for the reaction conditions. Reactor stirrers are divided into two groups depending on how they mix the reactor contents. These groups are further subdivided according to the medium viscosity for which they are designed.[20–22]

1) Radial stirrers have a greater mixing efficiency in the lateral direction within the reactor (Figure 13.4). For a low-viscosity medium, disk stirrers and impeller stirrers are commonly used, while for a high-viscosity medium, paddles and anchor stirrers are used.
2) Axial stirrers have a greater mixing efficiency in the vertical direction of the reactor (Figure 13.4). Common stirrers for low-viscosity contents

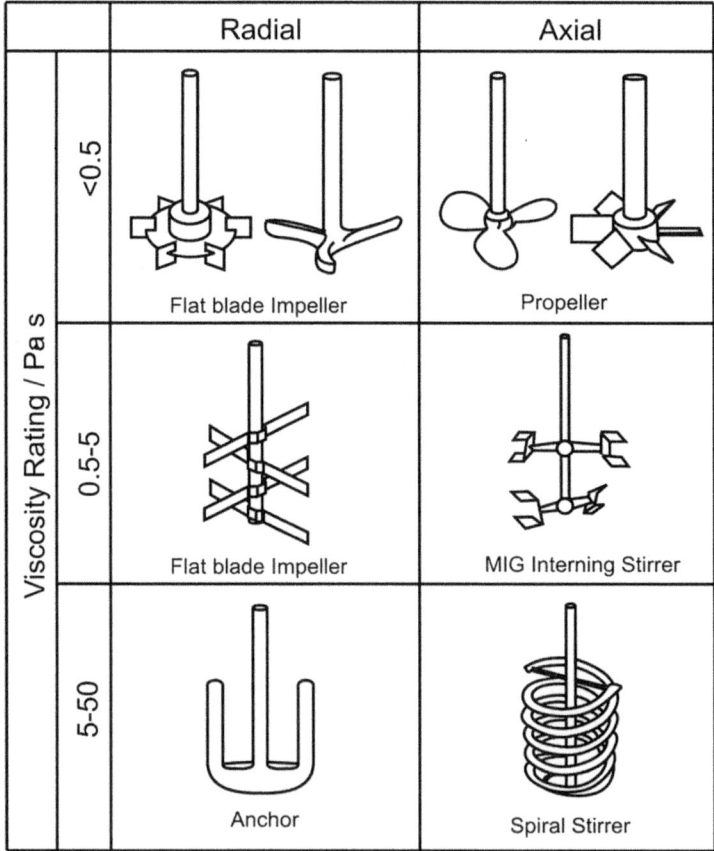

Figure 13.4   Stirrer choices for flow radial and axial flow patterns as described in Figure 13.5.
Adapted from Ref. 21.

are propeller and paddle stirrers, whereas for high viscosity, helical ribbon stirrers and MIG-stirrers are used.

Axial stirrers are used for reactions in which a solid must be suspended throughout the reactor or for gas introduction into through the surface aeration method (Figure 13.5). Radial stirrers are employed for exothermic reactions because the reaction heat is then more efficiently dissipated through the reactor walls.[20] The mixing efficiencies for both radial and axial mixing are displayed in Figure 13.5.

After the selection of the most adequate stirring device, the stirring rate should be determined. To determine the correct stirring rate, the reaction of interest should be conducted at a number of different stirring rates. This procedure is needed in order to determine the speed at which the mass-transfer problems are mitigated (*see also* Section 12.2.1.1). In general, the rate of the hydrogenation will increase with the stirring rate until a maximum is reached. At this point, the liquid phase is provided with enough H_2 indicating that the reaction is operating under the kinetic regime. It is noteworthy that the stirring rate should be selected to avoid the formation of a vortex. If a vortex is formed in the reactor, the gas/liquid interface area is small, and thus, the reaction rate will remain constant regardless of an

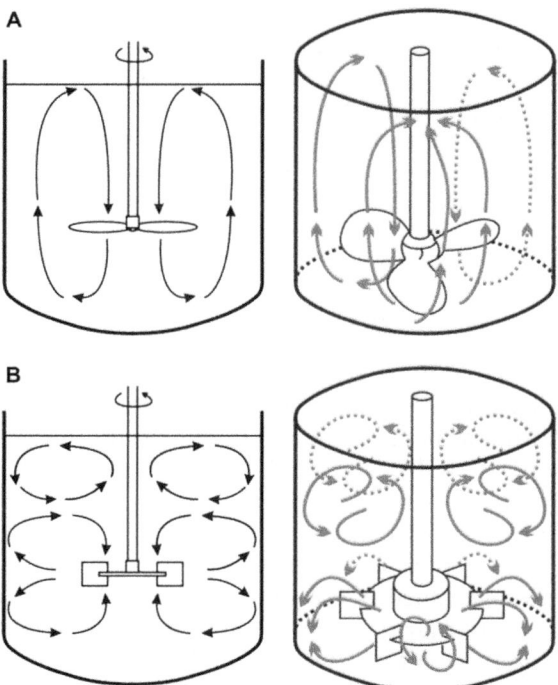

Figure 13.5 Flow patterns for different stirrers. (A) Axial stirrer; (B) Radial stirrer. Adapted from 21.

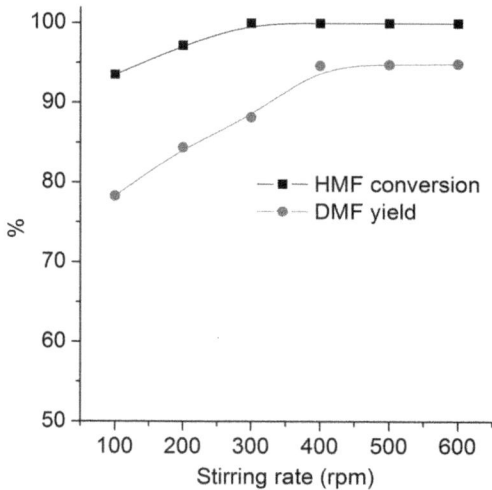

Figure 13.6   Dependence of HMF conversion and DMF yield with the agitation speed. Reaction conditions: substrate (2.5 wt% HMF relative to THF), THF (solvent, 20 mL), Ru/C catalyst (5 mol% relative to HMF), 2 MPa H_2, 473 K, 2 h.[23]

increase in the stirring rate.[21] This problem can be overcome by the introduction of a baffle into the reactor to perturb the flow.

Lin and coworkers[23] analyzed the effect of several different parameters (*e.g.* temperature, H_2 pressure, HMF concentration and agitation speed) on the conversion of HMF and DMF yield using a Ru/C catalyst (5 mol%). Figure 13.6 shows the dependence of HMF conversion and DMF yield with stirring speed. The results show that even when full conversion of HMF was obtained, further increasing the stirring speed improved the yield of DMF by 6.5%.[23]

13.5   Closing Remarks

Due to the severe conditions used in order to perform hydrogenation of biomass, an individual risk assessment prior to the reaction is mandatory. First, a suitable infrastructure, namely high-pressure boxes made of steel plates or ferroconcrete walls, is needed. The high-pressure boxes have to be able to both withstand a rupture of an autoclave, and allow an eventual controlled overpressure discharge. In order to avoid such a worst-case scenario, a proper selection of the reactor material and a suitable design in terms of pressure resistance is obligatory. Should these reactions be conducted in autoclaves of a larger scale ($\geq$ 500 mL), attention to the much more adiabatic behavior of the reactor, which could result in thermal runaways, is to be paid. The knowledge of reaction enthalpy, activation energy, and heat capacity of the system is important to insure that questions regarding sufficient dilution or stepwise addition of a key reaction component should be considered. Often, liquid-phase hydrogenations with molecular hydrogen are mass-transfer limited. Therefore, it is important to select the correct stirrer type, size and rotation speed to allow for optimization of gas-liquid mass transport and reaction heat transfer to the reactor walls. Through a full analysis of the gas–liquid mass transfer of the desired reaction, the reaction kinetics can be observed and important information on the reaction rates and mechanism can be determined.

Acknowledgements

This work was performed as part of the Cluster of Excellence "Tailor-Made Fuels from Biomass", funded by the German Federal and State governments to promote science and research at German universities.

References

1. D. P. Glasstone, *The effects of nuclear weapons*, US Department of Defense and the Energy Research and Development Administration, 1977.
2. N. Theyssen, K. Scovell and M. Poliakoff, In *Handbook of Green Chemistry*, Wiley-VCH Verlag GmbH & Co. KGaA, Editon edn., 2010.
3. H. J. Pasman and W. J. Rogers, *J. Loss Prevent. Proc.*, 2010, **23**, 697–704.
4. G. R. Astbury and S. J. Hawksworth, *Int. J. Hydrogen Energy*, 2007, **32**, 2178–2185.
5. National Institute of Standards and Technology, http://webbook.nist.gov/chemistry/fluid/.
6. F. Stoessel, *J. Loss Prevent. Proc.*, 1993, **6**, 79–85.
7. F. Stoessel, *Chimia*, 1996, **50**, 602–602.
8. F. Stoessel, *Chimia*, 1998, **52**, 691–693.
9. F. Stoessel, *Process Saf. Environ*, 2009, **87**, 105–112.
10. F. Stoessel, H. Fierz, P. Lerena and G. Kille, *Org. Process Res. Dev.*, 1997, **1**, 428–434.
11. HarsNet, *Reaction Calorimetry*, http://www.harsnet.net/harsbook/5.Reaction%20calorimetry.pdf.
12. HarsNet, *Adiabatic Calorimetry*, http://www.harsnet.net/harsbook/6.Adiabatic%20calorimetry.pdf.
13. M. Vasiliu, K. Guynn and D. A. Dixon, *J. Phys. Chem. C*, 2011, **115**, 15686–15702.
14. P. Microelectronics, *PG Series Gaskleen® Purifier Assemblies and Manifolds*, http://www.pall.com/pdfs/Microelectronics/MEPGGPEN.pdf.
15. L. Zhang, M. Wen, Z. Y. Li, J. Y. Zheng, X. X. Liu, Y. Z. Zhao, C. L. Zhou and Asme, *ASME 2012 Pressure Vessels and Piping Conference*, 2012, **2**, 533–545.
16. J. Zheng, X. Liu, P. Xu, P. Liu, Y. Zhao and J. Yang, *Int. J. Hydrogen Energy*, 2012, **37**, 1048–1057.

17. B. Panda, M. Sujata, M. Madan and S. K. Bhaumik, *Eng. Failure Anal.*, 2014, **36**, 379–389.
18. P. I. Company, *Safety in the Operation of Laboratory Reactors and Pressure Vessels*, http://www.parrinst.com/wp-content/uploads/downloads/2012/05/230M_Parr_Safety-Lab-Reactors.pdf.
19. D. Di Mondo, D. Ashok, F. Waldie, N. Schrier, M. Morrison and M. Schlaf, *ACS Catal.*, 2011, **1**, 355–364.
20. A. Behr and P. Neubert, *Applied Homogeneous Catalysis*, Wiley-VCH Verlag & Co. KGaA, Weinheim, Germany, 2012.
21. M. Baerns, A. Behr, A. Brehm, J. Gmehling, H. Hofmann, U. Onken, A. Renken, K. O. Hinrichsen and R. Palkovits, *Technische Chemie*, Wiley VCH Verlag GmbH, 2013.
22. H. A. Jakobsen, *Chemical Reactor Modeling: Multiphase Reactive Flows*, Springer, 2008.
23. L. Hu, X. Tang, J. Xu, Z. Wu, L. Lin and S. Liu, *Ind. Eng. Chem. Res.*, 2014, **53**, 3056–3064.
24. Harsnet, Measures, http://www.harsnet.net/harsmeth/final%20version/9_measures.html.

Subject Index